D1161741

Lecture Notes in Physics

Monographs

Springer
Berlin
Heidelberg
New York
Barcelona
Hong Kong
London
Milan
Paris
Singapore
Tokyo

The Editorial Policy for Monographs

The series Lecture Notes in Physics reports new developments in physical research and teaching - quickly, informally, and at a high level. The type of material considered for publication in the monograph Series includes monographs presenting original research or new angles in a classical field. The timeliness of a manuscript is more important than its form, which may be preliminary or tentative. Manuscripts should be reasonably self-contained. They will often present not only results of the author(s) but also related work by other people and will provide sufficient motivation, examples, and applications.
The manuscripts or a detailed description thereof should be submitted either to one of the series editors or to the managing editor. The proposal is then carefully refereed. A final decision concerning publication can often only be made on the basis of the complete manuscript, but otherwise the editors will try to make a preliminary decision as definite as they can on the basis of the available information.
Manuscripts should be no less than 100 and preferably no more than 400 pages in length. Final manuscripts should be in English. They should include a table of contents and an informative introduction accessible also to readers not particularly familiar with the topic treated. Authors are free to use the material in other publications. However, if extensive use is made elsewhere, the publisher should be informed. Authors receive jointly 30 complimentary copies of their book. They are entitled to purchase further copies of their book at a reduced rate. No reprints of individual contributions can be supplied. No royalty is paid on Lecture Notes in Physics volumes. Commitment to publish is made by letter of interest rather than by signing a formal contract. Springer-Verlag secures the copyright for each volume.

The Production Process

The books are hardbound, and quality paper appropriate to the needs of the author(s) is used. Publication time is about ten weeks. More than twenty years of experience guarantee authors the best possible service. To reach the goal of rapid publication at a low price the technique of photographic reproduction from a camera-ready manuscript was chosen. This process shifts the main responsibility for the technical quality considerably from the publisher to the author. We therefore urge all authors to observe very carefully our guidelines for the preparation of camera-ready manuscripts, which we will supply on request. This applies especially to the quality of figures and halftones submitted for publication. Figures should be submitted as originals or glossy prints, as very often Xerox copies are not suitable for reproduction. For the same reason, any writing within figures should not be smaller than 2.5 mm. It might be useful to look at some of the volumes already published or, especially if some atypical text is planned, to write to the Physics Editorial Department of Springer-Verlag direct. This avoids mistakes and time-consuming correspondence during the production period.
As a special service, we offer free of charge LaTeX and TeX macro packages to format the text according to Springer-Verlag's quality requirements. We strongly recommend authors to make use of this offer, as the result will be a book of considerably improved technical quality.
Manuscripts not meeting the technical standard of the series will have to be returned for improvement.
For further information please contact Springer-Verlag, Physics Editorial Department II, Tiergartenstrasse 17, D-69121 Heidelberg, Germany.

Jean Daillant   Alain Gibaud

X-Ray and Neutron Reflectivity: Principles and Applications

Springer

Author

Jean Daillant
Service de Physique de l'Etat Condensé
CEA Saclay
F-91191 Gif-sur-Yvette Cedex, France

Alain Gibaud
Laboratoire de Physique de l'Etat Condensé, UPRES A 6087
Université du Maine, Faculté des sciences
F-72085 Le Mans Cedex 9, France

Library of Congress Cataloging-in-Publication Data.

Die Deutsche Bibliothek - CIP-Einheitsaufnahme

X-ray and neutron reflectivity : principles and applications / Jean Daillant ; Alain Gibaud (ed.). - Berlin ; Heidelberg ; New York ; Barcelona ; Hong Kong ; London ; Milan ; Paris ; Singapore ; Tokyo : Springer, 1999
(Lecture notes in physics : N.s. M, Monographs ; 58)
ISBN 3-540-66195-6

ISSN 0940-7677 (Lecture Notes in Physics. Monographs)
ISBN 3-540-66195-6 Springer-Verlag Berlin Heidelberg New York

Typesetting: Camera-ready by the authors/editors
Cover design: *design & production*, Heidelberg

SPIN: 10644458 55/3144/du - 5 4 3 2 1 0 – Printed on acid-free paper

Foreword

The reflection of x-rays and neutrons from surfaces has existed as an experimental technique for almost fifty years. Nevertheless, it is only in the last decade that these methods have become enormously popular as probes of surfaces and interfaces. This appears to be due to the convergence of several different circumstances. These include the availability of more intense neutron and x-ray sources (so that reflectivity can be measured over many orders of magnitude and the much weaker surface diffuse scattering can now also be studied in some detail); the growing importance of thin films and multilayers in both technology and basic research; the realization of the important role which roughness plays in the properties of surfaces and interfaces; and finally the development of statistical models to characterize the topology of roughness, its dependence on growth processes and its characterization from surface scattering experiments. The ability of x-rays and neutrons to study surfaces over four to five orders of magnitude in length scale regardless of their environment, temperature, pressure, etc., and also their ability to probe buried interfaces often makes these probes the preferred choice for obtaining global statistical information about the microstructure of surfaces, often in a complementary manner to the local imaging microscopy techniques. This is witnessed by the veritable explosion of such studies in the published literature over the last few years. Thus these lectures will provide a useful resource for students and researchers alike, covering as they do in considerable detail most aspects of surface x-ray and neutron scattering from the basic interactions through the formal theories of scattering and finally to specific applications.

It is often assumed that neutrons and x-rays interact weakly with surfaces and in general interact weakly enough so that the simple kinematic theories of scattering are good enough approximations to describe the scattering. As most of us now appreciate, this is not always true, e.g. when the reflection is close to being total, or in the neighborhood of strong Bragg reflections (e.g. from multilayers). This necessitates the need for the full dynamical theory (which for specular reflectivity is fortunately available from the theory of optics), or for higher-order approximations, such as the distorted wave Born approximation to describe strong off-specular scattering. All these methods are discussed in detail in these lectures, as are also the ways in which the magnetic interaction between neutrons and magnetic moments can yield information on the magnetization densities of thin films and multilayers. I commend the organizers for having organized a group of expert lecturers to present this subject in a detailed but clear fashion, as the importance of the subject deserves.

S. K. Sinha
Advanced Photon Source
Argonne National Laboratory
December 1998

Contents

Part I. Principles

Part II. Applications

List of authors

- Dr. T. Baumbach
  Fraunhofer Institut Zerstörungsfreie Prüfverfahren, EADQ Dresden
  Krügerstraße 22
  D-01326 Dresden, Germany
  Present address: European Synchrotron Radiation Facility BP 220, F-38043, Grenoble Cedex France

- Dr. F. de Bergevin
  Laboratoire de Cristallographie associé à l'Université Joseph Fourier
  CNRS Bâtiment F
  25 avenue des martyrs, B.P. 166
  38042 Grenoble Cedex 09, France
  and
  European Synchrotron Radiation Facility B.P. 220, 38043 Grenoble Cedex, France

- Dr. J. Daillant
  Service de Physique de l'Etat Condensé, Orme des Merisiers
  CEA Saclay
  91191 Gif sur Yvette Cedex, France

- Dr. C. Fermon
  Service de Physique de l'Etat Condensé, Orme des Merisiers
  CEA Saclay
  91191 Gif sur Yvette Cedex, France

- Dr. J.M. Gay
  CRMC2 CNRS,
  Campus de Luminy, case 913
  13288 Marseille Cedex 9, France

- Prof. A. Gibaud
  Laboratoire de Physique de l'Etat Condensé, UPRESA 6087
  Université du Maine Faculté des sciences,
  72085 Le Mans Cedex 9, France

- Dr. L. Lapena
  CRMC2 CNRS,
  Campus de Luminy case 913,
  13288 Marseille Cedex 9, France

- Dr. A. Menelle
  Laboratoire Léon Brillouin CEA CNRS, CEA Saclay
  91191 Gif sur Yvette Cedex, France

- Dr. P. Mikulik
  Laboratory of Thin Films and Nanostructures
  Faculty of Science Masaryk University
  Kotlářská 2
  611 37 Brno, Czech Republic

- Dr. F. Ott
  Laboratoire Léon Brillouin CEA CNRS, CEA Saclay
  91191 Gif sur Yvette Cedex, France

- Prof. A. Pimpinelli
  LASMEA
  Université Blaise Pascal - Clermont 2
  Les Cézeaux
  63177 Aubière Cedex, France

- Dr. G. Reiter
  Institut de Chimie des Surfaces et Interfaces CNRS,
  15 rue Jean Starcky, B.P. 2488
  68057 Mulhouse, France

- Dr. A. Sentenac
  LOSCM/ENSPM
  Université de St Jérôme
  13397 Marseille Cedex 20, France

- Dr. G. Vignaud
  Université de Bretagne sud
  4 rue Jean Zay
  56100 Lorient, France

Acknowledgements

This book folllows a summer school on reflectivity held in Luminy, France, from June 9th to June 13th, 1997. The editors are particularly grateful to the Université du Maine (Le Mans, France), to the Direction des Sciences de la Matière of the Commissariat à l'Energie Atomique (C.E.A.), to the C.N.R.S. (Département Sciences Physiques et Mathématiques) and to the Région des Pays de la Loire for their help and sponsoring of this summer school. Many thanks are especially adressed to all of those who made this meeting possible: A. Radigois from the "délégation C.N.R.S. Bretagne-Pays de la Loire" who very kindly suggested the location of the school and who helped us through the admisnistrative tasks, J. Lemoine who made a wonderful job as the school secretary, and G. Ripault for his technical support at Luminy and a perfect organisation of social events. We are also indebted to Dr. D. Bonhomme for helping us during the preparation of the manuscript, to Drs. N. Cowlam and T. Waigh for reading some chapters of this book, and to the staff of the C.N.R.S. center of Luminy for their hospitality.

Introduction

In his paper entitled "On a New Kind of Ray, A Preliminary Communication" relating the discovery of x-rays, which was submitted to the Würzburg Physico-Medical Society on December 28, 1895, Röntgen stated the following about the refraction and reflection of the newly discovered rays: "The question as to the reflection of the X-ray may be regarded as settled, by the experiments mentioned in the preceding paragraph, in favor of the view that ***no noticeable regular reflection of the rays takes place from any of the substances examined.*** Other experiments, which I here omit, lead to the same conclusion.[1]"

This conclusion remained unquestioned until in 1922 Compton[2] pointed out that if the refractive index of a substance for x-rays is less than unity, it ought to be possible, according to the laws of optics, to obtain total external reflection from a smooth surface of it, since the x-rays, on entering the substance from the air, are going into a medium of smaller refractive index. This was the starting point for x-ray (and neutron) reflectivity. The demonstration that the reflection of x-rays on a surface was indeed obeying the laws of electromagnetism was pursued by Prins[3] and others who investigated the role of absorption on the sharpness of the limit of total reflection and showed that it was consistent with the Fresnel formulae. This work was continued by Kiessig[4] using nickel films evaporated on glass. Reflection on such thin films gives rise to fringes of equal inclination (the "Kiessig fringes" in the x-ray literature) which allow the measurement of thin film thicknesses, now the most important application of x-ray and neutron reflectivity. It was, however, not until 1954 that Parratt[5] suggested inverting the analysis and interpreting x-ray reflectivity as a function of angle of incidence via models of an *inhomogeneous* surface-density distribution. The method was then applied to several cases of solid or liquid[6] interfaces. Whereas Parratt noticed in his 1954 paper that "it is at first surprising that any experimental surface appears smooth to x-rays. One frequently hears that, for good reflection, a mirror surface must be smooth to within about one wavelength of the radiation involved..." it soon appeared that effects of surface roughness were important, the most

[1] A more complete citation of Röntgen's paper is given in an appendix to this introduction.

[2] A.H. Compton *Phil. Mag.* **45** 1121 (1923)

[3] J.A. Prins, *Zeit. f. Phys.* **47** 479 (1928); a very interesting account of these early developments is given in the famous book by R.W. James, "the optical principles of the diffraction of x-rays", Bell and sons, London, 1948.

[4] H. Kiessig *Ann. der Physik* **10** 715 and 769 (1931).

[5] L.G. Parratt, *Phys. Rev.* **95** 359 (1954).

[6] B.C. Lu and S.A. Rice, *J. Chem. Phys.* **68** 5558 (1978).

dramatic of them being the asymmetric surface reflection known as Yoneda wings[7]. These Yoneda wings were subsequently interpreted as diffuse scattering of the enhanced surface field for incidence or exit angle equal to the critical angle for total external reflection. The theoretical basis for the analysis of this surface diffuse scattering was established in particular through the pioneering work of Croce et al.[8] In a context where coatings, thin films and nanostructured materials are playing an increasingly important role for applications, the number of studies using x-ray or neutron reflectivity dramatically increased during the 90's, addressing vitually all kinds of interfaces: solid or liquid surfaces, buried solid-liquid or liquid-liquid interfaces, interfaces in thin films and multilayers[9]. Apart from the scientific and technological demand for more and more surface characterisation, at least two factors explain this blooming of x-ray and neutron reflectivity. First, the development of neutron reflectometers (Chap. 5) has been decisive, in particular for polymer physics owing to partial deuteration (Chap. 9), and an equally important contribution of neutron reflectivity can be expected for surface magnetism. Second, the use of 2nd and 3rd generation synchrotron sources has resulted in a sophistication of the technique now such that not only the thicknesses but also the morphologies and correlations within and between rough interfaces can be accurately characterised for in-plane distances ranging from atomic or molecular distances to hundreds of microns. In parallel more and more accurate methods have been developed for data analysis.

This book follows a summer school on reflectivity held in Luminy (France) in June 1997. It is organised into two parts, the first one being devoted to principles and the second one to the discussion of examples and applications. Organising the school and now editing the book, we had in mind that an increasing number of non-specialists are now using x-ray and neutron reflectometry and that the need for a proper introduction to the field was not yet fulfilled. It is also true that even if the principle of a reflectivity experiment is extremely simple (one just has to measure the intensity of a reflected beam), the technique is in fact really demanding. An important purpose of this book is therefore also to warn the beginners of experimental problems, often related to the experimental resolution, which are not necessarily apparent but may lead to serious misinterpretations. This is done in the second part of the book where specific aspects related to the nature of the samples are treated. An equally important purpose is also to share with the reader our enthusiasm for the many beautiful recent developments in reflectivity methods, and for the physics that can be can be done with it, and to give him/her the desire to do even more beautiful experiments.

---

[7] Y. Yoneda, *Phys. Rev.* **131** 2010 (1963).

[8] P. Croce, L. Névot, B. Pardo, *C. R. Acad. Sc. Paris* **274** 803 and 855 (1972).

[9] For a recent review see for example S. Dietrich and A. Haase, *Physics Reports* **260** 1 (1995) and the numerous examples cited in the different chapters of this book.

As strongly suggested by the short historical sketch given above, most of the revolutions in the use of x-rays (not only for interface studies) arise by considering new potentialities[10] related to their nature of electromagnetic waves, which was so controversial in the days of Röntgen. The book therefore starts with a panorama of the interaction of x-rays with matter, giving both a thorough treatment of the basic principles, and an overview of more advanced topics like magnetic or anisotropic scattering, not only to give a firm basis to the following developments but also to stimulate reflection on new experiments. Then, a rigourous presentation of the statistical aspects of wave scattering at rough surfaces is given. This point, obviously important for understanding the nature of surface scattering experiments, as well as for their interpretation, is generally ignored in the x-ray literature (this chapter has been written mainly by a researcher in optics). The basic statistical properties of surfaces are introduced first. Then an ideal scattering experiment is described, and the limitations of such a description, in particular the fact that the experimental resolution is always finite, are discussed. The finiteness of the resolution leads to the introduction of ensemble averages for the calculation of the scattered intensities and to a natural distinction between coherent (specular, equal to the average of the scattered field) and incoherent (diffuse, related to the mean-square deviation of the scattered field) scattering. These principles are immediately illustrated within the Born approximation in order to avoid all the mathematical complications resulting from the details of the interaction of an electromagnetic wave with matter. These more rigorous aspects of the scattering theory are treated in Chaps. 3 and 4 for specular and diffuse scattering. The matricial theory of the reflection of light in a smooth or rough stratified medium and its consequences are treated in Chap 3. This is used in Chap 4 for the treatment of diffuse scattering. The Croce approach to the distorted-wave Born approximation (DWBA) based on the use of Green functions is mainly used. This theory is currently the most popular for data analysis and is extensively used in the second part of the book, which is devoted to applications, in particular in Chap 8. However, other methods used in optics are also shortly reviewed. The general case of a stratified medium with interface roughness or density fluctuations is discussed using this DWBA, and different dynamical effects are discussed. Then, the theoretical aspects of a finite resolution function (the experimental aspects are treated in the second part of the book) are considered, as well as their implications for reflectivity experiments. The last chapter of this first part, principles, is devoted to neutron reflectometry whose specific aspects require a separated treatment. After an introduction to neutron-matter interactions, neutron reflectivity of non-magnetic materials is presented and the characteristics of the neutron spectrometers are given. Examples follow with

[10] It is our opinion that fully exploiting the spectroscopic capabilities of x-rays in reflectivity experiments would lead to most interesting developments.

a particular emphasis put on the newly developed methods of investigation of magnetic multilayers using polarised neutrons.

The second part of the book is devoted to examples of the physics that can be done using x-ray and neutron reflectivity. The first three chapters are related to solid surfaces and multilayers, whereas the last two chapters deal with soft condensed matter. In both cases, a statistical description of the surfaces and of their properties is given first (Chap 6 and beginning of Chap 9) and examples follow. In Chap 7, the complete characterisation of the roughness of a single solid surface is considered. The experimental geometry, diffractometers, resolution functions are introduced first. Then, examples are given and and the x-ray results are compared to the results obtained using complementary techniques like transmission electron microscopy and atomic force microscopy. More complicated cases of multilayers are discussed in Chap 8. The experimental setups are described and examples of reflectivity studies and non-specular scattering measurements are discussed with the aim of reviewing all the important situations that can be encountered. Examples include rough multilayers, stepped surfaces, interfaces in porous media, the role of roughness in diffraction experiments and multilayer gratings. Examples in soft condensed matter include liquid interfaces and polymers. This is a domain where the impact of reflectivity measurements has been very large because many of the very powerful complementary techniques which can be used with solid surfaces require high vacuum, and cannot be used for the characterisation of liquid interfaces. The specific aspects of liquid interface studies (mainly using x-rays) are discussed first. Experimental setups for the study of horizontal interfaces are described, and the implications of the specific features of liquid height-height correlation functions for reflectivity experiments are described. Examples include liquid-vapour interfaces, organic films at the air-water interfaces, liquid metal surfaces, and finally buried liquid-liquid interfaces. Finally, polymers at interfaces are considered in a last chapter. This is a domain where neutron reflectivity has made an invaluable contribution, in particular owing to the transparency of many materials to neutrons and to the possibility of contrast variation.

J. Daillant and A. Gibaud,

Saclay and Le Mans,
May 1999

Appendix: Röntgen's report on the reflection of x-rays.

"With reference to the general conditions here involved on the other hand, and to the importance of the question whether the X-rays can be refracted or not on passing from one medium into another, it is most fortunate that this subject may be investigated in still another way than with the aid of prisms. Finely divided bodies in sufficiently thick layers scatter the incident light and allow only a little of it to pass, owing to reflection and refraction; so that if powders are as transparent to X-rays as the same substances are in mass–equal amounts of material being presupposed–it follows at once that neither refraction nor regular reflection takes place to any sensible degree. Experiments were tried with finely powdered rock salt, with finely electrolytic silver-powder, and with zinc-dust, such as is used in chemical investigations. In all these cases no difference was detected between the transparency of the powder and that of the substance in mass, either by observation with the fluorescent screen or with the photographic plate... The question as to the reflection of the X-ray may be regarded as settled, by the experiments mentioned in the preceding paragraph, in favor of the view that no noticeable regular reflection of the rays takes place from any of the substances examined. Other experiments, which I here omit, lead to the same conclusion.
One observation in this connection should, however, be mentioned, as at first sight it seems to prove the opposite. I exposed to the X-rays a photographic plate which was protected from the light by black paper, and the glass side of which was turned towards the discharge-tube giving the X-rays. The sensitive film was covered, for the most part, with polished plates of platinum, lead, zinc, and aluminum, arranged in the form of a star. On the developing negative it was seen plainly that the darkening under the platinum, the lead and particularly the zinc, was stronger than under the other plates, the aluminum having exerted no action at all. It appears, therefore, that these metals reflect the rays. Since, however, other explanations of a stronger darkening are conceivable, in a second experiment, in order to be sure, I placed between the sensitive film and the metal plates a piece of thin aluminum-foil, which is opaque to ultraviolet rays, but it is very transparent to the X-rays. Since the same result substantially was again obtained, the reflection of the X-rays from the metals above named is proved. If we compare this fact with the observation already mentioned that powders are as transparent as coherent masses, and with the further fact that bodies with rough surfaces behave like polished bodies with reference to the passage of the X-rays, as shown as in the last experiment, we are led to the conclusion already stated that regular reflection does not take place, but that bodies behave toward X-rays as turbid media do towards light."

Part I

Principles

1 The Interaction of X-rays (and Neutrons) with Matter

François de Bergevin

Laboratoire de Cristallographie associé à l'Université Joseph Fourier, CNRS, Bâtiment F, 25 avenue des martyrs, B.P. 166, 38042 Grenoble Cedex 09, France *and* European Synchrotron Radiation Facility, B.P. 220, 38043 Grenoble Cedex, France

1.1 Introduction

The propagation of radiation is generally presented according to an optical formalism in which the properties of a medium are described by a refractive index. A knowledge of the refractive index is sufficient to predict what will happen at an interface, that is to establish the Snell-Descartes' laws and to calculate the Fresnel coefficients for reflection and transmission.

One of the objectives in this introduction will be to link the laws of propagation of radiation and in particular the refractive index, to the fundamental phenomena involved in the interaction of radiation with matter. The main process of interaction in the visible region of the electromagnetic spectrum is the polarisation of the molecules (at least for an insulator). At higher energies as with x-rays, it is generally sufficient to take into account the interactions with the atoms and at the highest x-ray energies only the electrons need be considered in the interaction process. It is the nuclei of the materials which interact with neutrons, which also have a second interaction with the electrons for those atoms which carry a magnetic moment.

The conventions and symbols which will be used in this book will be defined in Sect. 1.2. In the same section the basics of wave propagation will be revised and the different physical quantities which characterise the scattering of radiation will be introduced, together with the appropriate definitions and the properties of Green functions. In Sect. 1.3 the link between the atomic scattering and the model of a continuous medium represented by a refractive index will be established. Section 1.4 will be devoted to the interaction of x-ray radiation with matter. That will include the inelastic and elastic scattering, and the absorption. The scattering will be described as split into a non resonant and a resonant part. Together with the questions of resonance and absorption a discussion of the dispersion relations will be given. In Sect. 1.5, the case when the scattering depends on the anisotropy of the material will be briefly examined with reference to magnetic and to Templeton scattering. Neutron scattering will not be presented in detail in this chapter since it will appear in Chap. 5 of this book but we shall frequently refer to it.

In the present chapter, the ***bold italic*** font will be used to ***define words or expressions*** and the *emphasized sentences* will be in *italic.*

1.2 Generalities and Definitions

1.2.1 Conventions

Two conventions can be found in the literature to describe a propagating wave, because complex quantities are not observed and the imaginary part has an arbitrary sign. In optics and quantum mechanics a monochromatic plane wave is generally written as

$$A \propto e^{-i(\omega t - \mathbf{k}.\mathbf{r})}, \tag{1.1}$$

which is also the notation used in neutron scattering, even when doing crystallography. On the other hand, x-ray crystallographers are used to writing the plane wave as,

$$A \propto e^{+i(\omega t - \mathbf{k}.\mathbf{r})}. \tag{1.2}$$

The imaginary part of all complex quantities are the opposite of one another in these two notations. Since the observed real quantities may be calculated from imaginary numbers, it is very important to keep consistently a unique convention. The imaginary part f'' of the atomic scattering factor for example, used in x-ray crystallography is a positive number. This is correct provided that it is remembered that the complex scattering factor $(f + f' + if'')$ (f is the atomic form factor, also positive) is affected by a common minus sign, usually left as implicit. In optics, the opposite convention is commonly used and the most useful quantity is the refractive index. Its imaginary part which is associated with absorption is always positive. The number of alternative choices is increased with another convention concerning the sign of the scattering wave-vector transfer $\mathbf{q}$ or scattering vector, which can be written as

$$\mathbf{q} = \mathbf{k}_{\mathrm{sc}} - \mathbf{k}_{\mathrm{in}} \tag{1.3}$$

or

$$\mathbf{q} = \mathbf{k}_{\mathrm{in}} - \mathbf{k}_{\mathrm{sc}} \tag{1.4}$$

where $\mathbf{k}_{\mathrm{in}}$ and $\mathbf{k}_{\mathrm{sc}}$ are the incident and scattered wave vectors. In this book, the conventions (1.3) and (1.2) as used in crystallography, will be adopted. *Only one exception will be made, in the chapter devoted to neutrons (Chapter 5), in which convention (1.1) will be used.* The structure factor which describes the scattered amplitude in the Born approximation will therefore be written in all cases (except with neutrons) as,

$$f(\mathbf{q}) = \int \rho(\mathbf{r}) e^{i\mathbf{q}.\mathbf{r}} d\mathbf{r} \tag{1.5}$$

where $\rho(\mathbf{r})$ is the scattering density, which will be discussed below. The real part of the refractive index is generally less than 1 with x-ray radiation and the refractive index is usually written as,

$$n = 1 - \delta - i\beta \qquad \text{where } \delta \text{ and } \beta \text{ are positive.} \tag{1.6}$$

Indeed the imaginary part β, equal to $\lambda\mu/4\pi$, is essentially positive (λ is the wavelength and μ is the absorption coefficient, see (1.84) and section 1.4.6). Note that because the opposite convention is used, the sign of the imaginary part of n is opposite in visible optics.

The waves will be assumed to be monochromatic in most instances, with the temporal dependence $e^{i\omega t}$. To satisfy the international standard of units, or SI units, the electromagnetic equations will be written in the rationalised MKSA system of units. The Coulombian force in vacuum is in this system $qq'/4\pi\varepsilon_0 r^2$ with $\varepsilon_0\mu_0 = c^{-2}$, $\mu_0 = 4\pi 10^{-7}$.

1.2.2 Wave Equation

Propagation in a Vacuum The propagation of a radiation whether neutrons or x-rays, obeys a series of second order partial differential equations which can be presented in a common form. We will discuss first the case of propagation in a vacuum. Electromagnetic radiation can be represented by the 4-vector potential $A_\nu (\nu = 0, 1, 2, 3)$ defined by

$$A_0 = \Phi/c, \qquad (A_1, A_2, A_3) = \mathbf{A}, \tag{1.7}$$

where Φ is the scalar electric potential and $\mathbf{A}$ is the 3-vector potential. The 4-vector potential obeys in the Lorentz gauge and away from any charge

$$\Delta A_\nu = \varepsilon_0\mu_0 \frac{\partial^2 A_\nu}{\partial t^2}, \quad \left(\Delta = \sum_{x_i = x,y,z} \frac{\partial^2}{\partial x_i^2}, \; \varepsilon_0\mu_0 = \frac{1}{c^2} \right). \tag{1.8}$$

For a neutron of wave function Ψ, the equivalent form of (1.8) is the Schrödinger equation without any potential

$$-\frac{\hbar^2}{2m}\Delta\Psi = i\frac{\hbar\partial\Psi}{\partial t} \tag{1.9}$$

(using the convention of quantum mechanics for the sign of i, as discussed above). We shall consider essentially time independent problems and only monochromatic radiation which has frequency $\omega/2\pi$. The time variable then disappears from the equations, through use of the relations

$$\frac{1}{c^2}\frac{\partial^2}{\partial t^2} = -\frac{\omega^2}{c^2} = -k_0^2 \qquad \text{(electromagnetic field)} \tag{1.10}$$

$$i\frac{\hbar\partial}{\partial t} = \hbar\omega = \frac{\hbar^2}{2m}k_0^2 \qquad \text{(Schrödinger equation).} \tag{1.11}$$

k_0 is the wave vector in a vacuum and $\hbar\omega$ is the energy. In both cases, writing the generic field or wave function as A, yields the ***Helmholtz equation***,

$$\left(\Delta + k_0^2\right) A = 0. \tag{1.12}$$

The solutions to this equation are plane waves with the wave vector k_0.

In optics this equation is more usually expressed in terms of the electric and magnetic fields **E** and **H**, or the electric displacement and the magnetic induction $\mathbf{D}, \mathbf{B}$ rather than the vector potential A_ν. **E** is related to the potential through

$$\mathbf{E} = -\,\mathbf{grad}\,\phi - \frac{\partial \mathbf{A}}{\partial t} = -c\,\mathbf{grad}\,A_0 - \frac{\partial \mathbf{A}}{\partial t}. \tag{1.13}$$

If the gauge is so that A_0=0, **E** reduces to $-(\partial \mathbf{A}/\partial t)$. If furthermore the radiation is monochromatic, then

$$\mathbf{E} = -i\omega \mathbf{A}. \tag{1.14}$$

For a free field those conditions may usually be satisfied. Therefore, **E** and **A** being proportional to each other, most of the discussion subsequent to equation (1.12) applies to **E** as well. Nevertheless, in the presence of electric charges, all the properties of the electromagnetic field cannot be described with the generic field written as a scalar. These particular vector or tensor properties will be addressed when necessary.

Propagation in a Medium Equation (1.12) still applies in a modified form even when the radiation propagates in a homogeneous medium rather than a vacuum. All media are inhomogeneous, at least at the atomic scale, so for the moment the homogeneity will be taken as a provisional assumption whose justification will be discussed in Sect. 1.3. We also assume the isotropy of the medium, which is not the case for all materials.

In the case of the electromagnetic radiation the medium is characterised by permeabilities ε and μ that replace ε_0 and μ_0 in (1.8), although μ can usually be kept unchanged. Though the static magnetic susceptibility can take different values in various materials, we are concerned here with its value at the optical frequencies and above which is not significantly different from μ_0. In a medium equation (1.12) can be written as either,

$$\left(\Delta + k^2\right) A = 0 \qquad \left(k = nk_0, \quad n^2 = \epsilon\mu/\epsilon_0\mu_0 \simeq \epsilon/\epsilon_0\right) \tag{1.15}$$

or,

$$\left(\Delta + k_0^2 - U\right) A = 0 \qquad \left(U = k_0^2\left(1 - n^2\right)\right). \tag{1.16}$$

The first form shows that the wave vector has changed by a factor n, which is the refractive index. The second form is similar to the Schrödinger equation in the presence of a potential. Indeed in the case of the Schrödinger equation, the material can be characterised by a potential V and the equation becomes,

$$\left[-\frac{\hbar^2}{2m}\left(\Delta + k_0^2\right) + V\right]\Psi = 0 \tag{1.17}$$

which is equivalent to the previous equation, with

$$U = \frac{2m}{\hbar^2}V, \tag{1.18}$$

and again we may define a refractive index

$$n^2 = 1 - U/k_0^2 = 1 - V/\hbar\omega. \tag{1.19}$$

It is important to realise that describing the propagation in the medium by a Helmholtz equation, with just a simple change of the wave vector by a factor n or with the input of a potential U, is really just a convenience. In reality, each atom or molecule produces its own perturbation to the radiation and the overall result is not just a simple addition of those perturbations. It happens in most cases that the Helmholtz equation can be retained in the form indicated above. How n or U depends on the atomic or molecular scattering has to be established. Before addressing this question we have to give some further definitions for the intensity, current, and flux of the radiation, and to introduce the formalism of scattering length, cross-section and Green functions which help to handle the scattering phenomena.

1.2.3 Intensity, Current and Flux

The square of the modulus of the field amplitude, i.e. $|A|^2$, defines the ***intensity*** of the radiation, which is used to represent either the probability of finding a quantum of radiation in a given volume or the density of energy transported by the radiation. $|A|^2$ is also used when combined with the wave-vector direction to measure the flux density. These definitions are trivially correct in vacuum but need to be revised in a material.

The ***flux*** across a given surface is the amount of radiation, measured as an energy or a number of particles, which crosses this surface per unit time; this is a scalar quantity. The ***flux density*** or ***current density*** that we shall also call the ***flow*** is a vector. For instance, the electromagnetic energy flux density (flow) is designed by $\mathbf{S}$; the energy flux in an elementary surface $d\sigma$ is then $\mathbf{S}.d\sigma$. The ***density of energy*** u is connected to the flow by a relation which expresses the energy conservation. The amount of energy which enters a given closed volume must be equal to the variation of the energy inside that volume:

$$\frac{\partial S_x}{\partial x} + \frac{\partial S_y}{\partial y} + \frac{\partial S_z}{\partial z} + \frac{\partial u}{\partial t} = 0. \tag{1.20}$$

This equation is no longer valid when the medium is absorbing.

Note that equation (1.20) can also be written in terms of the number of particles instead of the energy; for instance this is appropriate for the case of neutrons or for electromagnetic radiation if it is quantised. The same formalism stands for the flux, the density of current and the density of particles. The dimension of the density of flux is the one of the relevant quantity (energy, number of particles or other) divided by dimension L^2T.

In the case of electromagnetic radiation, the quantities $\mathbf{E}, \mathbf{H}, \mathbf{D}$ and $\mathbf{B}$, can be used instead of $\mathbf{A}$ as discussed above and the dielectric and magnetic permeabilities, ε and μ, can be used to characterise the medium. The energy density is then given by,

$$u = \left(\varepsilon \mathbf{E}.\mathbf{E}^* + \mu \mathbf{H}.\mathbf{H}^*\right)/4. \tag{1.21}$$

For a plane wave defined by the unit vector $\hat{\mathbf{k}}$ along the wave vector,

$$\mathbf{H} = \sqrt{\varepsilon/\mu}\,\hat{\mathbf{k}} \times \mathbf{E} \tag{1.22}$$

and the energy density becomes,

$$u = \varepsilon \left|\mathbf{E}\right|^2/2. \tag{1.23}$$

The energy flow is then equal to the Poynting vector

$$\mathbf{S} = \mathbf{E} \times \mathbf{H}^*/2 = c\varepsilon\sqrt{\varepsilon_0\mu_0/\varepsilon\mu}\,\left|\mathbf{E}\right|^2\hat{\mathbf{k}}/2. \tag{1.24}$$

Note that these formulae giving u and $\mathbf{S}$ are written in terms of complex field quantities whose real part represents the physical field. The complex and the real formulations differ by a factor 1/2 in the expressions of second order in the fields.

The change in the wave vector length in going from a vacuum into a medium has been written above (1.15) in terms of the refractive index n

$$\mathbf{k} = n\mathbf{k}_0 \tag{1.25}$$

$$n = \sqrt{\varepsilon\mu/\varepsilon_0\mu_0}, \tag{1.26}$$

so that if $\mu \simeq \mu_0$ (1.15), we obtain,

$$u = n^2\left(\varepsilon_0\mu_0/\mu\right)\left|\mathbf{E}\right|^2/2 \simeq n^2\varepsilon_0\left|\mathbf{E}\right|^2/2 \tag{1.27}$$

$$\mathbf{S} = nc\left(\varepsilon_0\mu_0/\mu\right)\left|\mathbf{E}\right|^2\hat{\mathbf{k}}/2 \simeq nc\varepsilon_0\left|\mathbf{E}\right|^2\hat{\mathbf{k}}/2. \tag{1.28}$$

This shows that *the flux through a surface depends on both the amplitude* $\mathbf{E}$ *and also on the refractive index of the medium.*

A similar expression stands for neutrons (beware, in what follows as usual in neutron physics i has the opposite sign). Here the probability density ρ

and the current density $\mathbf{j}$ of particles are considered. The amplitude is the wave function Ψ.

$$\rho = |\Psi|^2 , \qquad \mathbf{j} = (\hbar i/2m)\,(\Psi \mathbf{grad}\Psi^* - \Psi^* \mathbf{grad}\Psi) . \tag{1.29}$$

For a plane wave , $\Psi_0 e^{i\mathbf{k}.\mathbf{r}}$, $\hat{\mathbf{k}}$ being the unit vector along $\mathbf{k}$

$$\rho = |\Psi_0|^2 , \qquad \mathbf{j} = (\hbar k/m)\,|\Psi_0|^2\,\hat{\mathbf{k}}. \tag{1.30}$$

Here too, the current depends on both Ψ_0 and on the medium which is characterised by a potential V and

$$\hbar^2\mathbf{k}^2/2m + V = \hbar\omega. \tag{1.31}$$

As in optics, it is possible to introduce a refractive index, which is (1.19)

$$n = \sqrt{\frac{\hbar\omega - V}{\hbar\omega}} \tag{1.32}$$

and which from k_0 gives the length of the wave vector $\mathbf{k}$. Then

$$\mathbf{j} = (n\hbar k_0/m)\,|\Psi_0|^2\,\hat{\mathbf{k}}. \tag{1.33}$$

The above formulae are valid when the medium is isotropic. When it is anisotropic the flow of energy and the current are affected. In the electromagnetic case the direction of the flow does not always coincide with the direction of the wave vector.

Exercise 1.2.1. A beam impinging on a surface gives rise to a reflected and a transmitted beam. The amplitudes of these beams are given by the Fresnel formulae (see section 3.1.2). As assumed above, the two media are not absorbing. Check the conservation of the flux, at least for the (s) polarisation.

Exercise 1.2.2. Let us consider a wave function Ψ, such as an evanescent wave $\Psi_0 e^{(ik_x x - k_z z)}$. Calculate the current density.

1.2.4 Scattering Length and Cross-Sections

Let us consider an isolated scattering object (molecule, atom, electron), fully immersed in the field of an incident wave. The object reemits part of the incident radiation. We start with the assumption that its dimensions are small compared to the wavelength so that the scattered amplitude is the same in all the directions; for an extended object instead, direction-dependent phase shifts would appear between the scattered amplitudes coming from different regions in the sample. When examining the scattered amplitude at large distances r from the object, simple arguments yield the following expression of the scattered amplitude (see also the appendix 1.A to this chapter)

$$A_{\rm sc} = -A_{\rm in}\, b\, \frac{e^{-ikr}}{r}. \tag{1.34}$$

Indeed this function which has the spherical symmetry (k and r are scalars), is proportional to the incident amplitude A_{in} and has locally the right wavelength $2\pi/k$; the decay as a function of the inverse of r guarantees the conservation of the total flux since the related intensity decays as the inverse of the surface of a sphere of radius r. The remaining coefficient b has the dimension of a length; this coefficient characterises the scattering power of the sample and is the so-called ***scattering length***. The notation b is rather used in the context of neutron scattering. Here we adopt it for x-rays as well. To be fully consistent with this notation we keep, as a mere convention, the minus sign in the definition of b. With this sign, the b value is positive for neutrons with most nuclei, and also for x-ray Thomson scattering. This length can have a complex value, since the wave can undergo a phase shift during the interaction process; we shall see that in our case if the sample is not absorbing then b is nearly real. A more rigorous justification of the expression (1.34) will be given in the next section.

To justify that b has the dimension of a length, we have considered the scattered flux. The ratio of this flux to the incident one per unit of surface (flux density, or current) has the dimension of a surface and is equal to

$$\sigma_{\text{scat,tot}} = 4\pi \left|b\right|^2 . \tag{1.35}$$

This is the so-called ***total scattering cross-section*** The scattered flux in the whole space is then equal to the one received by a surface equal to $\sigma_{\text{scat,tot}}$ which would be placed normal to the incident beam.

In general, with an extended object, the scattering depends on the direction of observation, defined by a unit vector $\hat{\mathbf{u}}$, so that b, which also depends on $\hat{\mathbf{u}}$, is written $b(\hat{\mathbf{u}})$. Therefore it is useful to define a cross-section for this particular direction that is called the ***differential scattering cross-section***

$$(d\sigma/d\Omega)\,(\hat{\mathbf{u}}) = \left|b(\hat{\mathbf{u}})\right|^2 \tag{1.36}$$

which is equal to the measured flux in the solid angle $d\Omega$ directed towards $\hat{\mathbf{u}}$, for a unit incident flux (Fig. 1.1). In this case the definition (1.35) is replaced by

$$\sigma_{\text{scat,tot}} = \iint \left|b(\hat{\mathbf{u}})\right|^2 d\Omega \tag{1.37}$$

where the integration is carried out over all the directions defined by $\hat{\mathbf{u}}$.

Any object (atom, molecule) also absorbs some part of the incident radiation without scattering it. Therefore one has to define the so-called ***cross-section of absorption***, σ_{abs}, equal to the ratio of the absorbed flux to the incident flux density. We have used in (1.35) and (1.37) a somewhat clumsy notation ($\sigma_{\text{scat,tot}}$) to recall that it is a scattering cross-section; indeed the ***total cross-section*** appellation, σ_{tot}, is also used to define the sum of the

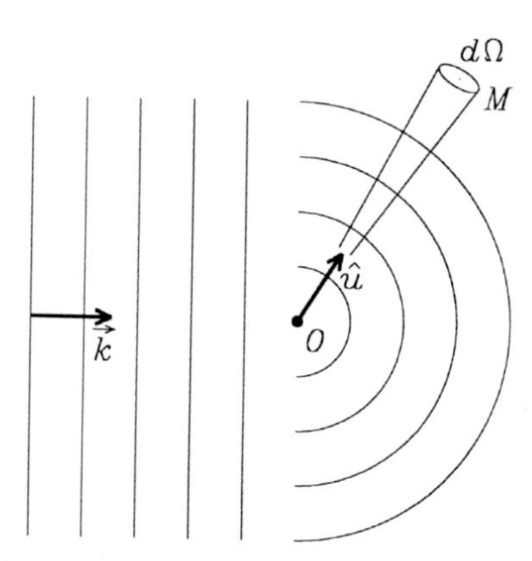

Fig. 1.1. Definition of the scattering length $b(\hat{\mathbf{u}})$ *and of the differential scattering cross-section* $(d\sigma/d\Omega)(\hat{\mathbf{u}})$. *The incident plane wave is* $A_{\text{in}}e^{-i\mathbf{k}.\mathbf{r}}$ *and the scattered wave* $A_{in}(b(\hat{\mathbf{u}})/OM)e^{-ikOM}$. *This last expression gives a well-defined flux in the cone OM, whatever the distance OM. The scattering length and the differential scattering cross-section in the direction* $\hat{\mathbf{u}}$ *are respectively* $b(\hat{\mathbf{u}})$ *and* $|b(\hat{\mathbf{u}})|^2$

cross-section concerning all the interaction processes (absorption, elastic and inelastic scattering); it is the whole relative flux picked up by the object,

$$\sigma_{\text{tot}} = \sigma_{\text{scat,tot}} + \sigma_{\text{abs}}. \tag{1.38}$$

1.2.5 The Use of Green Functions

The scattering amplitude b in equation (1.34) has not been introduced very rigorously and it is possible to define it more formally. The field scattered by a point like object obeys the wave equation (1.15) everywhere except at the center of the object, which is both the source and a singular point. The simplest mathematical singularity is the Dirac δ function. The ***Green function*** of equation (1.15), $G(\mathbf{r})$ is a solution of the equation

$$\left(\Delta + k^2\right) G(\mathbf{r}) = \delta(\mathbf{r}). \tag{1.39}$$

Physically, $G(\mathbf{r})$ represents the field emitted by the source normalised to unity. More generally, any partial derivative equation which is homogeneous in A such as

$$DA(x) = 0 \tag{1.40}$$

(here D represents a sum of differential operators with constant coefficients plus a constant term, and x is a scalar or a vector variable), admits Green functions G which satisfy

$$DG(x) = \delta(x). \tag{1.41}$$

A common application of Green functions is the resolution of non homogeneous partial derivative equations. For example, if $G(x)$ is a Green function and $A_0(x)$ is any of the solutions of the homogeneous equation, the equation

$$DA(x) = f(x) \tag{1.42}$$

admits the following solutions

$$A(x) = A_0(x) + \int G(x - x')f(x')dx'. \tag{1.43}$$

This can be shown by substitution into (1.42) and use of the equation,

$$f(x) = \int \delta(x - x')f(x')dx', \tag{1.44}$$

and by finally applying (1.41).

Let us now check that the diverging wave (1.34) (or the converging wave having the opposite sign for k) is indeed, to a certain coefficient, a Green function solution of (1.39). Due to the spherical symmetry, it is worth using the spherical coordinates $r, \hat{\mathbf{u}}$ ($\mathbf{r} = r\hat{\mathbf{u}}$; $\hat{\mathbf{u}}$ is defined by the polar angles θ, ϕ). The differential operators yield

$$\mathbf{grad} = \hat{\mathbf{u}}\frac{\partial}{\partial r} + \frac{1}{r}\mathbf{d}_1(\theta, \phi), \qquad \hat{\mathbf{u}}.\mathbf{d}_1 = 0 \tag{1.45}$$

$$\Delta = \frac{\partial^2}{\partial r^2} + \frac{2}{r}\frac{\partial}{\partial r} + \frac{1}{r^2}d_2(\theta, \phi), \tag{1.46}$$

where $\mathbf{d}_1$ and d_2 are differential operators relative to (θ, ϕ) and $\mathbf{d}_1$ is a vector perpendicular to $\hat{\mathbf{u}}$. For the moment, the above expressions are sufficient since we only use functions having the spherical symmetry and therefore $\mathbf{d}_1$ and d_2 vanish; we shall also use these expressions for less symmetrical functions, but in such a case we shall only consider the asymptotic behavior at large values of r where the exponent of $1/r$ is sufficient to make $\mathbf{d}_1$ and d_2 negligible. We then have exactly

$$\left(\Delta + k^2\right)\frac{e^{\pm ikr}}{r} = 0 \qquad \text{for } r \neq 0. \tag{1.47}$$

At $r = 0$, we must compare the singularity with $\delta(r)$. It is possible to integrate the left hand side of (1.47), inside a sphere of radius r_0 centered at the origin. Indeed from the definition of G, the integral of $(\Delta + k^2)\ G(r)$ must be equal to 1 when performed over the whole volume including the origin. This calculation is proposed in the exercise 1.2.3 and yields -4π. The Green function of the three dimensional Helmholtz equation is then

$$G_\pm(\mathbf{r}) = -\frac{1}{4\pi r}e^{\pm ikr}. \tag{1.48}$$

It is also useful to express the Green function in one dimension. Indeed in specular reflectivity some problems can be solved in one dimension. A similar calculation yields

$$G_{1d,\pm}(\mathbf{r}) = -\frac{\pm i}{2k} e^{\pm ikr}. \tag{1.49}$$

The Green function of Helmholtz equation in two dimensions can be expressed with the help of Bessel functions. The asymptotic form at large r is yet harmonic, with a $r^{-0.5}$ decay and an additional phase shift equal to $\pm(\pi/4)$.

Exercise 1.2.3. Calculate in three dimensions

$$\int_{r<r_0} (\Delta + k^2)(1/r)e^{\pm\, ikr}\, d\mathbf{r}.$$

Hints: The integral of the first term, Δ..., can be transformed into the integral of the gradient over the sphere of radius $r = r_0$; the integral of the second, k^2..., can be successively performed over spheres and then over r. Note that the independence of the result with respect to r_0 yields (1.47) and is sufficient to prove that this is a Green function.

1.2.6 Green Functions: the Case of the Electromagnetic Field

While the Green functions of the Helmholtz equation are valid for scalar fields as for instance the neutron wave function, the case of the electromagnetic field is more complicated. Not only the field is a vector (if the potential is used) but also, the simplest sources are vibrating dipoles which are represented by vectors and which cannot be described by a simple δ function.

The 4-vectors A_ν and j_ν represent respectively the potential, and the current-charge density, as follows

$$A_0 = \Phi/c, \qquad (A_1, A_2, A_3) = \mathbf{A} \tag{1.50}$$

$$j_0 = c\rho, \qquad (j_1, j_2, j_3) = \mathbf{j}. \tag{1.51}$$

Φ is the scalar electric potential as previously defined (1.7), ρ the charge density and $\mathbf{j}$ the electric current density. Since j_ν describes the charge motion it must fulfill the conservation relationship

$$div\,\mathbf{j} + \frac{1}{c}\frac{\partial j_0}{\partial t} = 0. \tag{1.52}$$

We shall have to integrate the current density over a volume,

$$\mathbf{J} = \int \mathbf{j}(\mathbf{r}) d\mathbf{r} \tag{1.53}$$

to get a 3-vector $\mathbf{J}$. If we consider a conductor in which there is a current, $\mathbf{J}$ is the product of the current intensity by the vector identified to a portion of the conductor. If we consider a moving charge (an electron for instance), $\mathbf{J}$ is the product of the charge by the velocity and if it is a vibrating dipole of amplitude $\mathbf{d}$ such as $\mathbf{d}e^{i\omega t}$, $\mathbf{J}(t)$ yields $i\omega\mathbf{d}e^{i\omega t}$. *Such a vibrating dipole, if infinitesimally small, is the simplest radiating point source.* It is characterised by the following current-charge density, which fulfills the conservation law:

$$j_0(\mathbf{r},t) = \frac{ic}{\omega} div\,(\mathbf{J}(t)\delta(\mathbf{r})) = \frac{ic}{\omega}\mathbf{J}(t).\mathbf{grad}\,\delta(\mathbf{r}) \tag{1.54}$$

$$\mathbf{j}(\mathbf{r},t) = \mathbf{J}(t)\delta(\mathbf{r}). \tag{1.55}$$

The charge density j_0/c has the form of the derivative in the direction $\mathbf{J}$, of the scalar function δ.[1] We write it into two different forms which both are useful.

In the presence of the current-charge density j_ν , the potential A_ν (written with help of the Lorentz gauge) verifies, instead of the four homogeneous equations (1.8), the inhomogeneous ones

$$\Delta A_\nu - \frac{1}{c^2}\frac{\partial^2 A_\nu}{\partial t^2} = -\frac{j_\nu}{\varepsilon_0 c^2}. \tag{1.56}$$

We take as j_ν the dipole just described. We then keep as the useful solutions A_ν those which have the same oscillating time dependence as j_ν. When ω is replaced by ck, (1.56) transforms into the inhomogeneous Helmholtz equations

$$\left(\Delta + k^2\right) A_0\,(\mathbf{r},t) = -\frac{i}{\varepsilon_0 c^2 k}\,\mathbf{J}(t).\mathbf{grad}\,\delta(\mathbf{r}) \tag{1.57}$$

$$\left(\Delta + k^2\right) \mathbf{A}\,(\mathbf{r},t) = -\frac{\mathbf{J}(t)}{\varepsilon_0 c^2}\,\delta(\mathbf{r}). \tag{1.58}$$

The solution of the second equation is a Green function $G_\pm(\mathbf{r})$ (1.48). The first one can be solved by the use of the method proposed in (1.43). The outgoing solution is

$$A_0\,(\mathbf{r},t) = -\frac{i}{\varepsilon_0 c^2 k}\,\mathbf{J}(t).\mathbf{grad}\,G_-(\mathbf{r}) \tag{1.59}$$

$$\mathbf{A}\,(\mathbf{r},t) = -\frac{\mathbf{J}(t)}{\varepsilon_0 c^2}\,G_-(\mathbf{r}), \tag{1.60}$$

where G_- is given by (1.48). Up to a constant factor $\mid \mathbf{J} \mid /\varepsilon_0 c^2$, (1.59) and (1.60) are *the equivalent for the electromagnetic potential of the Green*

[1] This idealised dipole, isolated in a vacuum, can be used to represent what happens at a microscopic scale in a dielectric material in the range of a few atoms (for x-rays and any material the relevant scale lies inside a single atom). Once the average has been made over a larger volume, these microscopic currents disappear from the equations. They are implictly accounted for through the dielectric constant and the new fields $\mathbf{D}$ and $\mathbf{H}$, otherwise equal to $\epsilon_0\mathbf{E}$ and $\mathbf{B}/\mu_0$. This is the point of view of chapter 4.

function for the scalar field. These particular expressions are due to both the vector character of the field and to the electric dipolar character of the source. Other kinds of sources exist that we shall not describe here, as for example magnetic dipoles, or multipoles of higher order. One can also imagine the scalar field of multipole sources.

For practical purpose we may need the electric field $\mathbf{E}$. Following (1.13)

$$\mathbf{E}(\mathbf{r},t) = \frac{i\omega}{\varepsilon_0 c^2}\left[\mathbf{J}(t)G_-(\mathbf{r}) + \frac{1}{k^2}\mathbf{grad}\,(\mathbf{J}(t).\mathbf{grad}\,G_-(\mathbf{r}))\right]. \tag{1.61}$$

The second derivative $\mathbf{grad}\,\mathbf{J}.\mathbf{grad}$ can be handled in two different ways. First, since we often consider the radiated field far from the source, we look for an asymptotic value valid when $kr \gg 1$. For this, the expression of the gradient (1.45) is used, but only the derivative according to r is kept, and in the derivative of $G_- \propto e^{-ikr}/r$, only the derivative of e^{-ikr} is calculated. All the other derivatives are of higher order in $1/kr$. Thus we can write ($\hat{\mathbf{u}}$ is the unit vector along $\mathbf{r}$)

$$\begin{aligned}\mathbf{grad}\,(\mathbf{J}(t).\mathbf{grad}\,G_-(\mathbf{r})) &\underset{kr\to\infty}{\sim} -\mathbf{grad}\,(\mathbf{J}(t).\hat{\mathbf{u}}\,ik\,G_-(\mathbf{r}))\\ &\underset{kr\to\infty}{\sim} -k^2(\mathbf{J}(t).\hat{\mathbf{u}})\hat{\mathbf{u}}\,G_-(\mathbf{r}),\end{aligned} \tag{1.62}$$

and one can recognize in this expression the projection of $\mathbf{J}$ on the vector $\mathbf{r}$. The asymptotic form of $\mathbf{E}(\mathbf{r},t)$ is,

$$\mathbf{E}(\mathbf{r},t) \underset{kr\to\infty}{\sim} -\left[\mathbf{J}(t) - (\mathbf{J}(t).\hat{\mathbf{u}})\hat{\mathbf{u}}\right]\frac{i\omega e^{-ikr}}{4\pi\varepsilon_0 c^2 r}, \tag{1.63}$$

i.e. *the scalar Green function multiplied by the component of the current normal to* $\hat{\mathbf{u}}$ *and by* $i\omega/\epsilon_0 c^2$.

An other way to transform expression (1.61), now without any approximation, relies on the alternative form $\mathbf{grad}\,div(\mathbf{J}G_-)$ (this equivalence is given by (1.54)) for the second derivative term. The following equation is also identically valid

$$\mathbf{grad}\,div \equiv \Delta + \mathbf{curl}\,\mathbf{curl}, \tag{1.64}$$

and since G_- is solution of Helmholtz equation away from the origin, Δ may be replaced by $-k^2$. As a result we have

$$\mathbf{E}(\mathbf{r},t) = \mathbf{curl}\,\mathbf{curl}\left(\mathbf{J}(t)\,\frac{ie^{-ikr}}{4\pi\varepsilon_0\omega r}\right) \quad \text{for } r \neq 0. \tag{1.65}$$

1.3 From the Scattering by an Object to the Propagation in a Medium

1.3.1 Introduction

The amplitude of a progressive plane wave in a vacuum, at some fixed time, is $e^{-i\mathbf{k}_0 \cdot \mathbf{r}}$. If the vacuum is replaced by an assembly of scattering objects, the field is modified by the scattered waves. We have said that, when the medium is sufficiently homogeneous, the result is still a plane wave. The wave in the medium has a different wavelength and the amplitude becomes $e^{-in\mathbf{k}_0 \cdot \mathbf{r}}$. We shall derive this result and show what is meant by the term "sufficiently homogeneous". In addition, we shall link the optical index n to the scattering lengths of the objects.

A way to approach that problem is to search directly for a partial derivative equation which is satisfied by the total field (incident field A_{in} plus scattered A_{sc}). From the definition of the scattering length b (1.34) and the Green function (1.39, 1.48), and from the Helmholtz equation (1.12) satisfied by A_{in}, it may be shown that the following equation is verified if b is independent of the scattering direction

$$[\Delta + k_0^2 - 4\pi\rho(\mathbf{r})b](A_{\text{in}} + A_{\text{sc}}) = 0. \tag{1.66}$$

Here $\rho(\mathbf{r})$ is the density of objects of scattering length b. This question is discussed in the Appendix 1.A to the present Chapter, about the Born approximation. If in this equation $\rho(\mathbf{r})b$ can be replaced by its space average ρb, we obtain for the refractive index n.

$$n^2 = 1 - \left(4\pi/k_0^2\right) \rho b. \tag{1.67}$$

The equation (1.66) is exact for a scalar field with b independent of the direction. It turns out that in the case of neutrons and x-rays, because they give small values of ρb, the average can be made safely in most cases and only the forward scattering is relevant (see next section). With these small values, the vector character of the electromagnetic radiation does not make any difference and the same formula (1.67) applies to x-rays as well as to neutrons. Also its last term is small enough to make equivalent the writing

$$n \simeq 1 - \left(2\pi/k_0^2\right) \rho b. \tag{1.68}$$

A discussion specific to the x-ray case is found in the treatise of Landau and Lifshitz [1], section 97.

Although the method that we have just sketched is straightforward and gives the right answer, *once the approximations are accepted as valid*, we shall discuss the problem along another line. It will give a more intuitive insight to the building up of the total field from the scattered field. It will also give more handles to grasp and understand the role of the smallness of ρb and of the homogeneity, and to discuss the approximations. For ρb to be considered as "small" we shall see that the Born approximation must apply to a scattering volume whose dimensions are at least of the order of the wavelength and that great simplifications then follow.

We start with the optical theorem (section 1.3.2), which exactly links the total cross-section of an isolated object to its forward scattering length. We then propose an extension of the optical theorem, valid under approximations which are discussed with some details. This extension yields the formula (1.68). The equivalence between those approximations and the Born approximation is commented (section 1.3.3). Finally (section 1.3.4) we shall discuss briefly the case of a strong interaction where these approximations are no longer valid.

1.3.2 The Optical Theorem and its Extensions

The Optical Theorem A relation, known as the optical theorem, exists between the total cross-section of an object and its forward scattering amplitude. It is worth examining this relation because it looks like a partial solution to the problem of finding the optical index of a medium made of such objects. Indeed the total cross-section of the objects (atoms or molecules) which constitute a medium, approximately yields the attenuation of that medium, then the attenuation is linked to the imaginary part of the index. What we call attenuation includes the absorption and the loss of radiation due to scattering out of the direction of propagation. The relation between the molecular total cross-section and the attenuation is only approximate because the scattering cross-sections of all the molecules cannot be simply summed, as can be the absorption.

The ***optical theorem*** exactly relates the total cross-section (absorption and scattering) σ_{tot} of an isolated object, with the imaginary part of the amplitude that this object scatters in the forward direction, i.e. $\mathcal{I}\mathrm{m}\,[b(0)]$.

$$\sigma_{\text{tot}} = 2\lambda\,\mathcal{I}\mathrm{m}\,[b(0)]\,. \tag{1.69}$$

Here b may depend on the scattering direction, but only its value at zero angle is relevant. The proof relies on the use of the Green theorem applied to a volume surrounding the object (see reference [2], section 9.14); we stress that no approximation is made. When the field, such as an electromagnetic field, has several components, $b(0)$ represents the scattering having the same polarisation as the incident wave.

We cannot make use of the optical theorem as it is. First because it gives no access to the real part of the index, and second because of the approximation made when going from the cross-section of the objects to the attenuation of the medium. Its validity can hardly be discussed directly. We present a similar relation instead, which yields the complete refractive index of a medium, with its real and imaginary part. Again it is related to the scattering in the forward direction. That calculation is not exact, but relies on the smallness of ρb. We shall call this condition, "the weak interaction". It is equivalent to saying that the index is close to 1. In fact, for x-rays and neutrons, the difference to 1 is of the order of 10^{-5}, or even less. This is not true for visible light, and in this case other formulations must be used. We write below the formulae for a scalar field. Under the condition that the interaction is weak, those will also be valid for the electromagnetic field. The demonstration follows Jackson [2].

The Amplitude Scattered by a Planar Assembly of Scattering Objects Let us consider a population of scattering objects homogeneously located in the surface of a plane P normal to the direction of propagation of the incident plane wave (Fig. 1.2). We shall consider the amplitude of the wave at a point M, far enough behind this plane but not at an infinite distance (like in Fresnel diffraction). The field is supposed to be a scalar.

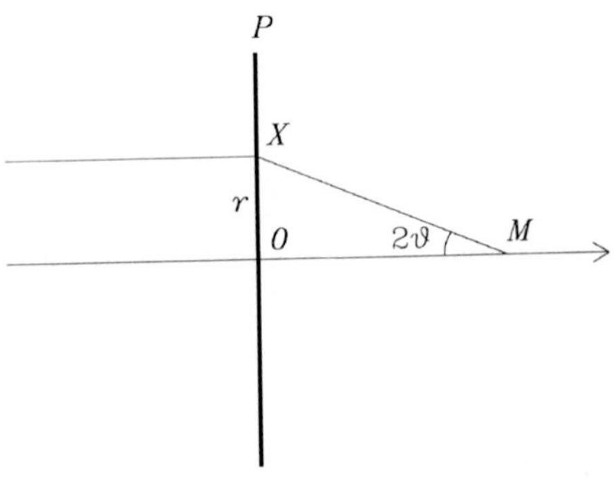

Fig. 1.2. A plane wave coming from the left encounters a plane P *containing an array of scattering objects. The axis OM is normal to the wave planes. The value of the field which is modified by the scattering objects, will be calculated at the point M*

Let ρ_s be the surface density of the scattering objects and $A(O)$ the incident amplitude of the wave at O, that is in the plane P. The objects located within the surface ds around a point X in the plane P will contribute an amplitude at the point M given by,

$$dA_X(M) = -A(O)\rho_s b(\widehat{XM})\,\frac{e^{-ik_0 XM}}{XM}\,ds, \tag{1.70}$$

where $b(XM)$ is the scattering length in the direction $\widehat{XM}$. To integrate this expression over the whole plane P, one first integrates around a ring of radius r and of thickness dr centered in O. The mean value of $b(\widehat{XM})$ around this ring is called $b(2\theta)$ (2θ being the angle OMX). The elementary amplitude scattered by this elementary ring is

$$dA_r\,(M) = -A(O)\rho_s b(2\theta) 2\pi r\,dr\,\frac{e^{-ik_0 XM}}{XM}. \tag{1.71}$$

This must be integrated over r but since

$$XM^2 = OM^2 + r^2, \qquad \text{whence} \quad XM\,dXM = r\,dr, \tag{1.72}$$

it is possible to integrate over XM instead of r and the amplitude $A_{\rm sc}(M)$ scattered by the entire plane P, appears as

$$A_{\rm sc}(M) = -A(O)\rho_s 2\pi \int_{OM}^{\infty} b\,(2\theta)\,e^{-ik_0 XM}\,dXM. \tag{1.73}$$

If the point M is far enough behind the plane P, the function to be integrated oscillates very quickly as soon as θ differs from 0; the integration easily gives the result (see below Eq. (1.78)), since b remains almost constant and equal to its value at $\theta = 0$. This can be rigorously proved by integrating by parts:

$$A_{\rm sc}(M) = -\,(2\pi i/k_0)\,A(O)\rho_s \left\{ \left[b\,(2\theta)\,e^{-ik_0 XM} \right]_{OM}^{\infty} - \int_{OM}^{\infty} e^{-ik_0 XM}\,\frac{db\,(2\theta)}{dXM}\,dXM \right\}. \tag{1.74}$$

If $b(2\theta)$ varies slowly with θ, it clearly appears that the second term is of the same order as the first one multiplied by λ/OM. This comes from

$$\frac{d}{dXM} = -\frac{OM}{XM^2}\,\frac{d}{d\cos 2\theta} \tag{1.75}$$

which allows to bound the second term by

$$\int_{OM}^{\infty} \frac{db\,(2\theta)}{OM\,d\cos 2\theta}\,\left(e^{-ik_0 XM}\right)\,dXM. \tag{1.76}$$

This integral which contains the factor $(1/k_0)OM$, is very small if $OM \gg \lambda$ and if $b(2\theta)$ presents a non singular extremum at the origin. The second term in (1.74) is thus negligible on this condition.

As for the first term, it quickly oscillates around zero when XM tends towards the infinite value of the upper bound so that one can make the following approximation

$$b\,(2\theta)\,e^{-ik_0 XM} \underset{XM \to \infty}{\sim} 0. \tag{1.77}$$

It is worth noting that to average those oscillations, the upper bound value for the radius r of the ring used in integrating over the plane P should be much larger than a characteristic length. This length is the so-called first Fresnel zone radius which is of the order of $(\lambda OM)^{1/2}$.

Finally the forward scattered amplitude becomes

$$A_{\mathrm{sc}}(M) = iA(O)\lambda\rho_s b(0)e^{-ik_0 OM}. \tag{1.78}$$

The forward scattered field adds to the incident field $A(O)e^{-ik_0 OM}$ in M and yields a total field $A(M)$

$$A(M) = A(O)e^{-ik_0 OM} + A_{\mathrm{sc}}(M) = A(O)\left(1 + i\lambda\rho_s b(0)\right)e^{-ik_0 OM}. \tag{1.79}$$

If we now consider instead of a plane a thin layer of thickness dx the above calculation remains valid provided the surface density ρ_s is related to the volume density ρ_v by

$$\rho_s = \rho_v\, dx. \tag{1.80}$$

The total field for such a layer becomes

$$A(M) = A(O)\left(1 + i\lambda\rho_v b(0)\, dx\right)\, e^{-ik_0 OM}. \tag{1.81}$$

It is possible to deduce the optical theorem from this relation but this will be presented later. Note that the amplitude in M is outphased by $\pi/2$ relative to the one scattered by a volume element; that phase difference results from the summation of amplitudes in Fresnel diffraction.

The Propagation of a Wave in a Homogenous Population of Scattering Objects Let us now consider the plane P as an infinitesimal small layer of thickness dx made of a medium of index n. The wave vector in the medium is nk_0. If the point O is located at the entrance of the layer, a plane wave which has crossed the thickness dx in the medium of index n has an amplitude at the point M given by

$$A(O)e^{-ink_0 dx}e^{-ik_0(OM-dx)} \approx A(O)\left(1 - i\left(n-1\right)k_0\, dx\right)e^{-ik_0 OM}. \tag{1.82}$$

The comparison of (1.82) with (1.81) shows that

$$n = 1 - \lambda^2\rho_v b(0)/2\pi = 1 - \left(2\pi/k_0^2\right)\rho_v b(0). \tag{1.83}$$

Equations (1.82) and (1.81) are schematically represented in the complex plane in Fig. 1.3. As shown in this figure, the imaginary part of $b(0)$ or $(n-1)$ modifies the absolute value of the field amplitude in M, whereas the real part modifies its phase.

Equation (1.83) links the scattering by elementary objects to the propagation in the medium which is considered to be continuous. It is an extension of the optical theorem. Indeed the imaginary part of n, i.e. β, describes the attenuation of the radiation in the medium and $2\beta k_0$ is the absorption coefficient μ:

$$|A(M)|^2 \underset{dx\to 0}{\sim} |A(O)|^2\left(1 - 2\beta k_0\, dx\right) = |A(O)|^2\left(1 - \sigma_{\mathrm{atten}}\rho_v\, dx\right), \tag{1.84}$$

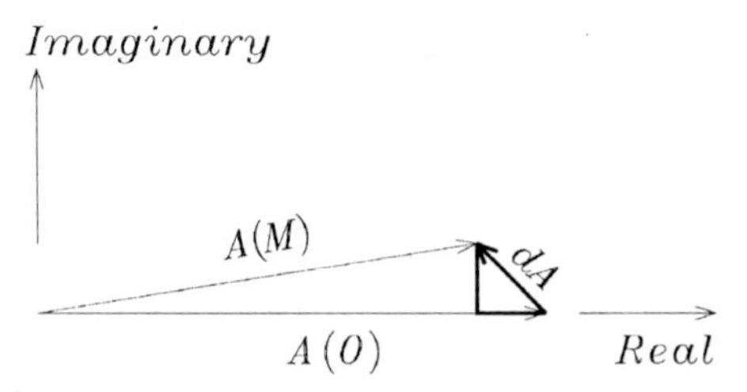

Fig. 1.3. Representation of the amplitude of the field in the complex plane. Up to the common factor e^{-ik_0OM}, *the total field* $A(M)$ *is the sum of the incident field* $A(O)$ *and of an infinitesimal field* $dA = i\lambda\rho_s b(0)A(O)$ *in the first calculation, and* $dA = -i(n-1)k_0\,dx A_0$ *in the second one. The component of the field* dA *associated with the real part of* $b(0)$ *or of* $(n-1)$ *is turned by* $\pi/2$ *from the incident field. This produces a phase shift of the total field. On the other hand, the imaginary part of* $b(0)$ *or of* $(n-1)$ *decreases the amplitude of the total field*

where σ_{atten} is the attenuation cross-section of these objects in this medium. It is "almost" the optical theorem (1.69). The "almost" means replacing σ_{tot}, the total cross-section of the isolated objects, by their attenuation cross-section in this particular medium, σ_{atten}.

All the derivations above consider the field as a scalar. Under the approximations made here ($XM \to \infty$) they can be extended to the electromagnetic field. Indeed, only the forward scattering, which is usually independent of polarisation and conserving it, is retained. Beyond the above approximations, one must take into account all the scattering directions and the scalar and vector fields display different behaviors.

About the Approximations We must now discuss the approximations which are made. The argument relies on the equality of the amplitudes calculated from the index (1.82) and from the scattering (1.81). On the one hand (1.82) is valid if

$$(n-1)\,k_0\,dx \ll 1$$

and on the other hand (1.81) holds if

$$OM \gg \lambda.$$

We are going to show that in the case of a medium of finite thickness x these two conditions can hold simultaneously under some restrictions on n or $b\,(0)$. An arbitrary thickness x of the material may be divided into thin layers of thickness dx. Let O and M be the points taken at the entrance and at the exit of layer j, such as $dx = OM$ (see Fig. 1.4).

Let us show that the amplitude at M can be deduced from the one at O by (1.82) as far as $\mid n-1 \mid k_0OM \ll 1$. Since OM does not appear into (1.81) except in the global phase factor, the expression of the total field

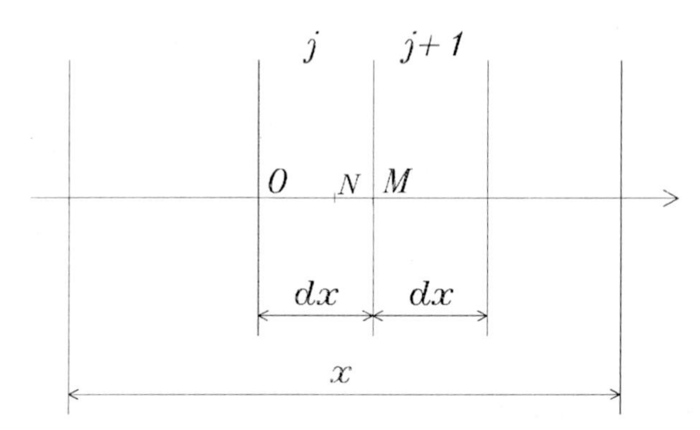

Fig. 1.4. The point M is located at the border of the two layers (j, j+1) of the material. We assume that the condition $L_e \gg OM \gg \lambda$ *(see text) is satisfied. Note that if $OM \gg \lambda$ then $NM \gg \lambda$ for nearly all N. Then the amplitude at M only comes from the layer j, and is given by (1.81) (with ρ_s the surface density of the layer). Since $L_e \gg OM$,* $(n-1)k_0\,dx$ is infinitesimally small, the approximation (1.82) does apply, and the material has an index given by (1.83)

holds whatever the value of OM even if $OM = dx$. Yet, for any point N located between the points O and M, the condition $NM \gg \lambda$ has to be verified. Although there are some points N very close to M which do not verify $NM \gg \lambda$, most of the points of the layer are at a distance from M larger than λ since the initial condition was $OM \gg \lambda$.

In order for the amplitude at M to be given by (1.81), it is also necessary for the back scattering coming from the layers $j+1$ located behind the point M to be negligible. The different points of that layer scatter towards M with different phase shifts. It is possible to show that the ratio of the sum of scattered amplitudes in the backward direction to the one in the forward direction by layer j is of the order of λ/OM. Therefore the condition $OM \gg \lambda$ is sufficient for equation (1.81) to be valid.

Equations (1.81) and (1.82) are simultaneously satisfied if $OM \gg \lambda$ and $|n-1|\,k_0 OM \ll 1$. The combination of these two inequalities yields

$$\lambda \ll OM \ll \frac{\lambda}{2\pi\,|n-1|}. \tag{1.85}$$

This makes obvious that a length L_e defined as

$$L_e = \frac{\lambda}{2\pi\,|n-1|} = \frac{V_a}{\lambda\,|b(0)|} \qquad \text{with} \quad V_a = 1/\rho_v\,, \tag{1.86}$$

must play an important role in the optical properties of the medium. It appears also in any scattering process. With reference to the dynamical theory of x-ray diffraction we shall call this length the ***extinction length***. We should remind ourselves that *this length must be much larger than the wavelength*, or equivalently that $| n - 1 |$ must be very small (actually it is 10^{-5} or less for x-rays and neutrons).

Another condition must also be discussed. We have replaced the sum over the atoms or molecules by integrals to make the calculation of the scattered amplitudes. This is allowed only if the intermolecular distances and more generally the dimensions of heterogeneity are smaller than the range of integration. In the longitudinal direction that range is OM and in the transverse direction the characteristic length for the integration, is the radius of the first Fresnel zone, which is of the order of $(OM\lambda)^{1/2}$. The volume V_{aver} which is large enough to represent on the average the material (V_{aver} is defined as a volume larger than the heterogeneity) must be less than $OM^2\lambda$. As the inequality $OM \ll L_e$ must stand, V_{aver} *must be very small compared to* $L_e^2\lambda = V_a^2/\lambda \mid b(0) \mid^2$, where V_a is the volume of the unit (namely the atom) of scattering length b. If this condition $V_{\text{aver}} \ll L_e^2\lambda$ is not fulfilled, the field in the material can fluctuate strongly around a value given by the ideal plane wave function and an important fraction of the radiation may be scattered out of the propagation direction.

In condensed matter and for x-rays of energy 10 keV or thermal neutrons, V_a is of the order of a few λ^3. For x-rays and for $Z \simeq 15$, we have $L_e/\lambda \simeq 10^4$. For neutrons this ratio is about ten times larger. The condition $L_e \gg \lambda$ is thus well satisfied. If V_a is of the order of λ^3 , the volume $L_e^2\lambda$ involved in the inequality $V_{\text{aver}} \ll L_e^2\lambda$ is of the order of $10^8\, V_a$ ($10^{10}\, V_a$ for neutrons). The last inequality can be easily checked in reasonably homogeneous materials. Provided these conditions are satisfied, the wave propagation according to the continuous medium field equations with the index given by (1.83) is consequently valid.

1.3.3 The Extinction Length and the Born Approximation

The condition (1.85) ($L_e \gg \lambda$) shows that the extinction length plays a major role in the evaluation of the strength of the interaction of a radiation with matter. When the radiation has travelled a distance L_e, it begins to undergo a measurable phase shift of exactly one radian because of the crossed material; the scattering produced then becomes substantial. The results may be qualitatively different when the thickness of the material which has been crossed is smaller or bigger than L_e. For x-rays of energy 10 keV, the extinction length is of the order of a micron ($| n - 1 | < 10^{-5}$), and it is one order of magnitude larger for neutrons.

The approximation which has been made to relate n with b is connected to the first Born approximation. We have used a single scattering to produce the plane wave propagating in the medium. In addition the extinction length allows us to decide whether the Born approximation is valid for a given situation. When $L_e \gg \lambda$, the criterion is that the path travelled in the volume of the material giving rise to a coherent scattering must be less than L_e. The kinematical theory of diffraction by crystals (equivalent to the Born approximation) is commonly used because the volume of the perfect crystal (coherently scattering) is often smaller than one micron cube. The property expressed in equation (1.85), which tells us that the extinction length is much larger than the wavelength, also presents beneficial effects for the physics of x-rays and neutrons. It is associated with the fact that even if the kinematical theory is no longer valid, in perfect crystals, the dynamical theory remains calculable. In visible optics, where this condition is not valid, the diffraction equations are most often not exactly solvable.

In the domain of reflection in grazing incidence on a surface, the extinction length plays a major role. First it is related to the critical angle of total external reflection, discussed in chapter 3. Indeed the following relation stands

$$1/\left|\mathbf{q}_c\right| = L_e \sin\theta_c/2\,(1+n) \approx L_e \sin\theta_c/4, \tag{1.87}$$

where $|\ \mathbf{q}_c\ |$ is the scattering wave vector transfer corresponding to the critical specular reflection at the critical angle θ_c. The left-hand side term represents (up to a factor $1\ /\ 4\pi$) a sort of wavelength perpendicular to the surface, and the right-hand side term (up to a factor $1\ /\ 4$) the extinction length projected on the perpendicular axis. The quasi equality of these two lengths is the sign that at the critical angle, the Born approximation is no longer valid. For less shallow angles, the perpendicular wavelength becomes smaller than the perpendicular extinction length and therefore the reflectivity becomes weak and calculable in this approximation. In the case of a rough surface, one must also compare the extinction length to the characteristic lengths of its waviness. If the waviness is longer or shorter than the extinction length, the losses in reflectivity and the scattering are different (see appendix 3.A).

1.3.4 When the Interaction Becomes Stronger

It is useful to know, even though this does not apply to neutrons or x-rays, the kind of propagation which arises when the interaction becomes stronger. In such a case, the representation by a continuous medium can still be retained, but the value of the index is no longer the one given above. In particular the index cannot be calculated easily without making the supposition that the scattering length $b(2\theta)$ is independent of the 2θ scattering direction. The formation of the index now implies that multiple scattering will be produced in all directions and not only in the forward direction. Also the scalar and vector fields do not have the same properties, since for the latter b depends on 2θ because of polarisation (but we know how to take it into account provided that there is no other anisotropy).

For a scalar field, when b is independent of the angle, the calculation that leads to the amplitude (1.78) scattered by a plane is exact, even if OM is not much larger than λ. Once the integration is made in the planes perpendicular to the propagation, it is then possible to work in one single dimension. Nevertheless the discussion which uses the decomposition of the material in layers (Fig. 1.4) must be revisited essentially because it is no longer possible to neglect the back scattering at M coming from the other layers which are located behind the point M. A calculation is proposed in exercise 1.3.3. As indicated in the introduction 1.3.1, it yields

$$n^2 = 1 - 2\lambda^2 \rho_v b / 2\pi = 1 - \left(4\pi / k_0^2\right) \rho_v b. \tag{1.88}$$

For vector fields (the case of the electromagnetic field), Clausius and Mossoti have found an expression which links the static polarisability of the medium to the molecular polarisability. A similar expression due to Lorenz and Lorentz, gives the refractive index. Using our notations, this is written (r_e is the Lorentz classical radius of the electron, defined in section 1.4.2),

$$\left(n^2 - 1\right) / \left(n^2 + 2\right) = -\left(4\pi / 3\right) k_0^2 \rho_v r_e. \tag{1.89}$$

These formulae apply when the extinction length and the wavelength are of the same order. The homogeneity must be verified at scales shorter than the wavelength. If it is not the case the propagation may be no longer possible; it is the phenomenon of localisation.

Exercise 1.3.1. A scalar plane wave, with the wave vector $\mathbf{k_0}$, enters a medium through a planar interface making the angle θ with $\mathbf{k_0}$. By dividing the medium in layers parallel to the interface, calculate the scattered amplitude at any point in the medium, as shown in (1.78). Find the direction of equiphase planes of the total amplitude and compare to Snell-Descartes's law. For which values of θ, is the approximation improper ?

Hint. One can show that the scattered amplitude at a point located at the back of an angled layer is given by expression (1.78) divided by $sin\ \theta$.

Exercise 1.3.2. In the same configuration as the one of the previous exercise, and assuming $b(2\theta)$ constant, find with the same method the amplitude reflected by the interface. Compare with the exact Fresnel expression given in chapter 3, section 3.1. In the section 3.3 in the same chapter the Born approximation is discussed as in this exercise. Note that the amplitude calculated here is the basckscattered one, considered as negligible in the discussion of the approximations at the end of section 1.3.2.

Hint. The expressions for the scattered amplitudes at two symmetrical points with respect to an infinitesimal layer are the same. Only b may change from $b(0)$ in one case to $b(2\theta)$ in the other (2θ is the angle between the reflected and incident wave vectors).

Notice. If the θ angle is big enough to allow $b(2\theta) \neq b(0)$, this calculation shows that the scalar Fresnel reflectivity expression is not exact.

Exercise 1.3.3. In a one dimensional medium, the dx element located at x', which receives the scalar amplitude A, scatters the wave in the two opposite directions

$$A\eta G_{1d}(x-x') = -A\rho b_{1d} dx e^{i|k_0(x-x')|} .$$

G_{1d} is the one dimension Green function, η a constant coefficient and ρb_{1d} the density of scattering power which in general is imaginary.

Find the relation between ρb_{1d} and the refractive index in this medium at one dimension.

Hints. One can consider an interface at $x = 0$ between the vacuum at negative x and the medium at positive x. A wave $A(x) = A_0 e^{-ik_0 x}$ comes from the vacuum and becomes $A'(x) = A' e^{-ink_0 x}$ in the medium. The field in the medium can be written into two ways:

- by the integral equation of the scattering

$$A'(x) = A(x) + \int_0^\infty A'(x')\eta G_{1d}(x-x')\, dx' ;$$

- by the transmission at the interface $A' = tA_0$ where t is the Fresnel transmission coefficient, which is $2/(n+1)$ (Chapter 3, section 3.1, Eqs. (3.18), (3.19)).

One can notice that the scattering is composed of two terms. The one in $-e^{-ik_0 x}$ is at the origin of the disappearance (so-called ***extinction***) of the incident wave (see the ***extinction theorem*** [3]).

1.4 X-Rays

1.4.1 General Considerations

The electromagnetic radiation interacts principally with the electrons, and very weakly with atomic nuclei (the ratio of the amplitudes is in the inverse of masses). The interaction is essentially between the electric field and the charge, but a much weaker interaction is also manifest between the electromagnetic field and the spin, or its associated magnetic moment.

A photon which meets an atom can undergo one of the three following events:

- ***elastic scattering***, with no change in energy;

- ***inelastic scattering***: part of the energy is transferred to the atom, the most frequently with the ejection of an electron (the so-called ***Compton effect***); however it may happen that the lost energy brings the atom in an excited state, without any ionisation (***Raman effect***);

- ***absorption***: all the energy is transferred to the atom and the photon vanishes. Another photon can be emitted, but with a lower energy: this is the so-called ***fluorescence***.

These mechanisms are described in many text books; the one of R. W. James [4] is particularly complete (except for the Raman effect which can be found in [5]).

To give an intuitive image, we shall begin with the classical mechanics theory which simply provides an exact result for the scattering by a free electron (Thomson scattering). When the electron is bound, this theory is still convenient enough. However the Compton scattering cannot be described by this classical theory. Also this theory does not describe correctly the motion of the electrons in the atom. Therefore we shall also review all the following processes in the frame of the quantum theory, i.e.:

- the elastic and inelastic scattering (mainly Compton), for a free electron or an electron bound to an atom, when the radiation energy is well above the atomic resonance;
- the photo-electric absorption by an atom;
- the dispersion correction brought to the elastic scattering by the atomic resonance.

Finally we shall discuss the general properties of dispersion which are independent of a particular interaction or radiation. One can show that the real and imaginary parts of the scattering are linked by the Kramers-Kronig relations which are extremely general and probe the response of nearly every system to some kind of excitation. The origin of these properties lies in the thermodynamical irreversibility that can be introduced through the principle of causality.

1.4.2 Classical Description: Thomson Scattering by a Free Electron

The scattering by a free electron is simple and presents the main characters of the scattering by an atom. We shall start with this case.

The electron undergoes an acceleration, which is due to the force exerted by the incident electric field

$$\mathbf{E}_{\text{in}}(t) = \mathbf{E}_0 e^{i\omega t}. \tag{1.90}$$

Let $\mathbf{z}$ be the electron position and $(-e)$ its charge, then

$$m\ddot{\mathbf{z}} = (-e)\,\mathbf{E}_0 e^{i\omega t}. \tag{1.91}$$

The electron exhibits oscillations of small amplitude, producing a localised current

$$\begin{aligned} \mathbf{j}(\mathbf{r},t) &= (-e)\,\dot{\mathbf{z}}\,\delta(\mathbf{r}) \\ &= \frac{(-e)^2\,\mathbf{E}_{\text{in}}(t)}{i\omega m}\,\delta(\mathbf{r}). \end{aligned} \tag{1.92}$$

The radiation of that vibrating current, similar to a dipole antenna, has been discussed in section 1.2.6. From the formulae (1.63) and (1.55), we have at large distances ($kr \gg 1$),

$$\mathbf{E}_{\text{sc}} \underset{kr\to\infty}{\sim} -\left[\mathbf{E}_{\text{in}} - (\mathbf{E}_{\text{in}}.\mathbf{r})\,\frac{\mathbf{r}}{\mathbf{r}^2}\right]\frac{(-e)^2 e^{-ikr}}{4\pi\varepsilon_0 mc^2 r}. \tag{1.93}$$

What is measured is the projection of the field on some polarisation direction given by the unit vector $\widehat{\mathbf{e}}_{\mathrm{sc}}$, and $\widehat{\mathbf{e}}_{\mathrm{in}}$ is the unit vector which describes the incident polarisation. These vectors are chosen so that $\widehat{\mathbf{e}}_{\mathrm{in}}$ is parallel or antiparallel to $\mathbf{E}_0$ and $\widehat{\mathbf{e}}_{\mathrm{sc}}$ normal to $\mathbf{r}$ (see Fig. 1.5)

$$\mathbf{E}_{\mathrm{in}} = (\mathbf{E}_{in}.\widehat{\mathbf{e}}_{in})\,\widehat{\mathbf{e}}_{\mathrm{in}} \qquad \text{and} \qquad \mathbf{r}.\widehat{\mathbf{e}}_{\mathrm{sc}} = 0. \tag{1.94}$$

In these conditions of polarisation the definition of the scattering length (1.34) can be adapted as follows

$$\mathbf{E}_{\mathrm{sc}}.\widehat{\mathbf{e}}_{\mathrm{sc}} = -\mathbf{E}_{\mathrm{in}}.\widehat{\mathbf{e}}_{\mathrm{in}}\; b(\widehat{\mathbf{e}}_{\mathrm{sc}}, \widehat{\mathbf{e}}_{\mathrm{in}})\frac{e^{-ikr}}{r}, \tag{1.95}$$

then we have

$$b(\widehat{\mathbf{e}}_{\mathrm{sc}}, \widehat{\mathbf{e}}_{\mathrm{in}}) = r_e \widehat{\mathbf{e}}_{\mathrm{sc}}.\widehat{\mathbf{e}}_{\mathrm{in}}, \tag{1.96}$$

where r_e is the ***Lorentz classical radius of the electron*** with charge e and mass m ($r_e = e^2/4\pi\varepsilon_0 mc^2 = 2.818 \times 10^{-15}$ m).[2] The charge of the electron appears twice, first in the movement and then for the emission of the radiation. Thus it appears as a square and b does not depend on its sign. *The scattered field is however opposite to the incident one* because of its relation with the current (by convention, a positive value of b corresponds to such a sign reversal). If the ingoing polarisation is normal or parallel to the plane of scattering, the outgoing one has the same orientation. These polarisation modes are called (s)-(s) (or (σ)-(σ)) when perpendicular to the plane of scattering and (p)-(p) (or (π)-(π)) when parallel. The polarisation factor of the scattering length is 1 in the former case and $\cos 2\theta$ (Fig. 1.5) in the latter. The process that we have described is the so-called ***Thomson scattering***.

1.4.3 Classical Description: Thomson Scattering by the Electrons of an Atom, Rayleigh Scattering

The simple result of the Thomson scattering is exact, even for the bound electrons of an atom, as far as the frequency of the x-rays is large compared to the characteristic atomic frequencies. Nevertheless it is necessary to take into account both the number of electrons and their position in the electronic cloud when calculating the scattering from an atom. Every point of the electronic cloud is considered to scatter independently from the others and the scattered amplitudes add coherently. As in any interference calculation within the Born approximation (see the Appendix 1.A), justified whenever

[2] A system of units which is often used to describe microscopic phenomena is the Gauss system. In this system we have $r_e = e^2/mc^2$.

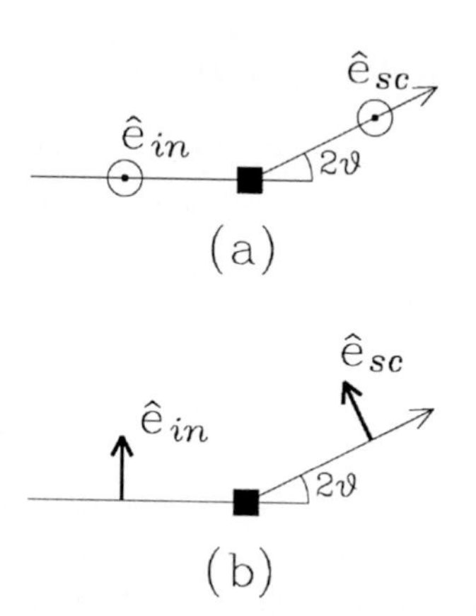

Fig. 1.5. Directions of incident and scattered polarisations for (a) the (s)-(s) or(σ)-(σ) *mode and (b) the (p)-(p) or*(π)-(π) *mode. The associated amplitude polarisation factor is respectively* 1 *and* $\cos 2\theta$

the scattering is weak, one obtains the total atomic scattering length b_{at} by the Fourier transform of the electron density $\rho(r)$

$$b_{\mathrm{at}} = r_e \widehat{\mathbf{e}}_{\mathrm{sc}} . \widehat{\mathbf{e}}_{\mathrm{in}} f(\mathbf{q}), \qquad f(\mathbf{q}) = \int \rho(\mathbf{r}) \, e^{i\mathbf{q}.\mathbf{r}} d(\mathbf{r}) \tag{1.97}$$

(see the definition of $\mathbf{q}$ in (1.25)). The quantity $f(\mathbf{q})$ is called the ***atomic scattering factor*** or the ***atomic form factor***. The integral of $\rho(\mathbf{r})$ over all $\mathbf{r}$ values must be equal to the number of electrons in the atom:

$$f(0) = Z. \tag{1.98}$$

There is no safe explanation to support the validity of this interference calculation. The justification comes from the alternative quantum calculation which gives the same result.

The assumption that the frequency of the radiation is greater than the atomic frequency may not be valid especially for the inner electronic shells. The model can be improved by introducing the binding of the electron to the atom which is modelled by a restoring force of stiffness κ and a damping coefficient γ. The damping is the result of the radiation which is emitted by the electron, or of the energy transferred to other electrons. The equation of motion (1.91), still written for a single electron, now becomes

$$m\ddot{\mathbf{z}} + \gamma \dot{\mathbf{z}} + \kappa \mathbf{z} = (-e)\mathbf{E}_0 e^{i\omega t}. \tag{1.99}$$

One looks for a solution of the kind $(e^{i\omega t})$ which must satisfy

$$(-m\omega^2 + i\gamma\omega + m\omega_0^2)\mathbf{z} = (-e)\mathbf{E}_0 e^{i\omega t}, \tag{1.100}$$

where $\kappa/m = \omega_0^2$. The current $(-e)\dot{\mathbf{z}}$ is then

$$\mathbf{j}(\mathbf{r},t) = -\frac{i\omega(-e)^2\mathbf{E}_{\text{in}}(t)\delta(\mathbf{r})}{m\left(\omega^2-\omega_0^2\right)-i\gamma\omega}. \tag{1.101}$$

As shown for the Thomson scattering above, this yields the following scattering length

$$b = r_e\frac{\omega^2}{\omega^2-\omega_0^2-i\gamma\omega/m}\widehat{\mathbf{e}}_{\text{sc}}.\widehat{\mathbf{e}}_{\text{in}}. \tag{1.102}$$

We shall now discuss how this expression is modified for different energies when only one electron and one resonance are considered although this discussion could have been more general. Actually it happens that $\omega, \omega_0 >> \gamma/m$ and we just have to compare ω with ω_0. For high energy x-rays and not too heavy atoms we have $\omega > \omega_0$ or even $\omega >> \omega_0$. Within these approximations (1.102) is just reduced to Thomson's expression. If on the other hand $\omega << \omega_0$, then b becomes,

$$b = -r_e\frac{\omega^2}{\omega_0^2}\widehat{\mathbf{e}}_{\text{sc}}.\widehat{\mathbf{e}}_{\text{in}}. \tag{1.103}$$

This is the so-called ***Rayleigh scattering***, originally proposed to explain the scattering of visible light produced by gasses or small particles.
Three important features of this kind of scattering should be noticed:

- the polarisation factor is the same as for the x-ray Thomson scattering;
- the scattered amplitude is proportional to the square of the frequency, and the cross-section is thus proportional to the fourth power of the frequency;
- the sign of the scattering length is opposite to the one of the Thomson scattering.

The second point explains the blue color of the sky (the highest frequency in the visible spectrum), which from the first point may appear to be highly polarised. The change of sign noted in the third point is important, since it corresponds to a sign change of $(n-1)$. We shall comment this further when we will dispose from a more quantitative theory.

Again for x-rays, the scattering length (1.102), when summed over all the atomic electrons becomes similar to the one of Thomson (1.97) but with real and imaginary corrections:

$$b_{\text{at}} = r_e\left(f+f'+if''\right)\widehat{\mathbf{e}}_{\text{sc}}.\widehat{\mathbf{e}}_{\text{in}}, \tag{1.104}$$

where f is the Thomson scattering, whereas f' and f'',which are real, give the correction due to resonance. This correction is the so-called ***dispersion correction*** or ***anomalous scattering*** .[3] One must take into account as in

[3] Originally, it was in optics that the anomalous dispersion was introduced. In the vicinity of resonances, the dispersion is opposite to the usual behavior for

the pure Thomson scattering the sum over all the electrons and their spatial distribution, but this discussion is difficult and uncertain in the classical theory. We shall see that in the quantum theory f' and f'' only slightly depend on $\mathbf{q}$ and have an energy dependence that we shall discuss.

Though crude, the classical model allows the calculation of the absorption as proposed in the exercise 1.4.1. In fact, one rather gets the total cross-section, including absorption and scattering. This result is very realistic, since it agrees with the prediction of the optical theorem discussed in section 1.3.2.

To summarize, the classical model although simple describes most of the phenomena and provides exact values for a certain number of physical quantities. Nevertheless the values of the resonance frequencies and of the damping coefficients are not calculable within this framework and are left arbitrary. In addition, it does not give much indications about the $\mathbf{q}$ dependence of the scattering factor at resonance but more important it does not describe the scattering when an electron is ejected (Compton effect). Although it is possible to give a classical description of such an effect by considering the reaction on the scattering of a vibrating electron, only the quantum approach is correct. Therefore the only coherent and completely exact description is given by the quantum theory of the interaction between the radiation and atoms.

Exercise 1.4.1. Calculate the total cross-section of an atom which exhibits only one resonance characterised by ω_0 and γ. We assume that the power taken by an atom from the radiation is the same as the one dissipated by the damping force $\gamma\dot{\mathbf{u}}$ (do not forget that when complex numbers are used to describe the oscillation of real variables, the answer is twice the one obtained with real numbers). The initial power of the radiation is given in section 1.2.3. Check that the optical theorem (section 1.3.2) is satisfied.

1.4.4 Quantum Description: a General Expression for Scattering and Absorption

In this description we shall assume that the radiation is quantised as photons. The scattering and absorption probabilities are then the squared modulus of the probability amplitudes. The amplitudes are transformed into scattering lengths and the probabilities into the scattering cross-section. The amplitudes themselves are derived from a perturbative calculation based on the interaction Hamiltonian between the radiation and the electrons.

which it is observed that the index of refraction varies in the same sense as the energy. By extension one refers to “anomalous scattering”. In French the two adjectives “anormale” and “anomale” are used. “Anormale” means that it does not follow the rule and “anomale” means different from other individuals from the same species. Since the normal behavior of the dispersion does not constitute a law in itself but only a usual behavior, the second expression seems to be more appropriate. We acknowledge B. Pardo for his comments.

The expression of the Hamiltonian of one electron in the radiation field contains the following term (we leave aside some other terms such as the potential of the atom)

$$(1/2m)\left(\mathbf{p}-e\mathbf{A}/c\right)^2=\mathbf{p}^2/2m+\left(e^2/2mc^2\right)\mathbf{A}^2-(e/mc)\mathbf{A}.\mathbf{p}. \qquad (1.105)$$

The $\mathbf{p}$ and $\mathbf{A}$ operators are the momentum of the electron and the vector potential of the radiation. The first term of the right-hand side gives the kinetic energy of the electron and the two others the energy of interaction. In this expression, the spin has been neglected which is permitted when the energy of the radiation is weak compared to the rest mass energy of the electron which is 511 keV. A perturbation calculation made at the lowest order on the two interaction terms yields the scattered amplitude. This approximation is sufficient because the strength of the interaction, measured by the ratio of the coefficient e^2/mc^2 (or $4\pi\varepsilon_0 r_e$) to the quantum size of the electron h/mc (or λ_c defined further) is small. The perturbation terms are sketched in Fig. 1.6. The smallest order of the perturbation is the first order for the term in $\mathbf{A}^2$ and the second order for the term in $\mathbf{A}.\mathbf{p}$. These two terms give rise respectively to one and to two terms in the scattering length (with our convention for the sign of imaginaries, unusual in quantum mechanics):

$$\begin{aligned}
b_{\mathrm{at}} &= r_e < s|\widehat{\mathbf{e}}_{\mathrm{sc}}^{*}\, e^{+i\mathbf{k}_{sc}.\mathbf{r}}.\widehat{\mathbf{e}}_{\mathrm{in}}\, e^{-i\mathbf{k}_{in}.\mathbf{r}}|i> \\
&- r_e \sum_c \frac{< s|\widehat{\mathbf{e}}_{\mathrm{sc}}^{*}.\mathbf{p}\, e^{+i\mathbf{k}_{sc}.\mathbf{r}}|c>< c|\widehat{\mathbf{e}}_{\mathrm{in}}.\mathbf{p}\, e^{-i\mathbf{k}_{in}.\mathbf{r}}|i>}{m(E_c-E_i-\hbar\omega_{\mathrm{in}}+i\Gamma_c/2)} \\
&- r_e \sum_c \frac{< s|\widehat{\mathbf{e}}_{\mathrm{sc}}^{*}.\mathbf{p}\, e^{-i\mathbf{k}_{sc}.\mathbf{r}}|c>< c|\widehat{\mathbf{e}}_{\mathrm{in}}.\mathbf{p}\, e^{+i\mathbf{k}_{in}.\mathbf{r}}|i>}{m(E_c-E_i+\hbar\omega_{\mathrm{sc}})} \\
&= b_{\mathrm{Th}}+b_{\mathrm{disp1}}+b_{\mathrm{disp2}}.
\end{aligned} \qquad (1.106)$$

Here $|\ i>$ (respectively $|\ s>$) stands for the initial (respectively after scattering) electron states. These two states are identical for elastic scattering and different for inelastic scattering. $\mathbf{r}$ is the position operator of the electron. In the last two rows a sum is made over all the excited states $|\ c>$ of this electron (bound or continuum states). E_c-E_i represents the energy of excitation and $\hbar\omega_{\mathrm{in}}$ ($\hbar\omega_{\mathrm{sc}}$) is the energy of the incident (scattered) photon. In elastic scattering $\omega_{\mathrm{sc}}=\omega_{\mathrm{in}}$. Γ_c is the width of the excited level $|\ c>$ and $\hbar/\Gamma_c$ its life time. The polarisation vectors may be complex so they can represent elliptical polarisation states.[4] The following discussion will show that the first term represents the Thomson scattering found in the classical theory. The two last terms, b_{disp1} and b_{disp2} define the dispersive part of the scattering.

[4] In most instance in this book, only linear polarisations are considered and no complex conjugate is indicated. In the case of anisotropic scattering, section 1.5, the circular polarisation may be required.

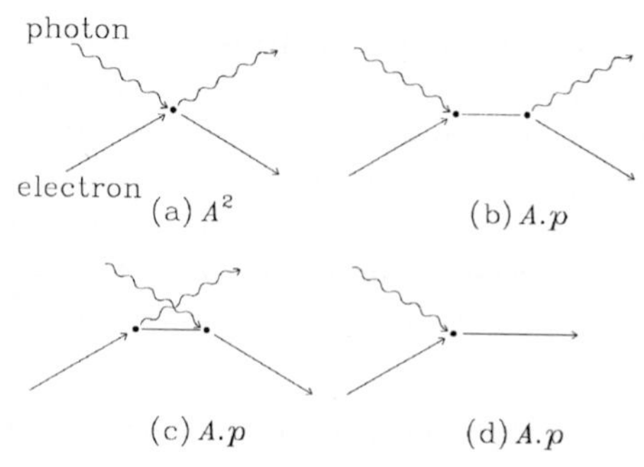

Fig. 1.6. These diagrams are the symbol of the amplitudes which are in formulae (1.106) and (1.107). A point represents a matrix element and a line the electron or the photon in the initial or final state of the matrix element. For instance in (b) where two matrix elements are represented, the initial and final states display one electron and one photon, and the intermediate state only one electron. In the formulae as written in the text, the photonic states are not made explicit, but their contribution $(\widehat{\mathbf{e}}, \mathbf{k})$ *is present through the* $(\widehat{\mathbf{e}}\, e^{-i\mathbf{k}.\mathbf{r}})$ *terms. For any of the four amplitudes, (a)* b_{Th}*, (b)* b_{disp1}*, (c)* b_{disp2}*, and (d) the absorption, the Hamiltonian term is indicated*

The absorption cross-section is also derived from the interaction Hamiltonian, once again at the lowest order of perturbation, Ref. [6] section 44,

$$\sigma_{\mathrm{abs}}\left(\hbar\omega_{\mathrm{in}}\right) = \frac{2\pi\hbar c r_e}{m} \sum_c \frac{\hbar\omega_{\mathrm{in}}\,\Gamma_c}{(E_c - E_i)^2} \frac{\left| < c|\widehat{\mathbf{e}}_{\mathrm{in}}.\mathbf{p}\, e^{-i\mathbf{k}_{\mathrm{in}}\mathbf{r}}|i > \right|^2}{(E_c - E_i - \hbar\omega_{in})^2 + \Gamma_c^2/4}. \quad (1.107)$$

In this process the photon completely disappears. The $\mathbf{A}^2$ term in the Hamiltonian does not contribute and therefore only the $\mathbf{A}.\mathbf{p}$ term is used. Every term of the sum corresponds to the excitation towards a $\mid c >$ state. The numerator suggests that the electric field transfers some momentum to the electron, and changes the $\mid i >$ level into the $\mid c >$ level. In a similar way, in (1.106), one can say that the scattering $b_{\mathrm{disp1}} + b_{\mathrm{disp2}}$ is obtained by excitation $\mid i > \rightarrow \mid c >$, then desexcitation $\mid c > \rightarrow \mid s >$. This order is reversed in b_{disp2} , since the $\mid c >$ state, which is virtual, is destroyed before being created (Fig. 1.6).

To calculate the scattering as well as the absorption, one generally uses the ***dipolar approximation***, that is to say one replaces the factors $e^{i\mathbf{k}.\mathbf{r}}$ by one, supposing the wavelength much bigger than the atomic dimensions. This approximation which is excellent in the visible spectrum, is still good for x-rays because the electronic levels which are excited are usually very much localised. Under certain conditions however, this approximation is not sufficient and the next term in the expansion of the exponential ($i\mathbf{k}.\mathbf{r}$, the ***quadrupolar term***) must be included.

When the energy of a photon is sensibly larger than all the excitation thresholds of the atom ($\omega \gg \omega_c$), the first term in (1.106) which represents the Thomson scattering becomes preponderant. In the extremes of light atom and very high energies, the scattering cross-section given by this first term is even greater than the absorption cross-section (1.107). We shall start the discussion of the Thomson scattering b_{Th} to show that it can be separated into the elastic and inelastic (Compton) scattering. Then we shall describe the absorption spectrum which comes from (1.107). Finally we shall discuss the dispersive $b_{\mathrm{disp1}} + b_{\mathrm{disp2}}$ scattering, in relation with absorption.

1.4.5 Quantum Description: Elastic and Compton Scattering

For a free electron and in the classical Thomson scattering, the backward move of the electron is ignored. Compton performed a kinematical calculation which took into account the momentum and the energy carried by the radiation quantised as photons. For an electron initially at rest, the conservation of these two quantities implies that the photon releases an energy such that the wavelength after the scattering process λ_{sc} becomes larger than the initial one λ_{in} , and satisfies the equation

$$\lambda_{\mathrm{sc}} = \lambda_{\mathrm{in}} + \lambda_c (1 - \cos 2\theta) \qquad \lambda_c = 2\pi\hbar/mc = 0.002426 \text{ nm} \tag{1.108}$$

where 2θ is the angle between the incident and scattered beams and λ_c is the ***Compton wavelength*** of the electron.

In the present calculation, we are doing non relativistic approximations which are not valid if the photon energy becomes close to the rest energy of the electron. Neglected relativistic effects are the influence of the spin and a factor which diminishes the Compton scattering cross-section.

When the electron is bound to an atom two processes are possible: the radiation may be elastically scattered with the conservation of the electron state (the momentum being transferred to the atom which is assumed to have an infinite mass), or inelastically with the ejection of the electron. One must determine the respective probabilities of these two processes. We start first with the case of an atom which has only one electron.

Let us evaluate the elastic, then the total scattering. The inelastic scattering will be obtained by subtraction. Keeping only b_{Th} from (1.106) we have,

$$b_{\mathrm{at}} = r_e \widehat{\mathbf{e}}^*_{\mathrm{sc}} . \widehat{\mathbf{e}}_{\mathrm{in}} \, f_{si}, \qquad f_{si} = < s | e^{i\mathbf{q}.\mathbf{r}} | i >, \tag{1.109}$$

where $\mathbf{q}$ is equal to $\mathbf{k}_{\mathrm{sc}} - \mathbf{k}_{\mathrm{in}}$ (1.3). For elastic scattering, $| \, i >= | \, s >$. Let $\psi(\mathbf{r})$ and $\rho(\mathbf{r})$ be the wave function and the electron density then

$$f_{si} = f_{ii} = \int \psi^*(\mathbf{r}) \psi(\mathbf{r}) e^{i\mathbf{q}.\mathbf{r}} d\mathbf{r} = \int \rho(\mathbf{r}) e^{i\mathbf{q}.\mathbf{r}} d\mathbf{r} \tag{1.110}$$

We have derived here more rigorously, the form factor which we previously determined by the classical theory (for the atom having one electron). The calculation is completed by the evaluation of the total scattering cross-section, elastic plus inelastic. This total is obtained by summing the square modulus of the scattering factor over all the final states of the electron,

$$\sum_{|s\rangle} \left|\langle s| e^{i\mathbf{q}.\mathbf{r}} |i\rangle\right|^2 = \sum_{|s\rangle} \langle i| e^{-i\mathbf{q}.\mathbf{r}} |s\rangle \langle s| e^{i\mathbf{q}.\mathbf{r}} |i\rangle . \tag{1.111}$$

Since the sum over the final states is made over all the possible states, these ones form a complete set and satisfy the closure relation

$$\sum_{|s\rangle} |s\rangle \langle s| = \text{unitoperator.} \tag{1.112}$$

The final states disappear from expression (1.111) which becomes equal to unity. The inelastic cross-section is obtained by subtraction and finally we have,

$$(d\sigma/d\Omega)_{\text{elas+inel}} = (r_e \widehat{\mathbf{e}}^*_{\text{sc}}.\widehat{\mathbf{e}}_{\text{in}})^2 \tag{1.113}$$

$$(d\sigma/d\Omega)_{\text{elas}} = (r_e \widehat{\mathbf{e}}^*_{\text{sc}}.\widehat{\mathbf{e}}_{\text{in}})^2 |f_{ii}|^2 \tag{1.114}$$

$$(d\sigma/d\Omega)_{inel} = (r_e \widehat{\mathbf{e}}^*_{\text{sc}}.\widehat{\mathbf{e}}_{\text{in}})^2 \left(1 - |f_{ii}|^2\right) . \tag{1.115}$$

This calculation prompts two remarks. We first observe that in the sum (1.111), the terms which do not conserve the momentum seem to play no part: they cancel the matrix element $< s \mid e^{i\mathbf{q}.\mathbf{r}} \mid i >$. However these terms must be included in the sum to enable the use of the closure relation (1.112). Next, some information about the conditions in which this sum is performed should be given. To be correct we must sum over all the final states of the radiation, with the scattering direction $\hat{\mathbf{u}}$ kept fixed (this is a result of the definition of the differential cross-section) and with the energy conservation obeyed. Instead of (1.111), the exact expression is (E_s and E_i being the energies of the electron states)

$$\sum_{|s\rangle} \int_{\mathbf{k}_{\text{sc}}/|\mathbf{k}_{\text{sc}}|=\hat{\mathbf{u}}} \left|\langle s| e^{i\mathbf{q}.\mathbf{r}} |i\rangle\right|^2 \delta(E_s - E_i - \hbar c\,|\mathbf{k}_{\text{in}}| + \hbar c\,|\mathbf{k}_{\text{sc}}|) d\mathbf{k}_{\text{sc}} . \tag{1.116}$$

Let us note that this expression imposes the two conditions used by Compton which are the energy conservation as shown by the δ function, and the conservation of momentum in the matrix element. Performing the integral over $\mathbf{k}_{\text{sc}}$ one gets back the sum (1.111) over the final states of the electron, with the condition $\mathbf{k}_{\text{sc}}/ \mid \mathbf{k}_{sc} \mid = \hat{\mathbf{u}}$. One can see that being in the δ function $\mathbf{k}_{\text{sc}}$ depends on $< s \mid$, and $\mathbf{q}$ consequently has the same dependence. For the discussion to be consistent, this dependence must be neglected otherwise the closure relation (1.112) could not be used in (1.111). The approximation is very good but the small dependence of $\mid \mathbf{k}_{\text{sc}} \mid$ on the final state of the electron, which, through the momentum conservation, is also a dependence on the electron initial momentum, can be used to measure the momentum distribution inside the atom or inside the solid. This application of Compton scattering will not be developed further in this book.

The inelastic scattering for which we have calculated the cross-section is frequently considered as the Compton scattering. This is not completely correct since the total scattering cross-section also includes the ***Raman scattering***. In such a case the final state $< s \mid$ of the electron is not a free plane wave but a bound excited state [5]. To be fully complete we must also consider another inelastic scattering process, the so-called ***resonant Raman scattering***. This process does not appear in the above calculation, but rather in the development of the second term in (1.106), b_{disp1}, when one assumes $< s \mid \neq < i \mid$; it is obvious at energies close to an excitation edge. It is thus more associated with absorption and fluorescence than with Compton scattering. However far from resonances, the dominating inelastic process is usually the Compton scattering.

The calculation that we have just carried out has to be changed for an atom having more than one electron. The electronic states are multi-electron states and each interaction operator is replaced by the sum of operators acting each on one electron. For an atom having two electrons,

$$|i\rangle \quad \rightarrow \quad (1/\sqrt{2})\, |\Psi_1(\mathbf{r}_1)\Psi_2(\mathbf{r}_2) - \Psi_1(\mathbf{r}_2)\Psi_2(\mathbf{r}_1)\rangle \tag{1.117}$$

$$e^{i\mathbf{q}.\mathbf{r}} \quad \rightarrow \quad e^{i\mathbf{q}.\mathbf{r}_1} + e^{i\mathbf{q}.\mathbf{r}_2}. \tag{1.118}$$

Expression (1.117) is the Slater's determinant which represents the antisymmetric state with respect to the permutation of the electrons. The elastic scattering factor becomes

$$f = \langle i|\, e^{i\mathbf{q}.\mathbf{r}_1} + e^{i\mathbf{q}.\mathbf{r}_2}\, |i\rangle\,. \tag{1.119}$$

With $\mid i >$ given by (1.117) and using the orthogonality between Ψ_1 and Ψ_2, this yields

$$f = f_{11} + f_{22} \qquad \text{where} \qquad f_{jl} = \int \Psi_j^*(\mathbf{r})\Psi_l(\mathbf{r})e^{i\mathbf{q}.\mathbf{r}}d\mathbf{r}. \tag{1.120}$$

To obtain the total cross-section one must sum the amplitude squares over all the final states as in (1.111). Although it is not necessary to know them, we explicitly write them for more complete view

$$|s\rangle = (1/\sqrt{2})\, |\Psi_1(\mathbf{r}_1)\Psi_2(\mathbf{r}_2) - \Psi_1(\mathbf{r}_2)\Psi_2(\mathbf{r}_1)\rangle \tag{1.121}$$

$$|s\rangle = (1/\sqrt{2})\, |\Psi_x(\mathbf{r}_1)\Psi_2(\mathbf{r}_2) - \Psi_x(\mathbf{r}_2)\Psi_2(\mathbf{r}_1)\rangle \qquad x \neq 1,2 \tag{1.122}$$

$$|s\rangle = (1/\sqrt{2})\, |\Psi_1(\mathbf{r}_1)\Psi_x(\mathbf{r}_2) - \Psi_1(\mathbf{r}_2)\Psi_x(\mathbf{r}_1)\rangle \qquad x \neq 1,2 \tag{1.123}$$

$$|s\rangle = (1/\sqrt{2})\, |\Psi_x(\mathbf{r}_1)\Psi_y(\mathbf{r}_2) - \Psi_x(\mathbf{r}_2)\Psi_y(\mathbf{r}_1)\rangle \qquad x,y \neq 1,2. \tag{1.124}$$

The last state corresponds to a two electrons excitation and gives rise to a zero amplitude, but once again, it must be included to use the closure relation. With this latter, the total cross-section is proportional to

$$\langle i| \left(e^{-i\mathbf{q}.\mathbf{r}_1}+e^{-i\mathbf{q}.\mathbf{r}_2}\right)\left(e^{i\mathbf{q}.\mathbf{r}_1}+e^{i\mathbf{q}.\mathbf{r}_2}\right) |i\rangle \,. \tag{1.125}$$

Substituting $|i\rangle$ by (1.117) and using the definition (1.120), this expression becomes

$$2 + f_{11}f_{22}^{*} + f_{11}^{*}f_{22} - |f_{12}|^2 - |f_{21}|^2 \,. \tag{1.126}$$

It is easy to extend this calculation to any number of electrons Z. The elastic and inelastic scattering cross-sections become,

$$(d\sigma/d\Omega)_{\text{elas+inel}} = (r_e\widehat{\mathbf{e}}_{\text{sc}}^{*}.\widehat{\mathbf{e}}_{\text{in}})^2 \left(Z + \sum_{1\leq j\neq l\leq Z} f_{jj}^{*}f_{ll} - \sum_{1\leq j\neq l\leq Z} |f_{jl}|^2\right) \tag{1.127}$$

$$(d\sigma/d\Omega)_{\text{elas}} = (r_e\widehat{\mathbf{e}}_{\text{sc}}^{*}.\widehat{\mathbf{e}}_{\text{in}})^2 \left|\sum_{1\leq j\leq Z} f_{jj}\right|^2 \tag{1.128}$$

$$(d\sigma/d\Omega)_{\text{inel}} = (r_e\widehat{\mathbf{e}}_{\text{sc}}^{*}.\widehat{\mathbf{e}}_{\text{in}})^2 \left(Z - \sum_{1\leq j\leq Z} |f_{jj}|^2 - \sum_{1\leq j\neq l\leq Z} |f_{jl}|^2\right). \tag{1.129}$$

One can see that the elastic scattering factor is written as the Fourier transform of the electron density (which is the sum of the densities of all the wave functions), c.f. (1.110). When the terms $|\, f_{jl}\,|^2$ are ignored, *the case of the many electrons atom is naively deduced from the one having one electron: on one hand the elastic scattering lengths and on the other hand the inelastic cross-sections of all the electrons are added.* The $|\, f_{jl}\,|^2$ terms constitute in fact a modest correction. They are called the exchange terms since they come from matrix elements in which electrons have been interchanged as for instance in the case of two electrons,

$$\langle \Psi_1(\mathbf{r}_1)\Psi_2(\mathbf{r}_2)| \ldots. |\Psi_1(\mathbf{r}_2)\Psi_2(\mathbf{r}_1)\rangle \,. \tag{1.130}$$

They are subtracted because electrons are fermions.

The general evolution of the cross-sections is presented as a function of $|\mathbf{q}|$ in Fig. 1.7.

To end this section we now discuss how easily the Compton scattering is observed. It is suitable to rewrite the change of wavelength (1.108) in the following way as a function of the scattering vector transfer $\mathbf{q}$

$$|\mathbf{k}_{\text{in}}| - |\mathbf{k}_{\text{sc}}| = (\lambda_c/4\pi)\,\mathbf{q}^2 + (\textit{higher order in } \mathbf{q}). \tag{1.131}$$

A radiation of wavelength λ_c has an energy of 511 keV, i.e. the mass energy of the electron at rest. For a radiation of energy 10 keV scattered at an angle

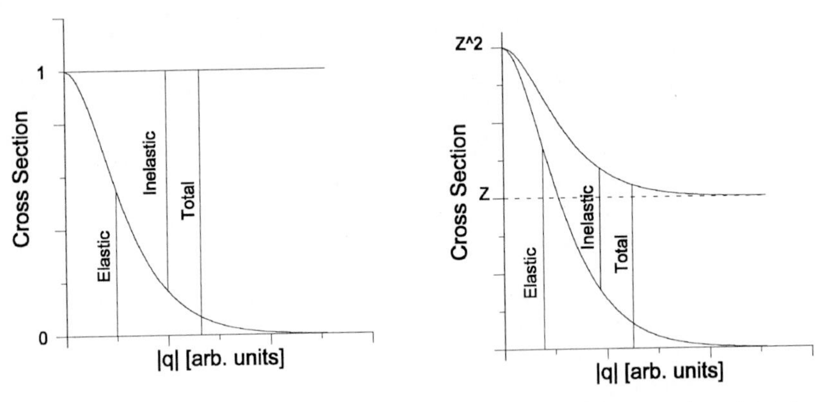

Fig. 1.7. Schematic representation of elastic, inelastic (Compton) and total cross-sections as a function of $|\mathbf{q}|$, (a) for an atom having one electron, (b) for an atom having any number of electrons in units of r_e^2

of say one degree (typical of a grazing incidence surface experiment), the wavelength change is very weak since it depends on the square of $\mathbf{q}$. Things are different in the range of medium and large values of scattering angles where this change is easily measurable, for instance by using an analyzer crystal. At energies greater than about 100 keV and medium angles of scattering this change is appreciable.

The Compton cross-section varies in a similar way as shown in (1.115) (1.129) and in Fig. 1.7. For radiation of 10 keV, it is negligible at small angles, but this is not true at wider angles. At higher energies, some tens of keV, the Compton scattering achieves its highest value already at medium angles. At those energies and for light elements it dominates the other processes. Indeed its proportion is larger for small Z scatterers. The table 1.1[5] gives some values of the total scattering cross-sections (integrated over the whole angular space); the elastic cross-section is condensed in the forward direction in a cone which becomes narrower when the energy increases.

1.4.6 Resonances: Absorption, Photoelectric Effect

In the interaction process, part of the radiation disappears instead of being scattered. As shown in (1.107), the energy is transferred to an electron which is excited to an empty upper state $| c >$. Most frequently it is expelled from the atom; this is the so-called ***photoelectric absorption***. After a delay of about h/Γ_c, the atom de-excites, according to various processes which can be radiative or not. The most obvious process in a diffraction or scattering experiment is the emission of ***fluorescence radiation***. It corresponds to the

[5] More cross-section values can be found in the International Tables for x-ray crystallography [7], vol. III and IV

fall of a second electron of the atom into the level vacated by the first one. Its energy is necessarily lower than the energy of excitation. The *fluorescence yield*, i.e. the fraction of excited atoms which are de-excited in this way, depends on the elements and on the levels; for the K level of copper, the fluorescence yield is 0.5. Let us note that for a given excited level $| c >$, the cross-section varies with the energy and exhibits a Lorentzian behavior with a FWHM Γ_c. In the x-ray domain, Γ_c lies between about a bit less than 0.5 eV and a bit more than 5 eV.

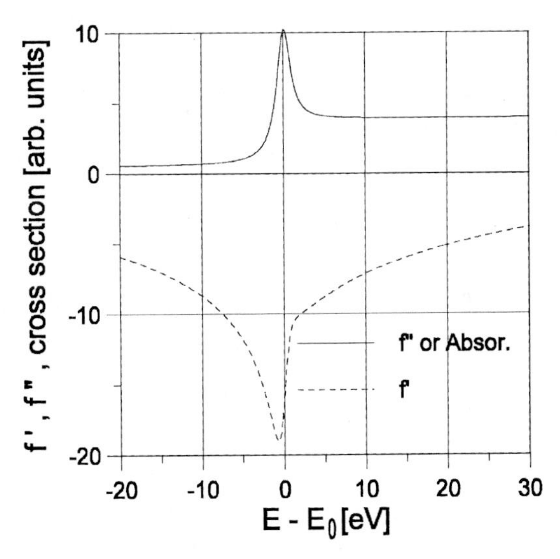

Fig. 1.8. Schematics of an absorption edge, with a white line ($\Gamma = 2$ eV). *The solid line shows the variations of both the absorption cross-section and of the imaginary part of dispersion correction (see next section), and the dotted line the real part of this correction in arbitrary units. The origin of energy, E_0, is taken at the edge. This schematic figure is not intended to show the real details of these curves in the vicinity and above the edge. On this short interval of energy, the E^{-3} decay has been neglected*

The important transitions for x-rays are those of the inner electrons which belong to the K, L, . . . shells. The transitions may bring the excited electron towards the continuum of the free states; their spectral signature is then characterised by an ***absorption edge***, located at the excitation energy, since any level above the edge is equally accessible (Fig. 1.8). They can also arise towards the first free bound levels. These states may have a large enough density to give rise to one (or several) well-defined absorption peaks superimposed to the edge, the so-called ***white line***(s). The white line(s) spectrum is not an exact image of the density of free states of the atom, molecules or condensed system. The observed spectrum corresponds to a system which has lost a core electron and is deformed by the electric charge of the core hole. Peaks can paradoxically then appear below the edge, and the white lines can be narrower and more intense than the corresponding levels of the ground state of the system. In condensed matter, the absorption above the edge exhibits oscillations, the so-called EXAFS (Extended x-ray Absorption Fine Structure), which are interpreted as arising from interference effects in the wave function of the ejected electron. These interference effects are due to the scattering of the ejected electron by the neighbouring atoms.

In short, one can say that the absorption varies as $E^{-3}$and as Z^4. This does not take into account the discontinuities at the edges. The K edges produce a discontinuity of the absorption by a factor of about 5 to 10. Figure 1.9 shows for copper a discontinuity of a factor 7 at the K edge and of about the same amount for the three L edges together; one can see that the decay in between the edges is a bit slower than E^{-3}. Table 1.1 gives some values of the absorption and scattering cross-sections.

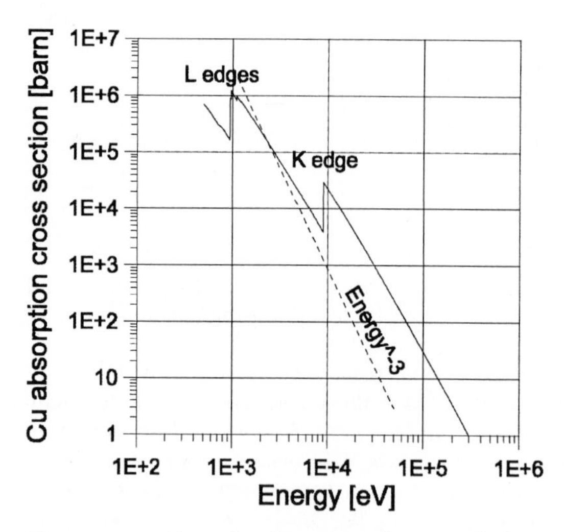

Fig. 1.9. Absorption cross-sections for the atom of copper in barns (10^{-22} mm^2), *after Cromer-Libermann (above 10 keV) and Henke (below 10 keV). The slope of the* E^{-3} *power law is presented by a dotted line*

In practice one frequently needs the absorption coefficient μ rather than the cross-section; this coefficient is defined by the fact that the transmission through a thickness t is given by $e^{-\mu t}$. It is also equal to $4\pi\beta/\lambda$ (β the imaginary part of the refractive index). For a homogeneous material made of a single element, μ depends on the cross-section σ and on the atomic volume V through

$$\mu = \sigma/V. \tag{1.132}$$

Table 1.1. Cross-sections of some elements as a function of energy (scattering cross-sections are integrated over a solid angle 4π). *In each case, are displayed the Compton scattering cross-section / the elastic scattering cross-section / the photoelectric absorption cross-section, in barns (i.e.* 10^{-28} m^2). *The irregularities in the evolution of the absorption are due to the presence of an edge close to the chosen energy. After [8]; see also the International Tables for x-ray Crystallography [7]*

Element (Z)	5 keV	10 keV	30 keV	100 keV
C(6)	2.1/5.8/371	2.7/3.2/39	3.3/0.67/1.1	2.9/0.07/0.02
Cu(29)	4.65/307/19500	8.2/153/22600	13/35.6/1090	13.3/4.5/30
Ag(47)	6.5/820/132000	11.5/459/20600	19/117/6420	20.9/15.8/224
Au(79)	8.22/2630/212000	15.3/1580/36100	27.8/432/8420	33.2/60.8/1590

To calculate the absorption coefficient of a material, it is sometimes useful to introduce the mass absorption coefficient, given by μ/ρ (ρ is the density) and commonly tabulated. This coefficient is characteristic of the element and independent of its density. If A is the molar mass and N Avogadro's number

$$\mu/\rho = N\sigma/A. \tag{1.133}$$

The absorption coefficient of a material composed of several elements i, each of them present with the partial density ρ_i, is simply given by

$$\mu = \sum_i \rho_i \left(\mu/\rho\right)_i . \tag{1.134}$$

1.4.7 Resonances: Dispersion and Anomalous Scattering

We return now to the case of the elastic scattering in which we had neglected the dispersive part $b_{\text{disp1}} + b_{\text{disp2}}$ in equation (1.106). Actually we shall take into account only b_{disp1}, which represents the second line of this expression. For a term of the sum over c to be appreciable it is necessary for its denominator to be small which never occurs in b_{disp2} .

We reproduce here the formula (1.104), which gives the separated Thomson and dispersive contributions.

$$b_{\text{at}} = r_e(f + f' + if'')\widehat{\text{e}}^*_{\text{sc}}.\widehat{\text{e}}_{\text{in}}. \tag{1.135}$$

This separation could appear artificial in the classical expression for b (1.102), but arises perfectly naturally in the quantum mechanical one (1.106). We have assumed that the polarisation contributes in b_{disp1} that is $f' + if''$, through the same polarisation factor as in the Thomson term. This is not true in every case, as discussed in Sect. 1.5.

Each of the terms $| \, c >$ of the sum (1.106) from which the dispersion correction $f' + if''$ arises, corresponds to an excitation energy (or commonly a resonance) $E_c - E_i$. The associated correction is

$$r_e(f_c' + i f_c'') \propto \frac{1}{x-i} = \frac{x}{1+x^2} + \frac{i}{1+x^2} \qquad (1.136)$$
$$\text{with } x = [\hbar\omega - (E_c - E_i)] \,/\, (\Gamma_c/2) \,.$$

The real and imaginary parts are presented in Fig. 1.10.

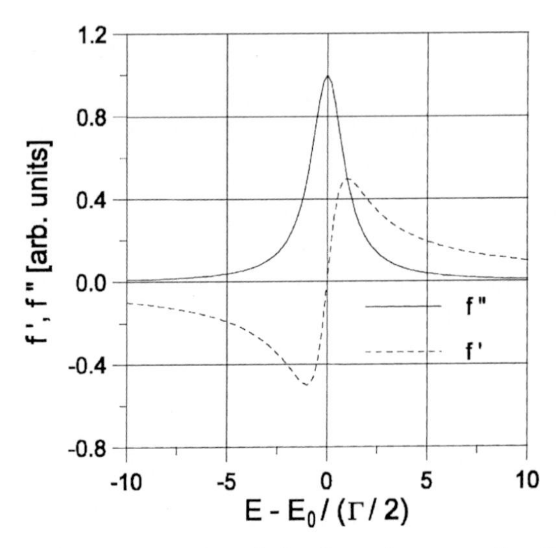

Fig. 1.10. Schematic representation of the dispersion correction for a single resonance at energy $E_c - E_i = E_0$. *f' and f'' are given by $b = r_e(f + f' + if")$*

The formulae (1.106) and (1.107) show a correspondence between the dispersion correction in terms of the scattering length and the absorption cross-section. Exactly at the resonance energy, we check the optical theorem ($\mathcal{I}$m is the imaginary part),

$$\sigma_{\text{abs}} = 2\lambda\, \mathcal{I}\text{m}[b\,(q=0)], \qquad (1.137)$$

discussed in Sect. 1.3.2.[6]

[6] We obtain here an expression for the *absorption* cross-section, while the optical theorem yields the same expression for the *total* cross-section. The error comes from our calculation of the scattering length, made in the first order Born approximation. The next order is required to obtain an imaginary part which expresses the intensity loss due to scattering (see appendix 1.A). The calculation at that order is made intricate because of some difficulties of the quantum theory of radiation (the renormalization of field theory). That error is negligible inasmuch as the absorption is the largest part of the cross-section, which is true up to moderate energies, but not at the highest.

The distribution of the resonance energies $E_c - E_i$, with the edges as main features, has been previously discussed about the absorption. Fig. 1.8 shows the comparison between the variations of the absorption cross-section close to an edge and the variations of the anomalous scattering.

Let us now look how the real part of the scattering factor, $f + f'$, varies when the energy changes from x-rays to near infrared, that is to say from several tenths of keV to one eV. The highest energies are far above the edges of most elements and the Thomson scattering factor f is dominant. For lower energies, a negative contribution f' appears at every edge and is more important below the edge than above because of the white lines. Low energy edges produce the most intense dispersion effects. Going to low energies, some edges for which $f + f'$ is negative are observed, and then a transition occurs towards 10 to 100 eV where $f + f'$ definitively changes its sign. In this range, the very intense absorption lines enormously reduce the propagation of light in matter, which makes it called ***vacuum ultraviolet*** radiation, because it propagates only in vacuum. When the sign of $f + f'$, which is also the sign of b, changes from positive to negative, the refractive index n goes from below to above the unit value (the link between scattering and the index is discussed in section 1.3).

1.4.8 Resonances: Dispersion Relations

The absorption cross-section is easily obtained directly by experiments as for example, the measurement of the transmission through a known thickness of a material. The imaginary part of the scattering length b is found at the same time. The real part of b however is more difficult to obtain accurately. Among the different methods, diffraction experiments but also reflectivity measurements have been used to extract the scattering length [9]. The drawback of such indirect methods can be overcome because it is possible to rebuild the real part of b if the imaginary part is known over the entire spectral range. Conversely the imaginary part can be deduced from the real one.

For a single resonance, if the variation $f''(E)$ as shown in (1.136) and Fig. 1.10 is known, then the energy, the amplitude and the width Γ of the resonance are determined, and $f'(E)$ can be obtained. Note however that the sign of Γ is left ambiguous in this procedure and must be given. The question is to know whether such a reconstruction of $f'(E)$ is still possible for the general case with several resonances. It will turn out to be possible, but not in the way it can be done for a single resonance whose shape is known. The key point is not the particular form of the function (1.136) but rather the well defined sign of Γ.

Let us start from the classical model, namely the expression (1.102) for the scattering length,

$$b = r_e \frac{\omega^2}{\omega^2 - \omega_0^2 - i\gamma\omega/m} \widehat{\mathbf{e}}_{\mathrm{sc}} . \widehat{\mathbf{e}}_{\mathrm{in}} . \tag{1.138}$$

The general case can be represented by summing many expressions of this kind corresponding to each different resonance ω_0 with a different damping constant γ. Since this model is defined by two independent functions of ω_0, a distribution of the resonance densities and a distribution of the damping constants, one could expect the real and imaginary parts of the scattering length to also constitute two independent functions. However some constraints are imposed because the damping constants γ are necessarily positive. Although these constraints seem to be weak, it is remarkable that they are sufficient to lead to a relation between the real and imaginary parts of b. We shall see that such a relation does not come from a particular scattering model; it is more general and concerns the response of any system to an excitation. For the proof, we return to the model with only one resonance (1.138) but this could be easily extended to the general case.

To prove the existence of a relation between the real and imaginary parts of b, it is necessary to make use of a mathematical trick, the analytical continuation of function b in the complex plane. The trick allows one to express some basic properties of complex functions. Indeed b is a complex function of the real variable ω. If such a function can be represented by a series expansion which converges for any real value of the variable, then it can also be defined for complex values of this variable. Hence, the series still converges in a domain of the complex plane. Inside this domain, the function that we shall call here $\phi(z)$ is analytic and follows Cauchy's theorem. This theorem ensures that for any closed contour C inside the domain of analyticity and for any point z inside the contour,

$$\phi(z) = \frac{1}{2\pi i} \int_C \frac{\phi(z')}{z' - z} dz' \qquad (z,\ z' \text{ complex, C taken in the positive sense}). \tag{1.139}$$

For this relation to be useful the integral must be taken only over the region where the function is known, i.e. the real axis. Let C be the real axis plus a curve which continuously approaches infinity in the lower half-plane for instance a semi-circle with a radius approaching infinity (Fig. 1.11, drawn with variable ω instead of z). We obtain the wanted relation provided that: (a) the function is analytic in all this half-plane, (b) it approaches zero when the modulus of the variable approaches infinity so that the integral taken over the semi-circle is zero. Under such conditions, relation (1.139) is expressed as an integral over z' real. These conditions however impose z to be inside the contour and therefore to have an imaginary part strictly negative though one would wish to have only real quantities. Nevertheless z can be on the real axis, but then the expression on the left-hand side is divided by two since z is at the border (a rigorous proof is available). Finally if P represents the principal part of the integral at the singularity $x' = x$, then

$$\phi(x) = -\frac{1}{\pi i} P \int_{-\infty}^{+\infty} \frac{\phi(x')}{x' - x} dx' \qquad (x, x' \text{ real}). \tag{1.140}$$

The real and imaginary parts of this relation can be written separately. This shows that *if $\phi(x)$ satisfies the above conditions (a) and (b), i.e. it is analytic in the lower half plane and tends to zero when $|x|$ goes to infinity, some integral relations exist on the real domain of x between its imaginary and real parts.*

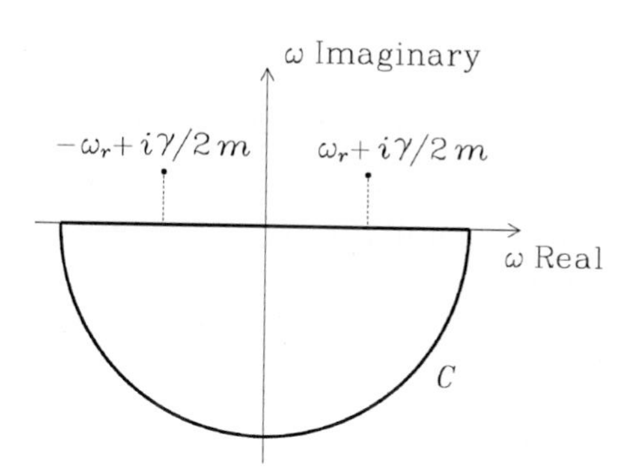

Fig. 1.11. Integration over a contour C *defined by the real axis and a semi-circle having its radius approaching infinity. If a function does not have any pole inside the contour it satisfies relation (1.139) (ω has the same role as the z variable). The poles of the scattering length $b(\omega)$ (1.138) have been represented. They are outside the contour*

The scattering length written in (1.138), which is a polynomial fraction of the variable ω, is analytic over any domain which does not contain its poles, i.e. the zeros of its denominator. These zeros, indicated in Fig. 1.11, are $i\gamma/2\pm\omega_r$ (ω_r depends on ω_0 and γ). The scattering length b satisfies condition (a) because the damping constant γ which is necessarily positive yields poles which are in the upper half plane. To satisfy condition (b) one could divide b by ω. The relations would then be valid in $b(\omega)/\omega$ (after replacing $\phi(x)$), but it is better to divide by ω^2 since we then get more general relations as we are going to comment. Let us notice that the division of b by ω^2 does not add any pole and does not change the domain of analyticity.

A relation such as (1.140) is not yet completely convenient because the physical domain does not extend over the entire real axis but only over its positive side (the variable is the radiation frequency). Integrating from 0 to ∞ is however sufficient since b verifies

$$b(-\omega) = b^*(\omega). \tag{1.141}$$

We should look if such a symmetry of the scattering length is attached to a particular model, or more general. A Fourier transform which transforms the above expression from ω to time space shows that this is simply the expression of a symmetry by time reversal. It is thus a general property which has however a limitation: this symmetry does not hold for magnetic moments so *the following expressions do not hold for magnetic scattering.* In that case, the equality (1.141) is written with a minus sign and different relations are obtained.[7] With this equality and a bit of algebra, one can rewrite the real and imaginary parts of (1.140). Replacing ϕ by $b(\omega)/\omega^2$ and x, x' by ω, ω' yields

$$\mathcal{R}\mathrm{e}[b(\omega)/\omega^2] = -\frac{2}{\pi} P \int_0^\infty \frac{\omega' \, \mathcal{I}\mathrm{m}[b(\omega')/\omega'^2]}{\omega'^2 - \omega^2} d\omega' \tag{1.142}$$

$$\mathcal{I}\mathrm{m}[b(\omega)/\omega^2] = \frac{2\omega}{\pi} P \int_0^\infty \frac{\mathcal{R}\mathrm{e}[b(\omega')/\omega'^2]}{\omega'^2 - \omega^2} d\omega' . \tag{1.143}$$

These relations are the so-called ***Kramers and Kronig or dispersion relations*** for the scattering length.

In this model, the proof we have given assumes all the poles of $b(\omega)/\omega^2$ to be above the real axis. We have inferred this from the positive value of the damping constant but *it can also be inferred from the principle of causality, which is of very general extent.* To understand the equivalence of these two hypotheses, positive value of the damping constant and principle of causality, it is worth returning to the resolution of the differential equation (1.99), which describes the movement of the electron in the incident field. We rewrite this equation by noting the displacement u instead of z to avoid any confusion with the variable z in the present section; for simplifying, u will be a scalar. The properties of u that we are going to discuss now are also the ones of the radiated field which is proportional to u.

$$m\ddot{u} + \gamma\dot{u} + m\omega_0^2 u = (-e)E_0 \, e^{i\omega t} . \tag{1.144}$$

A systematic method to solve such a differential equation with a right-hand side $f(t)$ consists in using the Green function of the equation. This method has been described in section 1.2.5. Let us recall that a solution of this kind of equation is given by

$$\begin{aligned} u(t) &= u_0(t) + \int_{-\infty}^{+\infty} G(t-t') f(t') \, dt' \\ &= \int_{-\infty}^{+\infty} G(t-t') f(t') \, dt' , \end{aligned} \tag{1.145}$$

where the Green function, $G(t)$, is solution of the equation with $\delta(t)$ instead of $f(t)$ in the right hand side; the solution $u_0(t)$ of the homogeneous equation (without the right hand side) becomes nearly zero after a certain amount of time due to damping. Writing the electron displacement $u(t)$ as in (1.145)

[7] In practice the same dispersion relations can be written for magnetic and non magnetic scattering lengths provided that the magnetic part is affected by a factor i.

allows the following physical interpretation to be given. The displacement u at a given time t is the result of the linear superposition of the excitation action f at any time t'; since the laws are invariant by time translation, the coefficient G only depends on the difference $t - t'$. $G(t)$ is obtained through its Fourier transform $g(\omega)$. We replace the right-hand side of equation (1.144) by $\delta(t)$, whose Fourier transform is one. The derivatives in the left-hand side transform into powers of ω, so we get

$$g(\omega) = -\frac{1}{m}\frac{1}{\omega^2 - \omega_0^2 - i\gamma\omega/m}, \tag{1.146}$$

which yields $G(t)$

$$G(t) = -\frac{1}{2\pi m}\int_{-\infty}^{+\infty} \frac{e^{i\omega t}}{\omega^2 - \omega_0^2 - i\gamma\omega/m} d\omega. \tag{1.147}$$

To calculate this integral, it is possible to integrate along a closed path in the complex plane: if the function does not have any pole inside the path of integration, its integral over it is zero. The poles of $g(\omega)$ are those of the scattering length that we have just discussed; the integral taken over the path of integration C (Fig. 1.11) is then zero. For $t < 0$ the integral over half the circle is also zero since the numerator is bound and the integral of $d\omega / \mid \omega \mid^2$ goes to zero when $\mid \omega \mid$ goes to infinity. Then $G(t) = 0$ for $t < 0$. It is important to mention that the proof depends on the position of the poles of $g(\omega)$, and on the positive sign of the damping constant. Conversely if $G(t)= 0$ for $t < 0$, it can be shown that $g(\omega)$ does not have any pole below the real axis and the Kramers-Kronig relations can be applied to $g(\omega)$ and $u(\omega)$. The condition $G(t)$ equal to zero at negative times constitutes the expression of *a causality principle, according to which an excitation given at a certain instant cannot produce any effect before this instant.* Making the dispersion relations to depend on this principle gives them a very general extent, beyond the cases where it is possible to clearly define some damping.

In our world most of the phenomena are irreversible and time is therefore asymmetric. The positive character of the damping and the principle of causality as discussed here, both constitute two manifestations of the irreversibility. We have proved that one or the other of these two principles yields the dispersion relations (1.143). But a question still remains. It is generally admitted that microscopic laws in physics are mainly symmetric with respect to time reversal. On the opposite, the irreversibility manifests itself in macroscopic phenomena and in statistical thermodynamics. We may be surprised that irreversibility is invoked in the scattering of radiation by an atom, which seems to be a rather elementary microscopic phenomenon. However one can also see that scattering involves some disorder. A plane wave travelling in vacuum, such as the incident wave constitutes a very unlikely state that can be considered as out of equilibrium. The ground state of a radiation is made of random waves in thermal equilibrium with neighbouring objects. Even considering a monochromatic radiation inside a perfectly reflecting box, it is only possible to find the initial wavelength that has been scattered into multiple disordered plane waves. One can see that the scattering of a plane wave by an atom is irreversible, a bit like the dilution of an alcohol droplet in a glass of water. The final state is the spherical wave moving away from the atom, superimposed to the incident wave which has a reduced amplitude. Therefore the unique incident plane wave has been changed into a superposition of plane waves travelling in all the directions. In addition, any of the plane components of the diverging wave can be associated to a particular movement of the atom since the momentum must be conserved. This is reminiscent of the dilution effect. If the scattering were reversible, one could produce the reverse operation: starting from a spherical wave converging towards an atom and from a plane wave, one could see the plane wave coming out with an increased amplitude. This would be difficult to realise and may be impossible. For this, one should correlate the different plane components of the converging wave to some particular movements of the atom. The difficulty is similar to the one which would be faced in an attempt to invert the dilution of the alcohol droplet by imposing to the molecules of the water-alcohol mixture some initial conditions such as the mixture would demix into two phases after a few instants. To what must be added the fact that we have simplified the problem not taking the absorption and the fluorescence in account. Multiple photons are then re-emitted for only one absorbed; in that case the radiation becomes still more disordered and this event is less reversible. As a matter of fact it appears that absorption and resonant scattering contribute much more to the dispersion than pure elastic scattering. From the above arguments one can be convinced that *even though the scattering looks like an elementary phenomenon, it is actually something irreversible, which has to obey the dispersion relations associated with irreversibility.*

1.5 X-Rays: Anisotropic Scattering

1.5.1 Introduction

In this section we present briefly some other types of x-ray scattering, observed essentially in crystalline materials. These are the magnetic scattering, which depends on the magnetic moment of the atom, and the Templeton anisotropic scattering, which depends on the neighbourhood of the atom in the crystal. A common feature to these two scattering effects is their anisotropy. The usual scattering amplitude which is described in the previous sections can be said isotropic because it depends on the incident and scattered polarisation directions through a unique factor, $\widehat{\mathrm{e}}_{\mathrm{sc}}.\widehat{\mathrm{e}}_{\mathrm{in}}$, independent of the orientation of the scattering object. The atomic scattering amplitudes which we discuss now can be said anisotropic because they depend on the orientation of the characteristic axes of the atom with respect to the incident and scattered polarisations. The characteristic axes may represent the magnetic moment direction of the atom if it exists, or the directions of the crystal field which eventually perturbs the state of that atom.

These scattering effects can take their origin from two different mechanisms. The first one is the interaction between the electromagnetic radiation and the spin of the electron. It produces some scattering, the so-called ***non resonant magnetic scattering***. This one is essentially independent of the binding of the electron in the atom, as is the Thomson scattering. A second type of anisotropic scattering arises as a part of the anomalous or resonant scattering, presented earlier in section 1.4.7. The atomic states $\mid i >$ and $\mid c >$ in (1.106), that is the initial state and the one in which the electron is promoted may be anisotropic. If that anisotropy originates from a magnetic moment, the resulting scattering is called ***resonant magnetic scattering***. If the anisotropy is some asphericity of the atom kept oriented by the peculiar symmetry of the material, it is the ***Templeton anisotropic scattering***.

X-ray magnetic scattering is a useful complement to neutron scattering. It can be used with some elements whose common isotopes strongly absorb thermal neutrons. The very good resolution (in all respects, position, angle and wavelength) of x-ray beams is an advantage for some studies. Since x-ray scattering depends on some characters of the magnetic moment in a way different from neutrons, it may raise some ambiguities left by neutron scattering experiments. One of these ambiguities is the ratio of the orbital to spin moment of the atom, because they contribute to neutron scattering exactly in the same way and cannot be discriminated from each other. x-ray amplitudes given by spin and orbital moment depend differently on the geometry of the experiment and they can be separated out. The resonant x-ray magnetic scattering is element dependent and eventually site dependent, which may give some useful information. It is also a spectroscopic method which probes the electronic state of the atom. The availability of small and brilliant x-ray beams compensates for the smallness of magnetic amplitudes in the study of thin films and multilayers. When the magnetic element has a very intense resonant magnetic scattering, even a single atomic layer can be probed.

Applications of Templeton anisotropic scattering are not yet fully developed. It can give some information on the orbital state of the atom in a crystal. Quite recently, a strong interest has developed about the so-called orbital ordering, where an alternating orientation of atomic orbitals can be detected with the help of this type of scattering.

In the present section we give a short description of the non resonant magnetic scattering, the resonant magnetic scattering and the Templeton anisotropic scattering. We also discuss the anisotropy of the optical index. The case of the magnetic neutron scattering is described in Chap. 5 of this book.

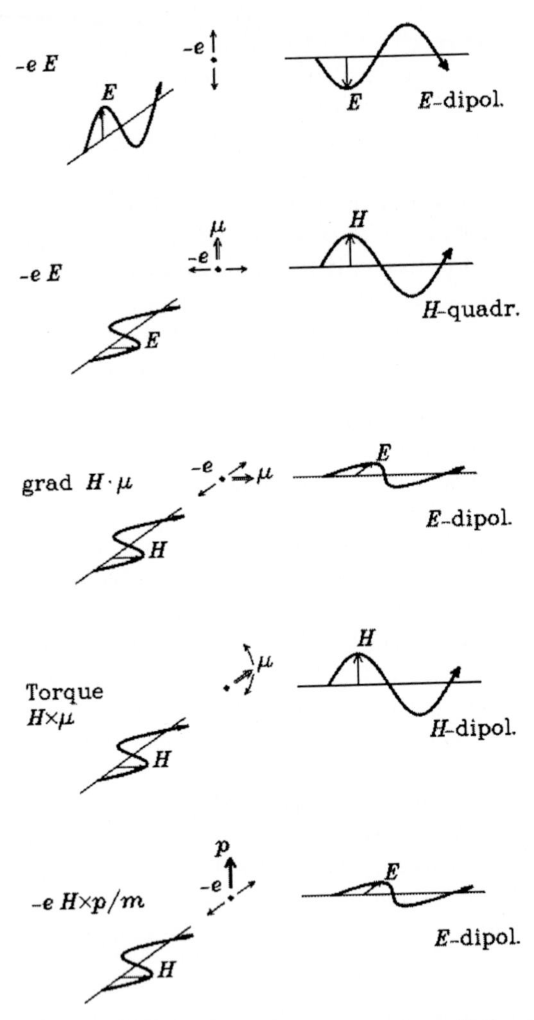

Fig. 1.12. The electron can scatter the electromagnetic radiation through a variety of processes. In each of them, the incident field moves the electron itself or its spin through a driving force on the left. The back and forth motion is indicated by a pair of thin opposite arrows. In this motion, the electron re-radiates through a mode indicated on the right. The first process is the well known Thomson scattering. Processes 2 to 4 describe the scattering by the spin, drawn as a double arrow. The process in the fifth line is a correction to Thomson scattering when the electron has a translation motion, indicated by the momentum $\mathbf{p}$. When integrated over the orbit of the electron in the atom, it gives rise to a scattering by the orbital moment

1.5.2 Non Resonant Magnetic Scattering

Similarly to the Thomson scattering, the non resonant magnetic scattering can be found either in the classical or quantum theory. The quantum calculation can be found in reference [10]. The spin of the electron is associated with a magnetic moment which, in a classical description, interacts with the magnetic component of the radiation. The Fig. 1.12 shows schematically how the interaction between the electromagnetic field and the electron, comprised of an electric charge and a magnetic moment can produce a magnetic dependent scattering. Having some interplay between spin and motion in space, and having some magnetic properties attached to an electrically charged particle are relativistic effects. That relativistic character introduces the scale factor

$$\left|\hbar \mathbf{q}\right| / 2\pi mc = 2\left(\lambda_c/\lambda\right)\sin\theta, \tag{1.148}$$

between the magnetic and Thomson scattering amplitudes of an electron. In a diffraction experiment that scale factor is typically of the order of 10^{-2}. Since only unpaired electrons, which are at most one or two tenths of all electrons of a magnetised atom, contribute to the magnetic scattering, the magnetic amplitude is in favorable cases 10^{-3} - 10^{-4} of the Thomson amplitude. The intensity of magnetic Bragg peaks of antiferromagnets is then affected by a factor of the order of 10^{-7}. The orbital moment contributes to the elastic scattering as well as the spin moment and with the same order of magnitude, but with a different dependence on wave vectors and polarisations. We write below the scattering length of an electron of spin **S** and orbital moment **L**

$$b_{mag} = -ir_e\left(\lambda_c/\lambda\right)\left[\left(\widehat{\mathbf{e}}_{\mathrm{sc}}^{*}.\overline{\overline{\mathbf{T}}}_S.\widehat{\mathbf{e}}_{\mathrm{in}}\right).\mathbf{S} + \left(\widehat{\mathbf{e}}_{\mathrm{sc}}^{*}.\overline{\overline{\mathbf{T}}}_L.\widehat{\mathbf{e}}_{\mathrm{in}}\right).\mathbf{L}\right]. \tag{1.149}$$

The tensors $\overline{\overline{\mathbf{T}}}_S$, $\overline{\overline{\mathbf{T}}}_L$ simply help to write these bilinear functions of the polarisations. Their elements are vectors. In their expression below, 2θ is the angle between $\widehat{\mathbf{k}}_{\mathrm{in}}$ and $\widehat{\mathbf{k}}_{\mathrm{sc}}$:

$$\overline{\overline{\mathbf{T}}}_S = \begin{matrix} \\ (s) \\ (p) \end{matrix} \begin{matrix} \quad(s) \qquad\qquad (p) \\ \begin{pmatrix} \widehat{\mathbf{k}}_{\mathrm{sc}} \times \widehat{\mathbf{k}}_{\mathrm{in}} & 2\widehat{\mathbf{k}}_{\mathrm{sc}}\sin^2\theta \\ -2\widehat{\mathbf{k}}_{\mathrm{in}}\sin^2\theta & \widehat{\mathbf{k}}_{\mathrm{sc}} \times \widehat{\mathbf{k}}_{\mathrm{in}} \end{pmatrix} \end{matrix} \tag{1.150}$$

$$\overline{\overline{\mathbf{T}}}_L = \begin{matrix} \\ (s) \\ (p) \end{matrix} \begin{matrix} \qquad(s) \qquad\qquad\qquad (p) \\ \begin{pmatrix} 0 & \left(\widehat{\mathbf{k}}_{\mathrm{sc}} + \widehat{\mathbf{k}}_{\mathrm{in}}\right)\sin^2\theta \\ -\left(\widehat{\mathbf{k}}_{\mathrm{sc}} + \widehat{\mathbf{k}}_{\mathrm{in}}\right)\sin^2\theta & 2\widehat{\mathbf{k}}_{\mathrm{sc}} \times \widehat{\mathbf{k}}_{\mathrm{in}}\sin^2\theta \end{pmatrix} \end{matrix}. \tag{1.151}$$

Remember that we use for i a sign opposite to the one used in quantum theory in the frame of which these equations are usually written.

Magnetic Compton scattering is also present but results only from the spin. Since $\hbar\mathbf{q}$ can be an important fraction of mc, the magnetic Compton amplitude can be significantly larger than the elastic one.

1.5.3 Resonant Magnetic Scattering

As explained in sections 1.4.4 the resonant, or dispersive, part of the scattering is based on the virtual excitation of an electron from a core level to an empty state, which can be just above the Fermi level. In the subsequent discussion in section 1.4.7, we have assumed that the polarisation factor was the same as for Thomson scattering, $\widehat{\mathbf{e}}_{\mathrm{sc}}.\widehat{\mathbf{e}}_{\mathrm{in}}$. This assumption in fact may be wrong. Let us write the numerator of a particular term in b_{disp1} (1.106), while making the dipolar approximation (the exponentials are reduced to 1)

$$< s|\widehat{\mathbf{e}}^*_{\mathrm{sc}}.\mathbf{p}|c >< c|\widehat{\mathbf{e}}_{\mathrm{in}}.\mathbf{p}|i > = \widehat{\mathbf{e}}^*_{\mathrm{sc}}.\overline{\overline{T}}_{\mathrm{res}}.\widehat{\mathbf{e}}_{\mathrm{in}}. \tag{1.152}$$

Again we express this bilinear function of $\widehat{\mathbf{e}}^*_{\mathrm{sc}}$, $\widehat{\mathbf{e}}_{\mathrm{in}}$ with a tensor $\overline{\overline{T}}_{\mathrm{res}}$. Instead of writing this tensor with the (s) and (p) polarisations as a basis, we may use a reference frame (x, y, z) attached to the medium, generally a crystal.

$$\overline{\overline{T}}_{\mathrm{res}} = \begin{matrix} \\ (x) \\ (y) \\ (z) \end{matrix} \begin{matrix} \begin{matrix} (x) & (y) & (z) \end{matrix} \\ \begin{pmatrix} a_1 & b_3 + ic_3 & b_2 + ic_2 \\ b_3 - ic_3 & a_2 & b_1 + ic_1 \\ b_2 - ic_2 & b_1 - ic_1 & a_3 \end{pmatrix} \end{matrix} . \tag{1.153}$$

With a_i, b_i, c_i being nine real coefficients, this is the most general tensor representing the expression (1.152). The actual structure of that tensor is determined by the symmetry of the scattering atom. The spherical symmetry is frequently a good approximation, though never completely exact in a crystal; then $\overline{\overline{T}}_{res}$ reduces to the unit matrix [11] and we recover the usual factor $\widehat{\mathbf{e}}^*_{\mathrm{sc}}.\widehat{\mathbf{e}}_{\mathrm{in}}$.

A case of lowering of the symmetry is the presence of a magnetic moment. We may observe that a time inversion, which should not change the scattering amplitude, exchanges the incident and scattering beams and reverses the magnetisation. This shows that the antisymmetrical part of $\overline{\overline{T}}_{res}$, that is the ic_i's, is of odd order in the magnetisation. That part has the form

$$\propto i\,(\widehat{\mathbf{e}}^*_{\mathrm{sc}} \times \widehat{\mathbf{e}}_{\mathrm{in}})\,.\widehat{\mathbf{z}}, \tag{1.154}$$

where $\widehat{\mathbf{z}}$ is just the direction of magnetisation, at least in the simplest cases.

These symmetry arguments should be completed by an explicit discussion of the physical process. The mechanism is described in reference [12] and shortly explained in Fig. 1.13 and caption. In the absence of any crystal field,

$$\begin{aligned} \widehat{\mathbf{e}}^*_{\mathrm{sc}}.\overline{\overline{T}}_{\mathrm{res}}.\widehat{\mathbf{e}}_{\mathrm{in}} \propto \widehat{\mathbf{e}}^*_{\mathrm{sc}}.\widehat{\mathbf{e}}_{\mathrm{in}}\,(F_{11} + F_{1-1}) + i\,(\widehat{\mathbf{e}}^*_{\mathrm{sc}} \times \widehat{\mathbf{e}}_{\mathrm{in}})\,.\widehat{\mathbf{z}}\,(F_{11} - F_{1-1}) \\ + (\widehat{\mathbf{e}}^*_{\mathrm{sc}}.\widehat{\mathbf{z}})\,(\widehat{\mathbf{e}}_{\mathrm{in}}.\widehat{\mathbf{z}})\,(2F_{10} - F_{11} - F_{1-1})\,, \end{aligned} \tag{1.155}$$

where F_{1-1}, F_{10} and F_{11} contain some transition probabilities. These transitions are described by two indices, the first one standing for the change in the orbital moment ΔL (1 in the dipolar term), and the second one for

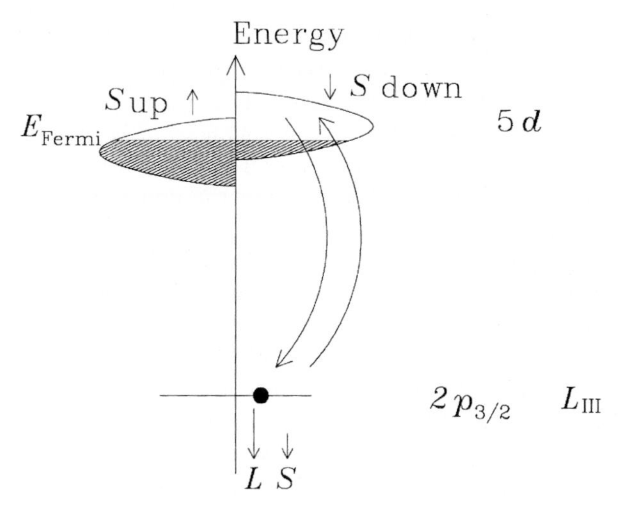

Fig. 1.13. Mechanism of the resonant magnetic scattering in the case of the L_{III} resonance of a third row transition element, such as platinum. Due to the magnetic moment the resonance occurs preferentially in the spin down ($S = -1/2$) half valence band, on the right. Because of the strong spin-orbit coupling in the core shell $2p$, the $2p_{3/2}$ level is completely separated out from the $2p_{1/2}$ and contributes alone to the resonance. Therefore the ($S = -1/2$) state involved in the resonance is coupled with a rather defined value of the $\widehat{z}$ component ($\widehat{z}$ the direction of magnetisation) of the orbital moment in the initial state of the electron. For a given polarisation of the radiation, this makes the amplitude to depend on the atom magnetisation direction, here the up direction

ΔL_z. The first term is the isotropic anomalous scattering, discussed in section 1.4.7. The second term is the just discussed antisymmetrical part. The third one depends on the axis along which the magnetisation is lying, but not on its sign; it is responsible for the magnetic linear dichroism. One should not forget that the above expression is to be multiplied by a resonance function of the energy showed in (1.136) and Fig. 1.10.

The spin orbit coupling is a key feature of this mechanism. In the example displayed in Fig. 1.13, the spin orbit coupling interaction is very large in the core state, but in some cases it may be present only in the excited state. In addition to the dipolar term written in the above formula, a quadrupolar one also exists. Though smaller, it cannot be neglected if it corresponds to a transition to a strongly magnetised atomic shell such as the $3d$ shell of transition elements, or $4f$ shell of lanthanides. It shows a quadrilinear dependence on $\widehat{\mathbf{e}}^*_{\text{sc}}, \widehat{\mathbf{e}}_{\text{in}}, \widehat{\mathbf{k}}_{\text{sc}}, \widehat{\mathbf{k}}_{\text{in}}$.

The order of magnitude of the resonant magnetic scattering may vary on a wide range. The K resonances, accessible for example in the $3d$ transition elements, have amplitudes which are comparable to the non resonant. Indeed the K shell has no orbital moment so that the effect relies on the spin orbit coupling in the valence shell, which is much less efficient. Furthermore the dipolar transition occurs to a weakly magnetised p valence shell, while the strongly magnetised d shell can give only a quadrupolar transition. The latter drawback limits also the $L_{II,III}$ resonances of lanthanides, but then the L shell is completely split by the spin orbit interaction. In that case the amplitude is typically ten times larger than the non resonant. The $L_{II,III}$ resonances of transition elements are favorable in all respects and enhance the amplitude by several orders of magnitude compared to non resonant. In the case of $3d$ elements, the long wavelength (of the order of 1.5 nm) can only fit long periodicities, and give diffraction on multilayer, or be used in reflectivity experiments. The L resonances of $5d$ transition elements arise in the 0.1 nm range but among those only the platinum group elements, and mainly the platinum itself, can take a magnetic moment. The $M_{IV,V}$ resonances of actinides offer the same favorable characters and are also quite effective, with amplitude enhancement by a factor of the order of one thousand. The wavelength, near 0.35 nm for the uranium allows for Bragg diffraction experiments.

1.5.4 Templeton Anisotropic Scattering

Even without any magnetic moment, the atom may show a low symmetry. Most often, this arises from the crystal field. A spontaneous orbital order (that is an orbital arrangement resulting mainly from the electrostatic interaction between orbitals of neighbouring atoms) is also expected in some materials. As a consequence the symmetrical part of (1.153), a_i, b_i, differs from the unit matrix. Again a quadrupolar term may exist. The Templeton scattering produces some change in the intensity of Bragg peaks at the absorption edges and this can be used to get more structural information. A striking feature is the occurrence of otherwise forbidden reflections [13–15]. When a reflection is forbidden because of a screw axis or glide plane, the amplitude cancels only if it is a scalar (that is independent on the orientation of the atom). For example with a screw axis, the atom rotates from one site to the next, so that the tensor amplitude in (1.152, 1.153) may not cancel. This breakdown of a crystallographic extinction rule should not be confused with the appearance of e.g. the 2 2 2 reflection in the diamond structure. In that case the structure factor is a scalar and the broken extinction rule is not a general rule for the space group; it applies only to a special position in the cell.

1.5.5 The Effect of an Anisotropy in the Index of Refraction

If the scattering is anisotropic, so may be the index of refraction and the optical properties. The non resonant magnetic scattering amplitude is zero in the forward direction. From the discussion in section 1.3 it cannot contribute to the refractive index. We shall therefore discuss only the consequences of resonant magnetic and of Templeton scattering. We give only some brief information on this question which could deserve quite a long development. The propagation of neutrons in magnetised materials and the associated reflectivity is examined in the chapter 5 of this book. It is different from, and somewhat simpler than the propagation of the electromagnetic radiation in an anisotropic medium, especially when the interaction is strong. The thermal neutron has a non relativistic motion which allows for a complete separation of the space and spin variables. The direction of propagation is in particular independent of the spin state. The electromagnetic radiation instead is fully relativistic, which intermixes the propagation and polarisation properties. In that case, the direction of propagation depends on the polarisation. Several unusual effects are consequently observed. For example the direction of a light ray may differ from the normal to the wave planes, or a refracted ray may lie outside of the plane of incidence.

Starting from the Helmholtz equation (1.15) the anisotropy of the medium modifies the dielectric constant ($\epsilon/\epsilon_0 \simeq n^2$) which becomes a tensor. Though we wrote the Helmholtz equation for the 4-vector A and the tensor should be of fourth order, all the useful coefficients are contained in a third order tensor acting in space, similar to (1.153). In the absence of magnetisation, we have the case of crystal optics, described in several textbooks, e.g. [3]. If the medium is magnetised, the antisymmetrical part of the tensor is non zero and some phenomena occur, such as the Faraday rotation of the polarisation plane or the magnetooptic Kerr effect (that is polarisation dependent and polarisation rotating reflectivity). The basic theory can be found in [16]. The theory as exposed in the textbooks is drawn in the dipolar approximation, which is legitimate in the range of the visible or near visible optics. I am not aware of a complete description of anisotropic optics including the quadrupolar terms. It seems reasonable in practice to use instead of such a full theory, some perturbative corrections since the quadrupolar resonance terms are always small. The incidence of the quadrupolar term is clear in an effect well observed with visible light and discovered nearly two centuries ago: it is the optical activity, that is the rotation of the polarisation plane in substances which lack a center of symmetry. Indeed the combination of a dipolar and a quadrupolar terms is required to produce such an effect. It is a small effect, since it corresponds to differences of the order of 10^{-4} between the indices of the two opposite circular polarisations. Yet it can be easily observed because the absorption of the visible light is still smaller and samples more than 10^5 wavelengths thick can be probed.

In the x-ray range, at energies of several keV and above, the refractive index differs from one by a small value and its anisotropic part is still smaller. Only a limited list of effects are observed and they are interpreted in simple terms. One of the most studied of those effects is the magnetic circular dichroism. In a ferro or ferrimagnet, where a net magnetisation is present, the refractive index changes, according to the helicity of circularly polarised x-rays being parallel or antiparallel to the magnetisation. Indeed the optical theorem (1.69) or its extension (1.83) yields the absorption atomic cross section or the dispersive part of the index of the medium, from the part of the scattering length written in (1.152, 1.155). For that we make $\widehat{\mathbf{k}}_{\mathrm{sc}}$ equal to $\widehat{\mathbf{k}}_{\mathrm{in}}$ and $\widehat{\mathbf{e}}_{\mathrm{sc}}$ equal to $\widehat{\mathbf{e}}_{\mathrm{in}}$. For a circular polarisation, the term (1.154) is real and reads

$$\pm\widehat{\mathbf{k}}_{\mathrm{in}}.\widehat{\mathbf{z}}. \tag{1.156}$$

The sign is switched from $-$ for the right handed helicity to $+$ for the left handed. This is to be multiplied by the complex resonance factor which we have left out from the formula. The difference in the real, δ, component of the index, between both helicities gives rise to the ***Faraday rotation*** of the polarisation plane. Similarly the difference in the imaginary, β, component gives rise to a difference in the absorption, called the ***magnetic circular dichroism***. Once a circularly polarised radiation is available, it is relatively easy to measure that change of absorption, usually by switching the magnetisation parallel or antiparallel to the beam. Similarly to the resonant scattering, the dichroism shows a spectrum in the region of the absorption edge.

At the L edges of the $3d$ elements, the resonances and their magnetic parts are quite large and the full optical theory, in the dipolar approximation, should be considered. Some reflectivity measurements have been done, e.g. [17].

1.A Appendix: the Born Approximation

Anne Sentenac, François de Bergevin, Jean Daillant, Alain Gibaud, and Guillaume Vignaud

In this appendix we give the Born development for the field scattered by a deterministic object.

In absence of any object, the field (scalar for simplicity) is solution of the homogeneous Helmholtz equation,

$$\left(\Delta + k_0^2\right) A_{\mathrm{in}}(\mathbf{r}) = 0. \tag{1.A1}$$

The object introduces a perturbation V on the differential operator, see Eqs. (1.16-17). In this case the field is solution of,

$$\left(\Delta + k_0^2 - V(\mathbf{r})\right) A(\mathbf{r}) = 0. \tag{1.A2}$$

The total field A can be written as the sum of an incident field $A_{\mathrm{in}}(\mathbf{r}) = A_{\mathrm{in}} e^{-i\mathbf{k}_{\mathrm{in}}.\mathbf{r}}$ (plane wave solution of the homogeneous equation) and a scattered field A_{sc} which satisfies the out-going wave boundary condition. Following section 1.2.5, we transform Eq. (1.A2) into an integral equation by introducing the Green function

$$G_-(\mathbf{r}) = -\frac{1}{4\pi}\frac{e^{-ik_0 r}}{r}, \tag{1.A3}$$

that satisfies outgoing wave boundary condition. We obtain

$$A(\mathbf{r}) = A_{\text{in}}(\mathbf{r}) + \int G_-(\mathbf{r}-\mathbf{r}')V(\mathbf{r}')A(\mathbf{r}')d\mathbf{r}'. \tag{1.A4}$$

Formally, one can write the solution of this integral equation in terms of a series in power of the convolution operator $[G_-V]$, [8]

$$A = A^{(0)} + A_{\text{sc}}^{(1)} + A_{\text{sc}}^{(2)} + \ldots \tag{1.A5}$$

with, $A^{(0)}(\mathbf{r}) = A_{\text{in}}(\mathbf{r})$,

$$A_{\text{sc}}^{(1)}(\mathbf{r}) = \int d^3r' G_-(\mathbf{r}-\mathbf{r}')V(\mathbf{r}')A_{\text{in}}(\mathbf{r}'),$$

$$A_{\text{sc}}^{(2)}(\mathbf{r}) = \int d^3r' \int d^3r'' G_-(\mathbf{r}-\mathbf{r}')V(\mathbf{r}')G_-(\mathbf{r}'-\mathbf{r}'')V(\mathbf{r}'')A_{\text{in}}(\mathbf{r}'').$$

When this series is convergent, one gets the exact value of the field. The main issue of such an expansion lies in its radius of convergence which is not easy to determine. Physically, the potential $V(\mathbf{r}')$ combined with the propagation operator G_- represents the action of the particule (or polarisation density) at $\mathbf{r}'$ on the incident wave, i. e. a scattering event. When the potential appears once, (in the first order term) the incident wave is singly scattered by the particules of the object. When it appears twice (in the second order term), one accounts for the double scattering events, etc. Eq. (1.A5) can also be viewed as a perturbative development in which the scattering event $[GV]$ is taken as a small parameter. The first Born approximation consists in stopping the development in Eq. (1.A5) to the first order in V (thus assuming the predominance of single scattering).

We now proceed by evaluating the scattered far-field and the scattering cross-section. We assume that the observation point $\mathbf{r}$ is far from all the points $\mathbf{r}'$ constituting the object (with respect to an arbitrary origin situated inside the object). In this case, one has

$$|\mathbf{r}-\mathbf{r}'| \approx r - \hat{\mathbf{u}}.\mathbf{r}', \tag{1.A6}$$

[8] In operator notation one can make an analogy with the Taylor expansion of $1/1-x = 1+x+x^2+\ldots$. Indeed, the field can be written as $A = A_{\text{in}}/1-[G_-V]$ which yields the series

$$A = A_{\text{in}} + [G_-V]A_{\text{in}} + [G_-V][G_-V]A_{\text{in}} + \ldots + [G_-V]^n A_{\text{in}} + \ldots,$$

with $[G_-V]f = \int G_-(\mathbf{r}-\mathbf{r}')V(\mathbf{r}')f(\mathbf{r}')d\mathbf{r}'$.

so that

$$G_-(\mathbf{r}-\mathbf{r}') = -\frac{1}{4\pi}\frac{e^{-ik_0|\mathbf{r}-\mathbf{r}'|}}{|\mathbf{r}-\mathbf{r}'|} \approx -\frac{1}{4\pi}\frac{e^{-ik_0 r}}{r}e^{ik_0\widehat{\mathbf{u}}.\mathbf{r}'}. \tag{1.A7}$$

Using this far-field approximation in Eq. (1.A4), one retrieves the expression given in section 1.2.4,

$$A_{\mathrm{sc}}(\mathbf{r}) = -A_{\mathrm{in}}b(\widehat{\mathbf{u}})\frac{e^{-ik_0 r}}{r}. \tag{1.A8}$$

Bearing in mind the Born development for the field, one can write the scattering length b in the form, $b(\widehat{\mathbf{u}}) = b^{(1)}(\widehat{\mathbf{u}}) + b^{(2)}(\widehat{\mathbf{u}}) + \dots$, with, for example

$$b^{(1)}(\widehat{\mathbf{u}}) = \frac{1}{4\pi}\int d^3r' V(\mathbf{r}')e^{i(k_0\widehat{\mathbf{u}}-\mathbf{k}_{\mathrm{in}}).\mathbf{r}'}. \tag{1.A9}$$

The calculation of the differential scattering cross-section, Eq. (1.36),

$$\frac{d\sigma}{d\Omega} = |b(\widehat{\mathbf{u}})|^2$$

is then straightforward.
It is worth noting that the perturbative development of the energy (which is proportional to the square of the field) starts at second order in V. Hence, to be consistent in our calculation, we should always develop the field up to the second order to account for all the possible terms (of order two) in the energy. A striking illustration of this remark is that the first Born approximation does not satisfy energy conservation. This can be readily shown by injecting the perturbative development of b in the optical theorem which is a direct consequence of the energy conservation. The optical theorem relates the total cross-section to the imaginary part of the forward scattered amplitude. One has, see section 1.3.2, $\sigma_{\mathrm{tot}} = 2\lambda\mathcal{I}m[b(\mathbf{k}_{\mathrm{in}})]$. If one disregards lossy media, the total cross-section is equal to the scattering cross-section,

$$\sigma_{\mathrm{sc}} = \int |b(\widehat{\mathbf{u}})|^2 d\Omega = 2\lambda\mathcal{I}m[b(\mathbf{k}_{\mathrm{in}})]. \tag{1.A10}$$

The expansion Eq. (1.A5) being a formally *exact* representation of the field, the optical theorem, written as a series, is verified at each order of the perturbative development. One gets, to the lowest order,

$$\int |b^{(1)}(\widehat{\mathbf{u}})|^2 d\Omega = 2\lambda\mathcal{I}m[b^{(1)}(\mathbf{k}_{\mathrm{in}}) + b^{(2)}(\mathbf{k}_{\mathrm{in}})]. \tag{1.A11}$$

This last equation shows clearly that if one wants the Born approximation to conserve energy, one should calculate the scattered amplitude up to second order *in the forward direction.*

References

1. L. Landau, E.M. Lifshitz, Course of theoretical physics vol. 8, Electrodynamics of continuous media, Pergamon Press, Oxford, 1960. L. Landau & E. Lifchitz. Physique théorique. Tome VIII, Electrodynamique des milieux continus. Ed. Mir, Moscou (1969).
2. J. D. Jackson. Classical Electrodynamics. 2nd Ed.. John Wiley & Sons (1975).
3. M. Born & E. Wolf. Principles of Optics. 4th Ed.. Oxford, New-York (1980).
4. R. W. James. The optical principles of the diffraction of X-rays. The cristalline state, vol II, Sir Lawrence Bragg ed.. G. Bell & Sons ltd, London (1962).
5. W. Schülke. Inelastic scattering by electronic excitations. In Handbook of synchrotron radiation, vol. 3, G. Brown & D. E. Moncton ed.. Elsevier Science Publisher (1991).
6. V.B. Berestetskii, E.M. Lifshithz, L.P. Pitaevskii, Course of theoretical physics vol. 4, Quantum electrodynamics, Pergamon Press, Oxford, 1982. E. Lifchitz & L. Pitayevski (L. Landau & E. Lifchitz). Physique théorique. Tome IV, Théorie quantique relativiste, Première partie. Ed. Mir, Moscou (1972).
7. International tables for x-ray crystallography; vol. III Physical and chemical tables; vol. IV Revised and supplementary tables to vol. II and III, The Kynoch Press, Birmingham, 1968.
8. E. Storm & H. I. Israel. Photon cross sections from 1 keV to 100 MeV for elements Z = 1 to Z = 100. *Nuclear Data Tables* **A7**, 565-681, (1970).
9. F. Stanglmeier, B. Lengeler, W. Weber, H. Göbel, M. Schuster, *Acta Cryst. A* **48**, 626 (1992).
10. M. Blume, *J. Appl. Phys.* **57**, 3615-3618, (1985).
11. D.H. Templeton and L.K.Templeton, *Acta Cryst. A* **36**, 237 (1980).
12. J. P. Hannon, G. T. Trammel, M. Blume & Doon Gibbs, *Phys. Rev. Lett.* **61**, 1245-1248, (1988).
13. V. M. Dmitrienko, *Acta Cryst. A* **39**, 29-35, (1983).
14. D.H. Templeton and L.K. Templeton, *Acta Cryst. A* **42**, 478 (1986).
15. A. Kirfel and A. Petcov, *Acta Cryst. A* **47**, 180 (1991).
16. A. V. Sokolov. Optical properties of metals. Blackie and Son Ldt, London (1967).
17. C. C. Kao, C. T. Chen, E. D. Johnson, J. B. Hastings, H. J. Lin, G. H. Ho, G. Meigs, J.-M. Brot, S. L. Hulbert, Y. U. Idzerda & C. Vettier, *Phys Rev. B* **50**, 9599-9602, (1994).

2 Statistical Aspects of Wave Scattering at Rough Surfaces

Anne Sentenac[1] and Jean Daillant[2]

[1] LOSCM/ENSPM, Université de St Jérôme, 13397 Marseille Cedex 20, France,
[2] Service de Physique de l'Etat Condensé, Orme des Merisiers, CEA Saclay, 91191 Gif sur Yvette Cedex, France

2.1 Introduction

The surface state of objects in any scattering experiment is, of necessity, rough. Irregularities are of the most varied nature and lengthscales, ranging from the atomic scale, where they are caused by the inner structure of the material, to the mesoscopic and macroscopic scale where they can be related to the defects in processing in the case of solid bodies or to fluctuations in the case of liquid surfaces (ocean waves, for example)

The problem of wave scattering at rough surfaces has thus been a subject of study in many research areas, such as medical ultrasonic, radar imaging, optics or solid-state physics [1], [2], [3], [4]. The main differences stem from the nature of the wavefield and the wavelength of the incident radiation (which determines the scales of roughness that have to be accounted for in the models). When tackling the issue of modelling a scattering experiment, the first difficulty is to describe the geometrical aspect of the surface. In this chapter, we are interested solely by surface states that are not well controlled so that the precise defining equation of the surface, $z = z(x, y)$ is unknown or of little interest. One has (or needs) only information on certain statistical properties of the surface, such as the height repartition or height to height correlations. In this probabilistic approach, the shape of the rough surface is described by a random function of space coordinates (and possibly time as well). The wave scattering problem is then viewed as a statistical problem consisting in finding the statistical characteristics of the scattered field (such as the mean value or field correlation functions), the statistical properties of the surface being given.

In the first section of this contribution we present the statistical techniques used to characterise rough surfaces. The second section is devoted to the description of a surface scattering experiment from a conceptual point of view. In the third section, we investigate to what extent the knowledge of the field statistics such as the mean field or field autocorrelation is relevant for interpreting the data of a scattering experiment which deals necessarily with deterministic rough samples. Finally, we derive in the fourth section a simple expression of the scattered field and scattered intensity from random rough surfaces under the Born approximation.

2.2 Description of Randomly Rough Surfaces

2.2.1 Introduction

Let us first consider the example of a liquid surface. The exact morphology of the surface is rapidly fluctuating with time and is not accessible inasmuch as the detector will integrate over many different surface shapes. However statistical information can be obtained and it provides an useful insight on the physical processes. Indeed, these fluctuations obey Boltzmann statistics and are characterised by a small number of relevant parameters such as the density of the liquid or its surface tension (see Chap. 9).

We now consider a set of surfaces of artificial origin (such as metallic optical mirrors) that have undergone similar technological treatments (like polishing and cleaning). Since it is impossible to reproduce all the microscopic factors affecting the surface state, these surfaces have complex and completely different defining equations $z = z(x, y)$. However, if the surface processing is well enough controlled, they will present some similarities, of statistical nature, that will distinguish them from surfaces that have received a totally different treatment.

In these two examples, we are faced with the issue of describing a set of real surfaces which present similar statistical properties and whose defining equations $z(x, y)$ are unknown or of small interest (see Fig. 2.2.1). It appears convenient [2] to approximate this set of surfaces by a statistical ensemble of surfaces that are realisations of a random continuous process of the plane coordinates $\mathbf{r}_\| = (x, y)$, whose statistical properties depend on some relevant parameters of the physical processes affecting the surface state (like the grain size of the polishing abrasive in the case of surfaces of artificial origin). It is likely that the characteristic functions $z(\mathbf{r}_\|)$ of the surfaces generated by the random process will be different from that of the real surfaces under study, but the statistical properties of both ensembles should be the same.

2.2.2 Height Probability Distributions

Generally speaking, a random rough surface is completely described statistically by the assignement of the n-point $(n \to \infty)$ height probability distribution $p_n(\mathbf{r}_{1\|}, z_1 ... \mathbf{r}_{n\|}, z_n)$ where $p_n(\mathbf{r}_{1\|}, z_1 ... \mathbf{r}_{n\|}, z_n) dz_1 ... dz_n$ is the probability for the surface points of plane coordinates $\mathbf{r}_{1\|}, ... \mathbf{r}_{n\|}$ of being at the height between $(z_1 ... z_n)$ and $(z_1 + dz_1 ... z_n + dz_n)$. However, in most cases, we restrict the description of the randomly rough surface to the assignement of the one and two-points distribution functions $p_1(\mathbf{r}_\|, z)$, and $p_2(\mathbf{r}_{1\|}, z_1; \mathbf{r}_{2\|}, z_2)$. Indeed, most scattering theories need solely this information.

From these probability functions, one can calculate the ensemble average of any functional of the random variables $(z_1 ... z_n)$ where $z_i = z(\mathbf{r}_{i\|\mathbf{r}_{i\|}})$,

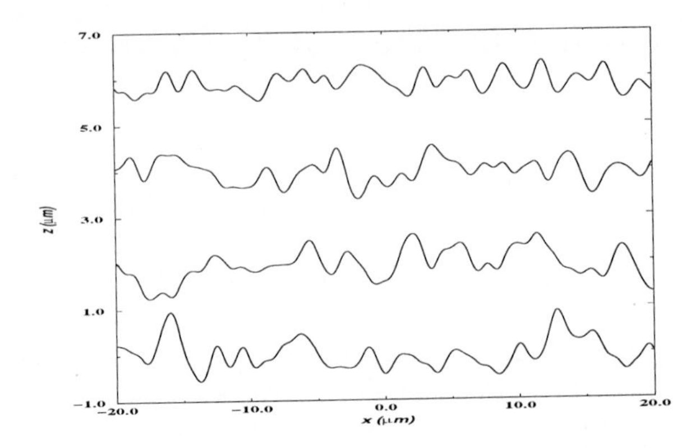

Fig. 2.1. Examples of various rough surfaces that present the same Gaussian statistical properties

through the integral,

$$\langle F\rangle(\mathbf{r}_{1\|}...\mathbf{r}_{n\|}) = \int_{-\infty}^{\infty} F(z_1...z_n)p_n(\mathbf{r}_{1\|}, z_1...\mathbf{r}_{n\|}, z_n)dz_1...dz_n. \quad (2.1)$$

The domain of integration covers all the possible values for $(z_1...z_n)$. This quantity is equivalent to an average of F calculated over an ensemble of surface realisations S_p,

$$\langle F\rangle(\mathbf{r}_{1\|}...\mathbf{r}_{n\|}) = \lim_{N\to\infty}\frac{1}{N}\sum_{p=1}^{N} F(z_1^p...z_n^p), \quad (2.2)$$

where z_j^p is the altitude of the p-th surface realisation at plane coordinates $\mathbf{r}_{j\|}$.
With this definition, one obtains in particular the mean height of the surface through

$$\langle z\rangle(\mathbf{r}_\|) = \int_{-\infty}^{\infty} z(\mathbf{r}_\|)p_1(\mathbf{r}_\|, z)dz. \quad (2.3)$$

The mean square height of the surface is given by

$$\langle z^2\rangle(\mathbf{r}_\|) = \int_{-\infty}^{\infty} z^2(\mathbf{r}_\|)p_1(\mathbf{r}_\|, z)dz. \quad (2.4)$$

The height-height correlation function C_{zz} is defined by

$$C_{zz}(\mathbf{r}_{1\|}, \mathbf{r}_{2\|}) = \langle z_1 z_2\rangle = \int_{-\infty}^{\infty} z_1 z_2 p_2(\mathbf{r}_{1\|}, z_1, \mathbf{r}_{2\|}, z_2)dz_1 dz_2, \quad (2.5)$$

where $z_j = z(\mathbf{r}_{j\|})$. It is also usual to introduce the pair-correlation function $g(\mathbf{r}_{1\|}, \mathbf{r}_{2\|})$ which averages the square of the difference height between two points of the surface,

$$g(\mathbf{r}_{1\|}, \mathbf{r}_{2\|}) = \langle (z_1 - z_2)^2 \rangle = \int_{-\infty}^{\infty} (z_1 - z_2)^2 p_2(\mathbf{r}_{1\|}, z_1, \mathbf{r}_{2\|}, z_2) dz_1 dz_2. \quad (2.6)$$

Note that $g(\mathbf{r}_{1\|}, \mathbf{r}_{2\|}) = 2\langle z^2 \rangle(\mathbf{r}_\|) - 2C_{zz}(\mathbf{r}_{1\|}, \mathbf{r}_{2\|})$.

2.2.3 Homogeneity and Ergodicity

Randomly rough surfaces have frequently the property that the character of the height fluctuations z does not change with the location on the surface. More precisely, if all the probability distribution functions p_i are invariant under any arbitrary translation of the spatial origin, the random process is called homogeneous. As a consequence, the ensemble average of the functional $F(z_1 ... z_n)$ will depend only on the vector difference, $\mathbf{r}_{j\|} - \mathbf{r}_{1\|}$ between one of the n space argument $\mathbf{r}_{1\|}$ and the $(n-1)$ remaining others $\mathbf{r}_{j\|}$, $j = 2...n$.

$$\langle F \rangle (\mathbf{r}_{1\|}, ..., \mathbf{r}_{n\|}) = \langle F \rangle (\mathbf{0}_\| ... \mathbf{r}_{n\|} - \mathbf{r}_{1\|}). \quad (2.7)$$

When the random process is isotropic (i.e has the same characteristics along any direction) the dependencies reduce to the distance $|\mathbf{r}_{j\|} - \mathbf{r}_{1\|}|$ between one of the space argument and the others. Hereafter we will only consider homogeneous isotropic random processes and we propose a simplified notation for the various functions already introduced.

The mean altitude $\langle z \rangle (\mathbf{r}_\|)$ does not depend on the $\mathbf{r}_\|$ position and one can find a reference plane surface such as $\langle z \rangle = 0$. The mean square deviation of the surface is also a constant and we define the root mean square (rms) height σ, as

$$\sigma^2 = \langle z^2 \rangle = \int_{-\infty}^{\infty} z^2 p_1(z) dz. \quad (2.8)$$

The rms height is often used to give an indication of the "degree of roughness", the larger σ the rougher the surface. Note that the arguments of the probability distribution are much simpler.
Similarly, the height-height correlation function can be written as,

$$C_{zz}(\mathbf{r}_{1\|}, \mathbf{r}_{2\|}) = \langle z(\mathbf{0}_\|) z(\mathbf{r}_\|) \rangle = C_{zz}(r_\|) = \int z_1 z_2 p_2(z_1, z_2, r_\|) dz_1 dz_2, \quad (2.9)$$

where $r_\| = |\mathbf{r}_\||$. We also introduce, with these simpler notations, the one point and two points characteristic functions,

$$\chi_1(s) = \int_{-\infty}^{\infty} p_1(z) e^{isz} dz, \quad (2.10)$$

$$\chi_2(s, s', r_{\parallel}) = \int_{-\infty}^{\infty} p_2(z, z', r_{\parallel}) e^{isz+is'z'} dz dz'. \quad (2.11)$$

One of the most important attributes of a homogeneous random process is its power spectrum, $P(\mathbf{q}_{\parallel})$ that gives an indication on the strength of the surface fluctuations associated with a particular wavelength. Roughly speaking, the rough surface is regarded as a superposition of gratings with different periods and heights. The power spectrum is a tool that relates the height to the period. We introduce the Fourier transform of the random variable z,

$$\tilde{z}(\mathbf{q}_{\parallel}) = \frac{1}{4\pi^2} \int z(\mathbf{r}_{\parallel}) e^{i\mathbf{q}_{\parallel} \cdot \mathbf{r}_{\parallel}} d\mathbf{r}_{\parallel}, \quad (2.12)$$

where $\mathbf{q}_{\parallel} = (q_x, q_y)$ is the in-plane wave-vector transfer. We define the spectrum as

$$P(\mathbf{q}_{\parallel}) = \langle |\tilde{z}(\mathbf{q}_{\parallel})|^2 \rangle = \langle \tilde{z}(\mathbf{q}_{\parallel}) \tilde{z}(-\mathbf{q}_{\parallel}) \rangle. \quad (2.13)$$

The Wiener-Kintchine theorem [5] states that the power spectrum is the Fourier transform of the correlation function,

$$P(\mathbf{q}_{\parallel}) = \frac{1}{4\pi^2} \int d\mathbf{r}_{\parallel} e^{i\mathbf{q}_{\parallel} \cdot \mathbf{r}_{\parallel}} \langle z(\mathbf{0}_{\parallel}) z(\mathbf{r}_{\parallel}) \rangle = \tilde{C}_{zz}(\mathbf{q}_{\parallel}). \quad (2.14)$$

More precisely, one shows that

$$\langle \tilde{z}^*(\mathbf{q}_{\parallel}) \tilde{z}(\mathbf{q}'_{\parallel}) \rangle = \langle \tilde{z}(-\mathbf{q}_{\parallel}) \tilde{z}(\mathbf{q}'_{\parallel}) \rangle = \tilde{C}_{zz}(\mathbf{q}_{\parallel}) \delta(\mathbf{q}_{\parallel} - \mathbf{q}'_{\parallel}). \quad (2.15)$$

The Fourier components of a homogeneous random variable are independent random variables, whose mean square dispersion is given by the Fourier transform of the correlation function. If the power spectrum decreases slowly with increasing $q_{\parallel}$, the roughness associated to small periods will remain important. Thus, whatever the length scale, the surface will present irregularities. In the real space, it implies that the correlation between the heights of two points on the surface will be small, whatever their separation. As a result, the correlation function will exhibit a singular behavior about 0 (discontinuity of the derivative for exemple). An illustration of the influence of the correlation function (or power spectrum) on the roughness aspect of the surface is presented in Fig. 2.2 and detailed in Sec. 2.2.4 in the special case of a Gaussian distribution of heights.

Until now we have been interested solely in ensemble average, which necessitates the knowledge of the complete set of rough surfaces generated by the homogeneous random process (or the probability distributions). However, sometimes only a single realisation S_p (with dimension L_x, L_y along Ox and Oy) of the random process is available and one defines the spatial average of any functional $F(z_1, ..., z_n)$ for this surface by,

$$\bar{F}_p(\mathbf{0}_{\parallel}, ..., \mathbf{r}_{n\parallel}) = \lim_{L_x \times L_y \to \infty} \frac{1}{L_x L_y} \int_{L_x \times L_y} d\mathbf{r}'_{\parallel} F[z(\mathbf{r}'_{\parallel}) ... z(\mathbf{r}'_{\parallel} + \mathbf{r}_{n\parallel})]. \quad (2.16)$$

It happens frequently that each realisation of the ensemble carries the same statistical information about the homogeneous random process as every other realisation. The spatial averages calculated for any realisation are then all equal and coincide with the ensemble average. The homogeneous random process is then said to be an ergodic process. In this case, the following particular relations hold,

$$\sigma^2 = \langle z^2 \rangle = \lim_{L_x, L_y \to \infty} \frac{1}{L_x L_y} \int_{L_x \times L_y} z^2(\mathbf{r}_\parallel) d\mathbf{r}_\parallel, \tag{2.17}$$

$$C_{zz}(r_\parallel) = \langle z(\mathbf{0}_\parallel) z(\mathbf{r}_\parallel) \rangle = \lim_{L_x, L_y \to \infty} \frac{1}{L_x L_y} \int_{L_x \times L_y} z(\mathbf{r}'_\parallel) z(\mathbf{r}'_\parallel + \mathbf{r}_\parallel) d\mathbf{r}'_\parallel. \tag{2.18}$$

One can show that Eqs. (2.17) and (2.18) will be satisfied if the correlation function $C_{zz}(r_\parallel)$ dies out sufficiently rapidly with increasing $r_\parallel$ (see for demonstration [5]). Indeed, this property implies that one realisation of the rough surface can be divided up into subsurfaces of smaller area that are uncorrelated so that an ensemble of surfaces can be constructed from a single realisation. Spatial averaging amounts then to ensemble averaging. If the random process is homogeneous and ergodic, all the realisations will look similar while differing in detail. This is exactly what we expect in order to describe liquid surfaces varying with time or set of surfaces of artificial origin. The fact that spatial averaging is equivalent to ensemble averaging when the surface contains enough correlation lengths to recover all the information about the random process is of crucial importance in statistical wave scattering theory.

2.2.4 The Gaussian Probability Distribution and Various Correlation Functions

In most theories, the height probability distribution is taken to be Gaussian. The Gaussian distribution plays a central role because it has an especially simple structure, and, because of the central limit theorem, it is a probability distribution that is encountered under a great variety of different conditions. If the height z of a surface is due to a large number of local independent events whose effects are cumulative, (like the passage of grain abrasive) the resulting altitude will obey nearly Gaussian statistics. This result is a manifestation of the central limit theorem which states that if a random variable X is the sum of N independent random variables x_i, it will have a Gaussian probability distribution in the limit of large N. Hereafter, we suppose that the average value of the Gaussian variate $z(\mathbf{r}_\parallel)$ is null, $\langle z \rangle = 0$. The Gaussian height distribution function is written as,

$$p_1(z) = \frac{1}{\sigma\sqrt{2\pi}} \exp\left(-\frac{z^2}{2\sigma^2}\right) \tag{2.19}$$

Gaussian variates have the remarkable property that the random process is entirely determined by the height probability distribution and the height-height correlation function C_{zz}. All higher order correlations are expressible in terms of second order correlation [5]. The two-points distribution function is given in this case by,

$$p_2(z, z', r_{\parallel}) = \frac{1}{2\pi\sqrt{\sigma^4 - C_{zz}^2(r_{\parallel})}} \exp - \left[\frac{\sigma^2(z^2 + z'^2) - 2zz'C_{zz}(r_{\parallel})}{2\sigma^4 - 2C_{zz}^2(r_{\parallel})}\right]. \tag{2.20}$$

Other useful results on the Gaussian variates are,

$$\chi_1(s) = \langle e^{isz} \rangle = e^{-s^2\sigma^2/2}, \tag{2.21}$$

$$\chi_2(s, s', r_{\parallel}) = \langle e^{i(sz - s'z')} \rangle = e^{-\sigma^2(s^2+s'^2)/2} e^{ss'C_{zz}(r_{\parallel})}. \tag{2.22}$$

The correlation function plays a fundamental role in the surface aspect. It provides an indication on the length scales over which height changes along the surface. It gives in particular the distance beyond which two points of the surface can be considered independent. If the surface is truly random, $C_{zz}(r_{\parallel})$ decays to zero with increasing $r_{\parallel}$. The simplest and often used form for the correlation function is also Gaussian,

$$C_{zz}(r_{\parallel}) = \sigma^2 \exp(-r_{\parallel}^2/\xi^2). \tag{2.23}$$

The correlation length ξ is the typical distance between two different irregularities (or bumps) on the surface. Beyond this distance, the heights are not correlated.

In certain scattering experiments, one can retrieve the behaviour of the correlation function for $r_{\parallel}$ close to zero. We have thus access to the small scale properties of the surface. We have seen that the regularity of the correlation function at zero mirrors the asymptotic behaviour of the power spectrum : the faster the high-frequency components of the surface decay to zero, the smoother the correlation function about zero. The Gaussian scheme whose variations about zero have the quadratic form $\sigma^2(1 - (r_{\parallel}/\xi)^2)$ is thus indicated solely for surfaces that present only one typical lateral length scale [6].

For surfaces with structures down to arbitrary small scales, one expects the correlation function to be more singular at zero. An example is the self affine rough surface for which,

$$g(r_{\parallel}) = A_0 r_{\parallel}^{2h} \tag{2.24}$$

where A_0 is a constant, or

$$C_{zz}(r_{\parallel}) = \sigma^2 \left(1 - \frac{r_{\parallel}^{2h}}{\xi^{2h}}\right), \tag{2.25}$$

with $0 < h < 1$. The roughness exponent or Hurst exponent h is the key parameter which describes the height fluctuations at the surface: small h values produce very rough surfaces while if h is close to 1 the surface is more regular. This exponent is associated to fractal surfaces with dimension $D = 3 - h$ as reported by Mandelbrodt [7]. The pair-correlation function given in Eq. (2.24) diverges for $r_{\|} \to \infty$. Hence, all the lengthscales along the vertical axis are represented and the roughness of the surface cannot be defined. We will see below that in that case, there is no specular reflection. However, very often, some physical processes limit the divergence of the correlation function, i.e. the roughness saturates at some in-plane cut-off ξ. Such surfaces are well described by the following correlation function,

$$C_{zz}(R) = \sigma^2 \exp(-\frac{R^{2h}}{\xi^{2h}}) \tag{2.26}$$

For liquid surfaces and surfaces close to the roughening transitions other functional forms described in Chaps. 6 and 9 are used.

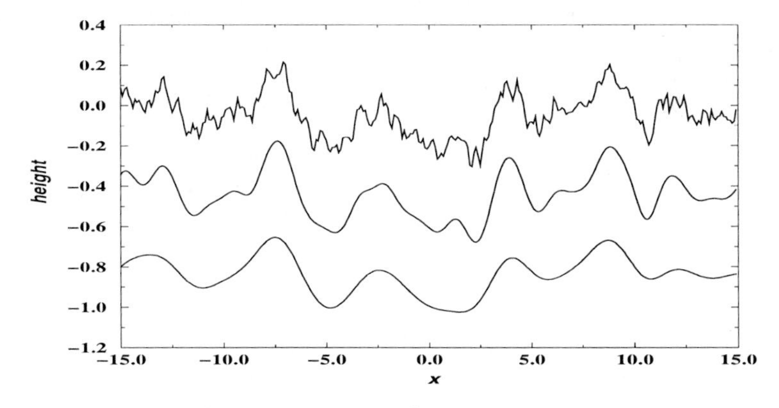

Fig. 2.2. Various rough surfaces with Gaussian height distribution but various correlation functions From bottom to top, $C_{zz}(R) = \sigma^2\xi^4/(\xi^2 + R^2)^2$, $C_{zz}(R) = \sigma^2 \exp(-\frac{R^2}{\xi^2})$, $C_{zz}(R) = \sigma^2 \exp(-\frac{R}{\xi})$

2.2.5 More Complicated Geometries: Multilayers and Volume Inhomogeneities

Up to now we have considered solely the statistical description of a rough surface separating two homogeneous media. The mathematical notions that have been introduced can be generalised to more complicated problems such as stacks of rough surfaces in multilayer components. In this case, one must also consider the correlation function between the different interfaces, $\langle z_i(\mathbf{0}_{\|}) z_j(\mathbf{r}_{\|})\rangle$

where z_i represents the height of the ith surface. A detailed description of the statistics of a rough multilayer is given in chap. 8, Sect. 8.2. One can also describe in a similar fashion the random fluctuations of the refractive index (or electronic density) ρ. In this case ρ is a random continuous variable of the three-dimensional space coordinates $(\mathbf{r}_{\|}, z)$. It will be introduced in Chap. 4 Sect. 4.5.

2.3 Description of a Surface Scattering Experiment, Coherence Domains

We have seen how to characterise, with statistical tools, the rough surface geometry. The next issue is to relate these statistics to the intensity scattered by the sample in a scattering experiment. In this section, we introduce the main theoretical results that describe the interaction between electromagnetic waves and surfaces. Attention is drawn on the notion of "coherence domains" which takes on particular importance in the modelling of scattering from random media. In this foreword, we present briefly the basic mechanisms that subtend this concept.

It can be shown (bear in mind the Huygens-Fresnel principle or see Chap. 4 Sect. 4.1.6) that a rough surface illuminated by an electromagnetic incident field acts as a collection of radiating secondary point sources. The superposition of the radiation of those sources yields the total diffracted field. If the secondary sources are coherently illuminated, the total diffracted field is the sum of the complex amplitudes of each secondary diffracted beam. In other words, one has to account for the phase difference in this superposition. As a result, an interference pattern is created. The coherence domain is the surface region in which all the radiating secondary sources interfere. It depends trivially on the nature of the illuminating beam (which can be partially coherent), but more importantly, it depends on the angular resolution of the detector. To illustrate this assertion, we consider the Young's holes experiment [8]. Light from a monochromatic point source (or a coherent beam) falls on two pinholes located in the sample plane (see Fig. 2.3). We study the transmitted radiation pattern on a screen parallel to the sample plane at a distance D. In this region, an interference pattern is formed. The periodicity Λ of the fringes, which is the signature of the coherence between the two secondary sources, depends on the separation d between the two pinholes, $\Lambda = \lambda D/d$. Suppose now that a detector is moved on the screen to record the diffrated intensity. As long as the detector width l is close to Λ, the modulation of the interference pattern will be detected. On the contrary, if $l > 10\Lambda$ the intensity measured by the detector is the average of the fringe intensities. We obtain a constant equal to the sum of the intensities scattered by each secondary sources. In this case, one may consider that, from the detector point of view, the sources radiate in an incoherent way. We see with this simple experiment that the coherence length is directly linked to the finite extent of the detector (equivalent to a finite angular resolution).[1]

We now turn to a more accurate description of a surface scattering experiment.

2.3.1 Scattering Geometry

We consider an ideal scattering experiment consisting in illuminating a rough sample with a (perfectly coherent monochromatic) beam directed along $\mathbf{k}_{\mathrm{in}}$ and detecting the flux of Poynting vector in an arbitrary small solid angle in the direction $\mathbf{k}_{\mathrm{sc}}$ with a point-like detector located in the far-field region.

The interaction of the beam with the material results in a wavevector transfer,

$$\mathbf{q} = \mathbf{k}_{\mathrm{sc}} - \mathbf{k}_{\mathrm{in}} . \tag{2.27}$$

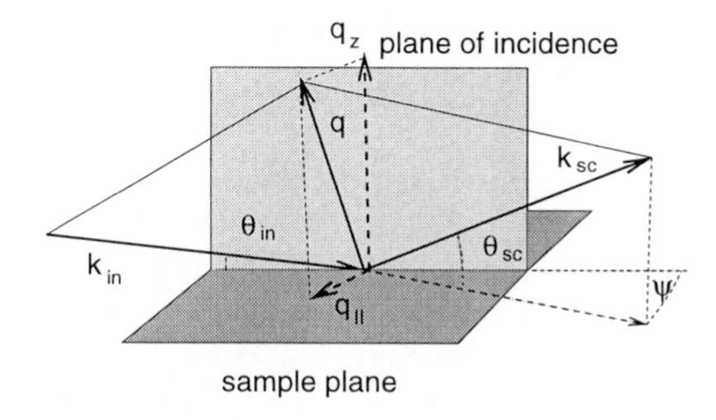

Fig. 2.3. Scattering geometry for interpreting surface scattering

Figure 2.3 shows the scattering geometry in the general case of a surface experiment. The plane of incidence contains the incident wave-vector $\mathbf{k}_{\mathrm{in}}$ and the normal to the surface Oz. In a reflectivity experiment, it is usual to work in the plane of incidence and thus to have ψ=0. Yet the case $\psi \neq 0$ is of special interest for surface diffraction experiments in grazing incidence geometry. When working in the plane of incidence it is also useful to distinguish the symmetric specular geometry for which $\theta_{\mathrm{in}} = \theta_{\mathrm{sc}}$ and the off-specular geometry for which $\theta_{\mathrm{in}} \neq \theta_{\mathrm{sc}}$. The following set of equations (2.28) gives the components of the wave-vector transfer with the notations introduced in Fig. 2.3.

[1] It is also obviously linked to the degree of coherence fixed by, for example, the incidence slit opening. However, for x-ray or neutron experiments the resolution is actually generally limited by the detector slits opening.

$$\begin{cases} q_x = k_0 \left(\cos\theta_{\rm sc} \cos\psi - \cos\theta_{\rm in} \right) \\ q_y = k_0 \left(\cos\theta_{\rm sc} \sin\psi \right) \\ q_z = k_0 \left(\sin\theta_{\rm sc} + \sin\theta_{\rm in} \right) \end{cases} \tag{2.28}$$

2.3.2 Scattering Cross-Section

In the ideal experimental setup presented in the previous section, one exactly measures the differential scattering cross-section as described in Fig. 1.1 of Chap. 1, (the isolated scattering object is the rough sample in this case). The vectorial electric field $\mathbf{E}$ is written as the sum,

$$\mathbf{E} = \mathbf{E}_{\rm in} + \mathbf{E}_{\rm sc} \tag{2.29}$$

of the incident plus scattered field. We are interested by the flux of the Poynting vector $\mathbf{S}$ through a surface dS located at the position $\mathbf{R}$ of the detector for a unit incident flux. The precise calculations of the differential scattering cross-section are detailed in Sect. 4.1.6. In this paragraph, we simply introduce the main steps of the derivation.
One assumes that the detector located at $\mathbf{R}$ is placed far from the sample (far-field approximation). We define the scattering direction by the vector $\mathbf{k}_{\rm sc}$ see Fig. 2.3,

$$\mathbf{k}_{\rm sc} = k_0 \widehat{\mathbf{u}} = k_0 \mathbf{R}/R. \tag{2.30}$$

It is shown in Sect. 4.1.6 that the scattered field can be viewed as the sum of the wavelets radiated by the electric dipoles induced in the material by the incident field, (these radiating electric dipoles are the coherent secondary sources presented in the introduction). The strength of the induced dipole located at $\mathbf{r}'$ in the sample, is given by the total field times the permittivity contrast at this point, $[k^2(\mathbf{r}') - k_0^2]\mathbf{E}(\mathbf{r}')$. Let us recall that for x-rays,

$$(k^2(\mathbf{r}') - k_0^2) = k_0^2[n^2(\mathbf{r}') - 1] = -4\pi r_e \rho_{el}(\mathbf{r}'), \tag{2.31}$$

where ρ_{el} is the local electron density and r_e the classical electron radius.[2] In the far-field region, the scattered field can be written as, see Eq. (4.19), (the

[2] If one is only interested in materials with low atomic numbers for which the x-ray frequency is much larger than all atomic frequencies, the electrons can be considered as free electrons plunged into an electric field $\mathbf{E}$. In this case, the movement of the electron is governed by $m_e d\mathbf{v}/dt = -e\mathbf{E}$, where m_e, $\mathbf{v}$, $-e$, are the mass, the velocity and the charge of the electron. We find $\mathbf{v} = (ie/m_e\omega)\mathbf{E}$ for a $e^{i\omega t}$ time dependence of the electric field. Thus the current density is $\mathbf{j} = -e\rho_{el}\mathbf{v} = -(ie^2\rho_{el}/m_e\omega)\mathbf{E}$ where ρ_{el} is the local electron density. Writing the Maxwell's equations in the form $curl\mathbf{H} = \mathbf{j} + \epsilon_0 \partial\mathbf{E}/\partial t = \partial\mathbf{D}/\partial t = n^2\epsilon_0\partial\mathbf{E}/\partial t$ (depending on whether the system is viewed as a set of electrons in a vacuum or as a material of refractive index n), one obtains by identification that $n = 1 - (e^2/2m_e\epsilon_0\omega^2)\rho_{el} = 1 - (\lambda^2/2\pi)r_e\rho_{el} \approx 1 - 10^{-6}$, with $r_e = (e^2/4\pi\epsilon_0 m_e c^2)$ the "classical electron radius". A complete and rigourous demonstration is given in Ref. [9]

far-field approximation and its validity domain are discussed in more details in Chap. 4).

$$\mathbf{E}_{\rm sc}(\mathbf{R}) = \frac{\exp(-ik_0 R)}{4\pi R} \int d\mathbf{r}' (k^2(\mathbf{r}') - k_0^2) \mathbf{E}_\perp(\mathbf{r}') e^{i\mathbf{k}_{\rm sc}.\mathbf{r}'}, \tag{2.32}$$

where

$$\mathbf{E}_\perp(\mathbf{r}') = \mathbf{E}(\mathbf{r}') - \widehat{\mathbf{u}}.\mathbf{E}(\mathbf{r}')\widehat{\mathbf{u}} \tag{2.33}$$

represents the component of the electric field that is orthogonal to the direction of propagation given by $\widehat{\mathbf{u}}$. Expression (2.32) shows that the scattered electric field $\mathbf{E}_{\rm sc}(\mathbf{R})$ can be approximated by a plane wave [8] with wavevector $\mathbf{k}_{\rm sc} = k_0\mathbf{R}/R = k_0\widehat{\mathbf{u}}$ and amplitude,

$$\mathbf{E}_{\rm sc}(\mathbf{k}_{\rm sc}) = \mathbf{E}_{\rm sc}(\mathbf{R}). \tag{2.34}$$

The Poynting vector is then readily obtained,

$$\mathbf{S} = \frac{1}{2\mu_0 c}|\mathbf{E}_{\rm sc}(\mathbf{R})|^2\widehat{\mathbf{u}}. \tag{2.35}$$

The flux of the Poynting vector for a unit incident flux (or normalized by the incident flux through a unit surface normal to the propagation direction) yields the differential scattering cross-section in the direction given by $\mathbf{k}_{\rm sc}$,

$$\frac{d\sigma}{d\Omega} = \frac{1}{16\pi^2|\mathbf{E}_{\rm in}|^2}\left|\int [k^2(\mathbf{r}') - k_0^2]\mathbf{E}_\perp(\mathbf{r}')e^{i\mathbf{k}_{\rm sc}.\mathbf{r}'}d\mathbf{r}'\right|^2. \tag{2.36}$$

Note that $d\sigma/d\Omega$ involves a double integration, that can be cast in the form,

$$\frac{d\sigma}{d\Omega} = \frac{1}{16\pi^2|\mathbf{E}_{\rm in}|^2}\int d\mathbf{r}\int d\mathbf{r}'(k^2(\mathbf{r}) - k_0^2)(k^2(\mathbf{r}+\mathbf{r}') - k_0^2)$$
$$\mathbf{E}_\perp(\mathbf{r}).\mathbf{E}_\perp^*(\mathbf{r}+\mathbf{r}')e^{i\mathbf{k}_{\rm sc}.\mathbf{r}'}, \tag{2.37}$$

where u^* stands for the conjugate of u. By integrating formally Eq. (2.32) over the vertical axis, one obtains a surface integral,

$$\mathbf{E}_{\rm sc}(\mathbf{R}) = \frac{\exp(-ik_0R)}{4\pi R}\int \mathcal{E}_\perp(\mathbf{r}'_{||}, k_{{\rm sc}z})e^{i\mathbf{k}_{{\rm sc}||}.\mathbf{r}'_{||}}d\mathbf{r}'_{||}, \tag{2.38}$$

with

$$\mathcal{E}_\perp(\mathbf{r}'_{||}, k_{{\rm sc}z}) = \int [k^2(\mathbf{r}') - k_0^2]e^{ik_{{\rm sc}z}z'}\mathbf{E}_\perp(\mathbf{r}')dz'. \tag{2.39}$$

We see that Eq. (2.39) is a 1D-Fourier transform, thus the variations of $\mathcal{E}_\perp$ with $k_{{\rm sc}z}$ are directly linked to the thickness of the sample. On the other hand, the variations of $\mathbf{E}_{\rm sc}$ with $\mathbf{k}_{{\rm sc}||}$ are related to the width of the illuminated area (i.e. the region for which $[k^2(\mathbf{r}') - k_0^2]\mathbf{E}$ is non-zero).

2.3.3 Coherence Domains

Up to now, we have considered an ideal experiment with a point-like detector. In reality, the detector has a finite size and one must integrate the differential scattering cross-section over the detector solid angle, $\Delta\Omega_{\mathrm{det}}$. Since the cross-section is defined as a function of wavevectors, it is more convenient to transform the integration over the solid angle $\Delta\Omega_{\mathrm{det}}$ centered about the direction $\mathbf{k}_{sc}$ into an integration in the (k_x, k_y) plane. The measured intensity (scattering cross-section convoluted with the resolution function) is then given by,

$$I = \frac{1}{16\pi^2}\frac{1}{|\mathbf{E}_{\mathrm{in}}|}\int d\mathbf{k}_{\|}\mathcal{R}(\mathbf{k}_{\|}) \times \int d\mathbf{r}_{\|}\int d\mathbf{r}'_{\|}\mathcal{E}_{\perp}^{*}(\mathbf{r}_{\|}+\mathbf{r}'_{\|},k_z).\mathcal{E}_{\perp}(\mathbf{r}_{\|},k_z)e^{i\mathbf{k}_{\|}.\mathbf{r}'_{\|}}, \tag{2.40}$$

where $\mathcal{R}(\mathbf{k}_{\|})$ is the detector acceptance in the (k_x, k_y) plane. The expression of $\mathcal{R}$ in the wavevector space is not easily obtained. In an x-ray experiment, it depends on the parameters (height, width) of the collecting slits. The reader is referred to section 4.7 for a detailed expression of $\mathcal{R}$ as a function of the detector shape. In this introductory chapter it is sufficient to take for $\mathcal{R}$ a Gaussian function centered about $\mathbf{k}_{\mathrm{sc}\|}$,

$$\mathcal{R}(k_{\mathrm{sc}x}, k_{\mathrm{sc}y}) = C\exp\left[-\frac{(k_x-k_{\mathrm{sc}x})^2}{2\Delta k_x^2}-\frac{(k_y-k_{\mathrm{sc}y})^2}{2\Delta k_y^2}\right]. \tag{2.41}$$

The variables Δk_x, Δk_y govern the angular aperture of the detector. If one assumes that the integrant does not vary significantly along $\mathbf{k}_z$ inside Δk_x Δk_y, [3] the resulting intensity is given by,

$$I = \frac{1}{16\pi^2}\frac{1}{|\mathbf{E}_{\mathrm{in}}|}\int\int d\mathbf{r}_{\|}d\mathbf{r}'_{\|}\mathcal{E}_{\perp}^{*}(\mathbf{r}_{\|}+\mathbf{r}'_{\|},k_{\mathrm{sc}z}).\mathcal{E}_{\perp}(\mathbf{r}_{\|},k_{\mathrm{sc}z})e^{i\mathbf{k}_{\mathrm{sc}\|}.\mathbf{r}'_{\|}}\tilde{\mathcal{R}}(\mathbf{r}'_{\|}) \tag{2.42}$$

where

$$\tilde{\mathcal{R}}(\mathbf{r}_{\|}) = 2\pi C\Delta k_x\Delta k_y e^{-\frac{1}{2}\Delta k_x^2x^2-\frac{1}{2}\Delta k_y^2y^2}. \tag{2.43}$$

We now examine Eq. (2.38) that gives the scattered field as the sum of the fields radiated by all the induced dipoles in the sample. We see that the

[3] This assumption is not straightforward. It is seen in Eq. (2.39) that the thicker the sample, the faster the variations of $\mathcal{E}_{\perp}$ with k_z. In an x-ray experiment, the sample under study is generally a thin film (a couple of microns) and we are interested by the structure along z of the material (multilayers). Hence, the size of the detector is chosen so that its angular resolution permits to resolve the interference pattern caused by the stack of layers. This amounts to saying that the k_z-modulation of $\mathcal{E}_{\perp}^{*}(\mathbf{r}_{\|}+\mathbf{r}'_{\|},k_z).\mathcal{E}_{\perp}(\mathbf{r}_{\|},k_z)$ is not averaged in the detector.

electric field radiated in the direction $\mathbf{k}_{sc}$ by the "effective" dipole placed at point $\mathbf{r}_{\|}$ is added coherently to the field radiated by another dipole placed at $\mathbf{r}_{\|} + \mathbf{r}'_{\|}$ whatever the distance between the points. The intensity, measured by an ideal experiment (coherent source and point-like detector), is given by a double integration of infinite extent which contains the incoherent term $|\mathcal{E}_{\perp}(\mathbf{r}_{\|}, k_{scz})|^2)$ and the cross-product (namely the interference term) $\mathcal{E}_{\perp}(\mathbf{r}_{\|}, k_{scz}).\mathcal{E}^*_{\perp}(\mathbf{r}_{\|} + \mathbf{r}'_{\|}, k_{scz})$. When the detector has a finite size, the double integration is modified by the introduction of the resolution function $\tilde{\mathcal{R}}$ which is the Fourier transform of the angular characteristic function of the detector. In our example, $\tilde{\mathcal{R}}$ is a Gaussian whose support in the (x, y) plane is roughly $1/[\Delta k_x \times \Delta k_y]$. This function limits the domain over which the contribution of the cross term to the total intensity is significant. This domain can be called the coherence domain S_{coh} due to the detector. The fields radiated by two points that belong to this domain will add coherently in the detector (the cross term value is important), while the fields coming from two points outside this domain will add incoherently (the cross term contribution is damped to zero). The resulting intensity can be seen as the incoherent sum of intensities that are scattered from various regions of the sample whose sizes coincide with the coherent domain given by the detector. This can be readily understood by rewriting Eq. (2.42) in the form [10],

$$I \propto \sum_{i=1,N} \int_{S_{\mathrm{coh}}} d\mathbf{r}_{\|} \int_{S_{coh}} d\mathbf{r}'_{\|}$$

$$\mathcal{E}^*_{\perp}(\mathbf{r_i}_{\|} + \mathbf{r}_{\|} + \mathbf{r}'_{\|}, k_{scz}).\mathcal{E}_{\perp}(\mathbf{r_i}_{\|} + \mathbf{r}_{\|}, k_{scz}) e^{i\mathbf{k}_{sc\|}.\mathbf{r}'_{\|}} \tilde{\mathcal{R}}(\mathbf{r}'_{\|}) \quad (2.44)$$

where $\mathbf{r}_i$ is the center of the different coherent regions S_{coh}. Hence, integrating the intensity over a certain solid angle is equivalent to summing the intensities (i. e. incoherent process) from various regions of the illuminated sample. This is the main result of this paragraph. *The finite angular resolution of the detector introduces coherence lengths beyond which two radiating sources can be considered incoherent (even though the incident beam is perfectly coherent).* Note that the plural is not fortuitous, indeed, the angular resolution of the detector can be different in the xOy and xOz plane, thus the coherent lengths vary along Ox, Oz and Oy. In a typical x-ray experiment (see Sect. 4.7.2, the sample is illuminated coherently over 5 mm^2 but the angular resolution of the detector yields coherence domains of solely a couple of square microns. More precisely, it is shown in Sect. 4.7.2 that a detection slit with height 100 μm width 1 cm placed at 1 meter of the sample with $\theta_{\mathrm{sc}} = 10$ $mrad$ limits the coherent length along Oz to 1 μm, the coherence length along Ox to 100 μm and that along Oy to 10 nm. Finally, in this introductory section, we have restricted our analysis solely to a detector of finite extent. In general, the incident source has also a finite angular resolution. However, coherence domains induced by the incident angular resolution is usually much bigger than that given by the detector angular resolution so that we do not consider it here. (The calculation scheme would be very similar). A more complete description of the resolution function of the experiment is given in Sect. 4.7.2.

2.4 Statistical Formulation of the Diffraction Problem

In this section, we point out, through various numerical simulations, the pertinence of a statistical description of the surface and of the scattered power for modelling a scattering experiment in which *the rough sample is necessarily deterministic*. The main steps of our analysis are as follow: Within the coherence domain, the field radiated by the induced dipoles (or secondary sources) of the sample interfere. We call ***speckle*** the complicated intensity pattern stemming from these interferences. The angular resolution of the detector yields an incoherent averaging of the speckle structures, (the intensities are added over a certain angular domain). This angular integration can be performed with an ensemble average by invoking

1. the ergodicity property of the rough surface, (i.e. we assume that the sample is one particular realisation of an ergodic random process) and
2. the equivalence between finite angular resolution and limited coherence domains.

It appears finally that the diffused intensity measured by the detector is adequatly modelled by the mean square of the electric field viewed as a function of the random variable z. Throughout this section, the numerical examples are given in the optical domain. The wavelength is about 1 μm and the perfectly coherent incident beam is directed along the Oz axis.

2.4.1 To What Extent is a Statistical Formulation of the Diffraction Problem Relevant?

In Sect. 2.3 it has been shown how to calculate formally the electromagnetic power measured by the detector in a scattering experiment. To obtain the differential scattering cross-section, one needs to know the permittivity contrast at each point of the sample, and the electric field at those points, Eq. (2.37). If the geometry of the sample is perfectly well known (i.e. deterministic like gratings), various techniques (such as the integral boundary method [11], [12]) permit to obtain without any approximation the field inside the sample. It is thus possible to simulate with accuracy the experimental results. In the case of scattering by gratings (i.e periodic surfaces) the good agreement between experimental results and calculations confirms the validity of the numerical simulations [12].

We study the scattered intensity from different rough deterministic surfaces s_n (e.g. those presented in Fig. 2.1) illuminated by a perfectly coherent beam. In this experiment, we suppose that the size of the coherence domains

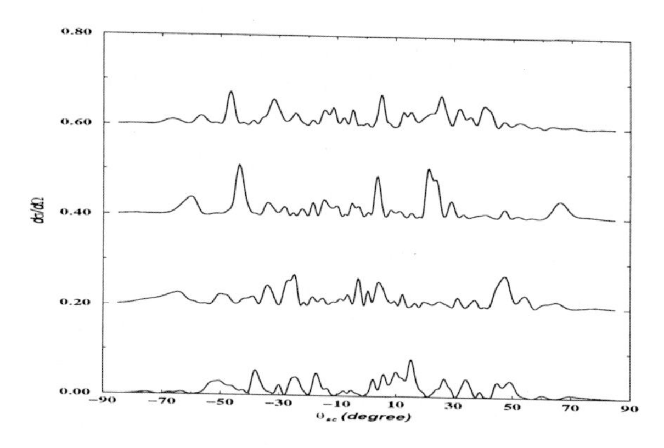

Fig. 2.4. Simulations of the differential scattering cross-section for the surfaces presented in Fig. 2.1. The illuminated area covers 40 μm which explains the large angular width of the speckle. The incident wavelength is 1 μm, the refractive index is $n = 1.5$. Normal incidence. The calculations are performed with a rigorous integral boundary method (no approximation in solving Eq. (2.37) other than the numerical discretizations) [13]

induced by the finite resolution of the detector is close to that of the illuminated area A. In other words, all the fringes of the interference pattern stemming from the coherent sum of the fields radiated by every illuminated point of the surface are resolved by the detector. We observe in Fig. 2.4 that the angular distribution of the intensity scattered by each surface presents a chaotic behavior. This phenomenon can be explained by recalling that the scattered field consists of many coherent wavelets, each arising from a different microscopic element of the rough surface, see Eq. (2.38). The random height position of these elements yields a random dephasing of the various coherent wavelets which results in a granular intensity pattern. This seemingly random angular intensity behavior, known as speckle effect, is obtained when the coherence domains include many correlation lengths of the surface, when the roughness is not negligible as compared to the wavelength (so that the random dephasing amplitude is important) and most importantly *when the size of the coherence domains is close to that of the illuminated area* so that the speckle is not averaged in the detector. To retrieve the precise angular behavior of the intensity, one needs an accurate deterministic description of the surface [14]. In Fig. 2.4 the surfaces s_n present totally different intensity patterns even though they have the same statistical properties. However, some similarities can be found in the curves plotted in Fig. 2.4. For example, the typical angular width of the spikes is the same for all surfaces. Indeed, in our numerical experiment it is linked to the width L of the illuminated area (which is here equivalent to the coherence domain). The smallest angular period of the fringes formed by the (farthest-off) coherent point-source pair on the surface determines the minimal angular width λ/L of the speckle spikes. This is clearly illustrated in Fig. 2.5, the larger the coherently illuminated area the thinner the angular speckle structures. In optics and radar imaging, sufficiently coherent incident beams (lasers) combined with detectors with fine angular resolution permits to study this phenomenon [14]. In x-ray experiments, the speckle effect can also be visualised in certain configurations. At grazing angles (e. g. $\theta_{\mathrm{sc}} = 1\ mrad$), the apparent resolution of the detector $\delta q_x = k_0 \theta \delta \theta$ (see Sect. 4.7.2) may be better than $10^{-7} k_0\ m^{-1}$. The size of the illuminated area being 5 mm, the speckle structures are resolved in the detector.

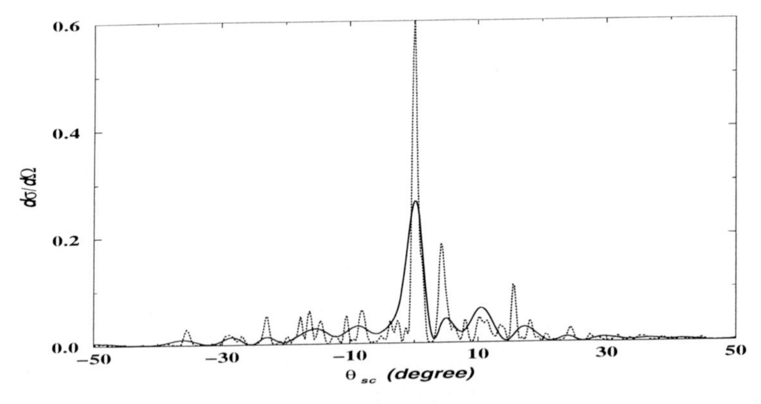

Fig. 2.5. Illustration of the dependence of the angular width of the speckle structures on the size of the illuminated area. Simulation of the intensity angular distribution for one rough surface illuminated in the first case over 60 μm and in second case over 30μm. The incident wavelength is 1 μm, the refractive index is $n = 1.5$, normal incidence

We now suppose that the illuminated area is increased enough so that the typical angular width of the speckle structures will be much smaller than the angular resolution of the detector. The detector integrates the intensity over a certain solid angle and, as a result, the fine structures disappear. One notices then that the smooth intensity patterns obtained for all the different surfaces s_n are quite similar. This is not surprising. Indeed, we have seen in the previous paragraph that the finite angular resolution of the detector is equivalent to the introduction of a coherence domain S_{coh} (that is smaller than the illuminated area A). The measured intensity can be considered the incoherent sum of intensities stemming from the different subsurfaces of size S_{coh} that constitute the sample. We now suppose that the illuminated area is big enough to cover many "coherent" subsurfaces, $A > 30 S_{\mathrm{coh}}$. Moreover, we suppose that the coherence domain is large enough so that each subsurface presents the same statistical properties $L_{\mathrm{coh}} > 30\xi$

where ξ is the correlation length and $L_{\rm coh}$ the coherence length. If the set of surfaces $\{s_n\}$ can be described by an ergodic stationary process, the ensemble of subsurfaces obtained from one particular realisation s_j will define the same random process with the same ensemble averaging as that created from any other realisation s_k. Consequently, the scattered intensity from one "big" surface s_j can be seen as the ensemble average of the "subsurface" $S_{\rm coh}$ scattered intensity which should be the same for all s_k. This assertion is supported by a comparison between two different numerical treatments of the same scattering experiment [13] [15].

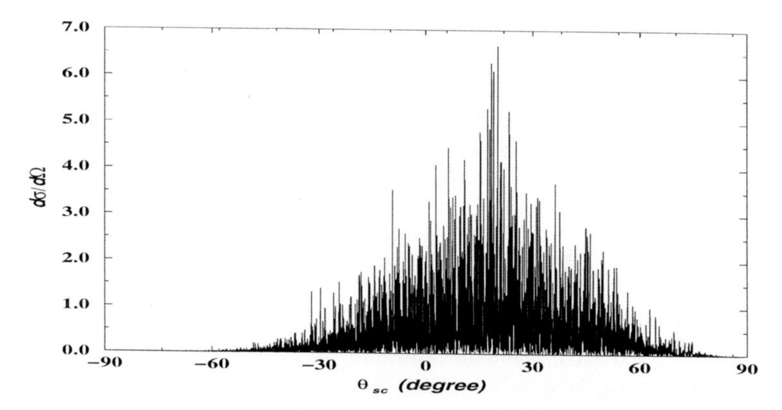

Fig. 2.6. Simulation of the differential scattering cross-section of a rough deterministic surface which is one realisation of a random process. The illuminated area covers $3mm$ (roughly several thousand of optical wavelengths). The statistics of the random process are: Gaussian height distribution with, $\sigma = 0.2\mu m$ and Gaussian correlation function with $\xi = 1\mu m$. The incident wavelength is $1\mu m$. Courtesy of prof. M. Saillard [13]

In Fig. 2.6 we have plotted the diffuse intensity obtained from a deterministic rough surface S_j illuminated by a perfectly coherent Gaussian beam, with a detector of infinite resolution. The rough surface is one realisation of a random process with Gaussian height distribution function and Gaussian correlation function with correlation length ξ. The incident beam is chosen wide enough so that the illuminated part of S_j is representative of the ergodic random process. In other words, S_j can be divided into many subsurfaces (with similar statistical properties) whose set describes accurately the random process. The total length of the illuminated spot is 5000ξ. It is seen in Fig. 2.6 that the scattered intensity exhibits a very thin speckle pattern. In general these fine structures are not visible. In Fig. 2.7 we have averaged the diffuse intensity over an angular width of 5 degree, corresponding to the angular resolution of a detector. We compare in Fig. 2.7 the *angular averaged pattern* with the *ensemble average* of the scattered intensity from subsurfaces

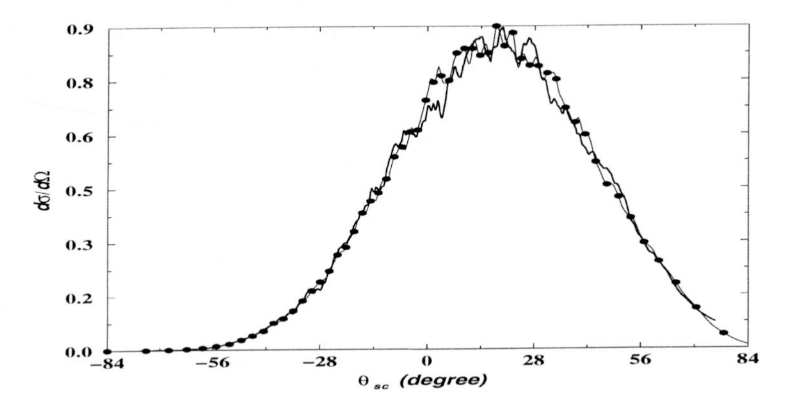

Fig. 2.7. Solid line: Angular average over 5 degrees of the differential scattering cross-section of the 'big surface' presented in Fig. 2.6, dotted line: ensemble average of the differential scattering cross-section of rough surfaces with the same statistics as the "big surface". Size of each realisation is $30\mu m$, no angular averaging. Courtesy of Prof. Saillard [13]

that are generated with the same random process as S_j but whose coherent illuminated domain is now restricted to 30ξ (i.e. to the coherence domain induced by the finite resolution of the detector). We obtain a perfect agreement between the two scattering patterns. In this example, we do not need any longer the precise value of the characteristic function $z(\mathbf{r}_{\|})$ but solely the statistical properties of the random process that describes conveniently these particular surfaces. The integration of the intensity over the solid angle $\Delta\Omega$ will then be replaced by the calculation of the ensemble average of the intensity. This ensemble averaging appears also naturally in the case of surfaces varying with time (such as liquid surfaces like ocean) by recording the intensity during a sufficiently long amount of time.

Each subsurface (either spread spatially via the coherence domains or temporally) generates an electric field $\mathbf{E}$. The latter can be viewed as a function of the random process z. The intensity measured by the detector is then related to the mean (in the ensemble averaging sense) square of the field, $\langle|\mathbf{E}|^2\rangle$. The purpose of most wave scattering theories is to evaluate the various moments of $\mathbf{E}$. More precisely, the random field can be divided into a mean and a fluctuating part,

$$\mathbf{E} = \langle\mathbf{E}\rangle + \delta\mathbf{E}. \tag{2.45}$$

We usually study separately the different contributions to the intensity.

2.4.2 Notions on Coherent (Specular) and Incoherent (Diffuse) Intensity

In the far-field, the scattered electric field $\mathbf{E}_{sc}$ behaves like a plane wave with wavevector $\mathbf{k}_{sc}$ and amplitude $\mathbf{E}(\mathbf{k}_{sc})$ see Eq. (2.32). It can be written as the sum of a mean part and a fluctuating part,

$$\mathbf{E}_{sc} = \langle \mathbf{E}_{sc} \rangle + \delta \mathbf{E}_{sc}. \tag{2.46}$$

The previous discussions have shown that the measured scattered intensity from a rough sample (whose deterministic surface profile is assumed to be one realisation of a given ergodic random process) can be evaluated with the ensemble average of the intensity $\langle |\mathbf{E}_{sc}(\mathbf{k}_{sc})|^2 \rangle$,

$$\langle |\mathbf{E}_{sc}|^2 \rangle = |\langle \mathbf{E}_{sc} \rangle|^2 + \langle |\delta \mathbf{E}_{sc}|^2 \rangle. \tag{2.47}$$

The first term on the right hand side of equation (2.47) is called the coherent intensity while the second term is known as the incoherent intensity. It is sometimes useful to tell the coherent and incoherent processes in the scattered intensity. In the following, we show that the coherent part is a Dirac function that contributes solely to the specular direction [4] if the randomly rough surface is statistically homogeneous in the (Oxy) plane.

In most approximate theories, the random rough surface is of infinite extent and illuminated by a plane wave. Suppose we know the scattered far-field $\mathbf{E}_{sc}$ from a rough surface of defining equation $z = z(\mathbf{r}_{\|})$. We now address the issue of how $\mathbf{E}_{sc}$ is modified when the whole surface is shifted horizontally by a vector $\mathbf{d}$. It is clear that such a shift will not modify the physical problem. However the incident wave amplitude acquires an additional phase factor $\exp(i\mathbf{k}_{in}.\mathbf{d})$ and similarly each scattered plane wave $\mathbf{E}_{sc}$ acquires, when returning to the primary coordinates, the phase factor $\exp(-i\mathbf{k}_{sc}.\mathbf{d})$. Thus we obtain,

$$\mathbf{E}_{sc}^{z(\mathbf{r}_{\|}-\mathbf{d})} = e^{-i(\mathbf{k}_{sc}-\mathbf{k}_{in}).\mathbf{d}} \mathbf{E}_{sc}^{z(\mathbf{r}_{\|})}. \tag{2.48}$$

We now suppose that the irregularities of the rough surface stem from a random spatially homogeneous process. In this case, the ensemble average is invariant under any translation in the (xOy) plane.

$$\langle \mathbf{E}_{sc}^{z(\mathbf{r}_{\|}-\mathbf{d})} \rangle = \langle \mathbf{E}_{sc}^{z(\mathbf{r}_{\|})} \rangle. \tag{2.49}$$

This equality is only possible if

$$\langle \mathbf{E}_{sc} \rangle = A\delta(\mathbf{k}_{sc\|} - \mathbf{k}_{in\|}). \tag{2.50}$$

Hence, when the illuminated domain (or coherence domain) is infinite, the coherent intensity is a Dirac distribution in the Fresnel reflection (or transmission) direction. For this reason it is also called specular intensity. Note that unlike the coherent term, the incoherent intensity is a function in the $\mathbf{k}_{sc\|}$ plane and its contribution in specular direction tends to zero as the detector acceptance is decreased. In real life, the incident beam is space-limited, the coherence domain is finite, thus the specular component becomes a function whose angular width is roughly given by λ/L_{coh}.

In many x-ray experiments, one is solely interested in the specularly reflected intensity. This configuration allows the determination of the z-dependent electron density profile and is often used for studying stratified interfaces (amphiphilic, or polymer adsorbed film). The modelisation of the coherent intensity requires the evaluation of the single integral Eq. (2.32) that gives the field amplitude while the incoherent intensity requires the evaluation of a double integral Eq. (2.37). It is thus much simpler to calculate only the coherent intensity and many elaborate theories have been devoted to this issue [4]. Chapter 3 of this book gives a thorough description of the main techniques developed for modelling the specular intensity from rough multilayers. However, it is important to bear in mind that the energy measured by the detector about the specular direction comes from both the coherent and incoherent processes inasmuch as the solid angle of collection is non zero. The incoherent part is not always negligible as compared to the coherent part especially when one moves away from the grazing angles. An estimation of both contributions is then needed to interpret the data.

2.5 Statistical Formulation of the Scattered Intensity Under the Born Approximation

In this last section, we illustrate the notions introduced previously with a simple and widely used model that permits to evaluate the scattering cross-section of random rough surfaces within a probabilistic framework. We discuss the relationship between the scattered intensity and the statistics of the surfaces. The main principles of the Born development have been introduced in Chap. 1, Appendix 1.A, and a complementary approach of the Born approximation is given in Chap. 4 with some insights on the electromagnetic properties of the scattered field.

2.5.1 The Differential Scattering Cross-Section

We start from Eq. (2.32) that gives the scattered far field as the sum of the fields radiated by the induced dipoles in the sample. The main difficulty of this integral is to evaluate the exact field $\mathbf{E}$ inside the scattering object. In the x-ray domain, the permittivity contrast is very small ($\approx 10^{-6}$) and one can assume that the incident field is not drastically perturbed by surrounding radiating dipoles. Hence, a popular assumption (known as the Born approximation) is to approximate $\mathbf{E}$ by $\mathbf{E}_{in}$. With this approximation the integrant is readily calculated. For an incident plane wave $\mathbf{E}_{\rm in} e^{-i\mathbf{k}_{\rm in}.\mathbf{r}}$, the differential scattering cross-section can be expressed as,

$$\frac{d\sigma}{d\Omega} = \frac{1}{16\pi^2} \frac{|\mathbf{E}_{\rm in\perp}|^2}{|\mathbf{E}_{\rm in}|^2} \int d\mathbf{r} \int d\mathbf{r}' [k^2(\mathbf{r}) - k_0^2][k^2(\mathbf{r}') - k_0^2] e^{i\mathbf{q}.(\mathbf{r}-\mathbf{r}')}, \quad (2.51)$$

where $\mathbf{E}_{\rm in\perp}$ is the projection of the incident electric field on the plane normal to the direction of observation of the differential cross-section. Denoting the unit vectors in direction $\mathbf{E}_{\rm in}$ and $\mathbf{E}_{\rm sc}$, $\widehat{\mathbf{e}}_{\rm in} = \mathbf{E}_{\rm in}/E_{in}$ and $(\widehat{\mathbf{e}}_{\rm sc})^2 = \mathbf{E}_{\rm sc}/E_{sc}$ respectively, we have $|\mathbf{E}_{\rm in\perp}| = E_{in}(\widehat{\mathbf{e}}_{\rm in}.\widehat{\mathbf{e}}_{\rm sc})^2$. In x-ray experiments, the incident field impiges on the surface at grazing angle and one studies the scattered intensity in the vicinity of the specular component. In this configuration, the orthogonal component of the incident field with respect to the scattered direction is close to the total incident amplitude. Yet, we retain the projection term $(\widehat{\mathbf{e}}_{\rm in}.\widehat{\mathbf{e}}_{\rm sc})^2$ in the differential scattering cross-section for completeness and coherence with the results of Chap. 1. Bearing in mind the value of the permittivity contrast as a function of the electronic density, Eq. (2.31), equation (2.51) simplifies to,

$$\frac{d\sigma}{d\Omega} = r_{\rm e}^2 (\widehat{\mathbf{e}}_{\rm in}.\widehat{\mathbf{e}}_{\rm sc})^2 \int d\mathbf{r} \int d\mathbf{r}' \rho_{el}(\mathbf{r}) \rho_{el}(\mathbf{r}') e^{i\mathbf{q}.(\mathbf{r}-\mathbf{r}')}, \quad (2.52)$$

with ρ_{el} the electron density and r_e the classical electron radius. [4] In the case of a rough interface separating two semi-infinite homogeneous media one gets,

$$\frac{d\sigma}{d\Omega} = r_e^2 \rho_{el}^2 (\widehat{\mathbf{e}}_{\rm in}.\widehat{\mathbf{e}}_{\rm sc})^2 \int_{-\infty}^{z(\mathbf{r}_\parallel)} dz \int_{-\infty}^{z(\mathbf{r}'_\parallel)} dz' \int d\mathbf{r}_\parallel \int d\mathbf{r}'_\parallel e^{i\mathbf{q}.(\mathbf{r}-\mathbf{r}')} \quad (2.53)$$

Integrating Eq. (2.53) over (z, z') (with the inclusion of a small absorption term to ensure the convergence at $-\infty$) yields,

$$\frac{d\sigma}{d\Omega} = \frac{\rho_{el}^2 r_e^2}{q_z^2} (\widehat{\mathbf{e}}_{\rm in}.\widehat{\mathbf{e}}_{\rm sc})^2 \int d\mathbf{r}_\parallel \int d\mathbf{r}'_\parallel e^{i\mathbf{q}_\parallel.(\mathbf{r}_\parallel - \mathbf{r}'_\parallel)} e^{iq_z[z(\mathbf{r}_\parallel) - z(\mathbf{r}'_\parallel)]}. \quad (2.54)$$

---

[4] One can make a general presentation of elastic scattering under the Born approximation from the scattering by an isolated object as presented in Chap. 1 Sect. 1.2.4 and appendix 1.A. The differential scattering cross-section can be cast in the form

$$\frac{d\sigma}{d\Omega} = \left| \sum_j b e^{i\mathbf{q}.\mathbf{r}_j} \right|^2 = \left| \int d\mathbf{r} \rho b e^{i\mathbf{q}.\mathbf{r}} \right|^2$$

where ρ is the density of scattering objects and b their scattered length as introduced in Eq. (1.34). The complex exponential is the result of the phase shift between waves scattered in the $\mathbf{k}_{\rm sc}$ direction by scatterers separated by a vector $\mathbf{r}$ as shown in figure 2.8. For neutrons, b is the scattering length which takes into account the strong interaction between the neutrons and the nuclei (we do not consider here magnetic materials); for x-rays, $b = r_e = (e^2/4\pi\epsilon_0 m_e c^2) = 2.810^{-15} m$ which is the classical radius of the electron.

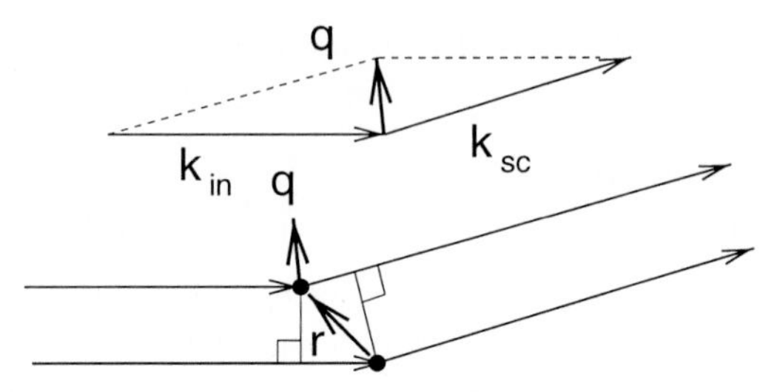

Fig. 2.8. Phase shift between the waves scattered by two points scatterers separated by a vector $\mathbf{r}$. The phase shift is $(\mathbf{k}_{sc} - \mathbf{k}_{in}).\mathbf{r} = \mathbf{q}.\mathbf{r}$

This equation concerns a-priori the scattering from any (deterministic or not) object. In this chapter, we are mostly interested by the scattering from surfaces whose surface profile z is unknown or of no interest. We have seen in the preceding sections that if z is described by a random homogeneous ergodic process, the intensity measured by the detector can be approximated by the ensemble average of the scattering cross-section. It amounts to replacing in Eq. (2.54) the integration over the surface by an ensemble average, $\int f(\mathbf{r}_{\|})d\mathbf{r}_{\|} = L_x L_y \langle f \rangle$, where L_x, L_y are the dimensions of the surface along Ox and Oy. One obtains,

$$\frac{d\sigma}{d\Omega} = \frac{\rho_{el}^2 r_e^2 L_x L_y}{q_z^2} (\widehat{\mathbf{e}}_{\text{in}}.\widehat{\mathbf{e}}_{\text{sc}})^2 \int d\mathbf{r}_{\|} e^{i\mathbf{q}_{\|}.\mathbf{r}_{\|}} \langle e^{iq_z[z(\mathbf{r}_{\|})-z(\mathbf{0}_{\|})]} \rangle . \tag{2.55}$$

Note that the expression (2.55) of the differential scattering cross-section accounts for both the coherent and incoherent processes. Hence, this integral does not converge in the function sense, it contains a Dirac distribution if the surface is infinite. This property will be illustrated with various examples in the following. If the probability density of z is Gaussian, we can write the differential cross-section as,

$$\frac{d\sigma}{d\Omega} = \frac{\rho_{el}^2 r_e^2 L_x L_y}{q_z^2} (\widehat{\mathbf{e}}_{\text{in}}.\widehat{\mathbf{e}}_{\text{sc}})^2 \int d\mathbf{r}_{\|} e^{i\mathbf{q}_{\|}.\mathbf{r}_{\|})} e^{-\frac{1}{2}q_z^2 \langle [z(\mathbf{r}_{\|})-z(\mathbf{0}_{\|})]^2 \rangle} . \tag{2.56}$$

We see that, under the Born approximation (where we neglect multiple scattering) the scattered intensity is related to the Fourier transform of the exponential of the pair-correlation function, $g(r_{\|}) = \langle [z(\mathbf{r}_{\|}) - z(\mathbf{0}_{\|})]^2 \rangle$. In the following we illustrate this result by studying the differential scattering cross-section for various pair-correlation functions. We start by the expression of the scattering differential cross-section in the case of a flat surface.

2.5.2 Ideally Flat Surfaces

For ideally flat surfaces $g(r_{\|})$ is zero everywhere at the surface and the scattering cross-section yields :

$$\frac{d\sigma}{d\Omega} = \frac{r_e^2 \rho_{el}^2 L_x L_y}{q_z^2} (\widehat{\mathbf{e}}_{\text{in}}.\widehat{\mathbf{e}}_{\text{sc}})^2 \int d\mathbf{r}_{\|} e^{i\mathbf{q}_{\|}.\mathbf{r}_{\|}}. \tag{2.57}$$

The integral is the Fourier transform of a constant so that, [5]

$$\frac{d\sigma}{d\Omega} = \frac{4\pi^2 r_e^2 \rho_{el}^2 L_x L_y}{q_z^2} (\widehat{\mathbf{e}}_{\text{in}}.\widehat{\mathbf{e}}_{\text{sc}})^2 \delta(\mathbf{q}_{\|}). \tag{2.58}$$

The scattered intensity is thus a Dirac distribution in the Fresnel reflection direction. As expected, for a perfectly flat surface, the reflectivity comes solely from a coherent process (Sect. 2.4.2), the incoherent scattering is null $\langle \delta E^2 \rangle = 0$. Note that the reflectivity decreases as a power law with q_z. We now turn to the more complicated problem of scattering from rough surfaces that are described statistically by an homogeneous ergodic random process.

2.5.3 Self-Affine Rough Surfaces

Surfaces Without Cut-Off We first consider self-affine rough surfaces with pair-correlation function g given by Eq. (2.24), $g(r_{\|}) = A_0 r_{\|}^{2h}$. With this pair-correlation function, the roughness cannot be determined since there is no saturation. The scattering cross-section is in this case,

$$\frac{d\sigma}{d\Omega} = \frac{r_e^2 \rho_{el}^2 L_x L_y}{q_z^2} (\widehat{\mathbf{e}}_{\text{in}}.\widehat{\mathbf{e}}_{\text{sc}})^2 \int d\mathbf{r}_{\|} e^{-\frac{q_z^2}{2} A R^{2h}} e^{i\mathbf{q}_{\|}.\mathbf{r}_{\|}}, \tag{2.59}$$

and can be expressed in polar coordinates as,

$$\frac{d\sigma}{d\Omega} = \frac{r_0^2 \rho_e^2 L_x L_y}{q_z^2} (\widehat{\mathbf{e}}_{\text{in}}.\widehat{\mathbf{e}}_{\text{sc}})^2 \int dr_{\|} e^{-\frac{q_z^2}{2} A R^{2h}} J_0(q_{\|} r_{\|}), \tag{2.60}$$

with $q_{\|}$ being the modulus of the in-plane scattering wave-vector, and J_0 the zeroth order Bessel function. The above integral has analytical solutions for $h = 0.5$ and $h = 1$ and has to be calculated numerically in other cases. For $h = 1$, the integration yields,

$$\frac{d\sigma}{d\Omega} = \frac{r_e^2 \rho_{el}^2 L_x L_y}{q_z^2} (\widehat{\mathbf{e}}_{\text{in}}.\widehat{\mathbf{e}}_{\text{sc}})^2 e^{-q_{\|}^2/q_z^4}, \tag{2.61}$$

and for $h = 0.5$,

$$\frac{d\sigma}{d\Omega} = (\widehat{\mathbf{e}}_{\text{in}}.\widehat{\mathbf{e}}_{\text{sc}})^2 \frac{r_e^2 \rho_{el}^2 L_x L_y}{q_z^2} \frac{\pi A}{\left(q_{\|}^2 + \left(\frac{A}{2}\right)^2 q_z^4\right)^{3/2}}. \tag{2.62}$$

The above expressions clearly show that for surfaces of this kind the scattering is purely diffuse (no Dirac distribution, no specular component).

[5] Let us recall that $\delta(\mathbf{q}_{\|}) = \frac{1}{4\pi^2} \int e^{-i\mathbf{q}_{\|}.\mathbf{r}_{\|}} d\mathbf{r}_{\|}$.

Surfaces with Cut-Off Rough surfaces are said to present a cut-off length when the correlation function $C_{zz}(\mathbf{r}_{\|})$ tends to zero when $r_{\|}$ increases, (for example see Eq. (2.26), when $C_{zz}(r_{\|}) = \sigma^2 \exp(-\frac{r_{\|}^{2h}}{\xi^{2h}})$, the cut-off is ξ). In this general case an analytical calculation is not possible and the scattering cross-section becomes,

$$\frac{d\sigma}{d\Omega} = \frac{r_e^2 \rho_{el}^2 L_x L_y}{q_z^2} e^{-q_z^2\sigma^2} (\widehat{\mathbf{e}}_{\text{in}}.\widehat{\mathbf{e}}_{\text{sc}})^2 \int d\mathbf{r}_{\|} e^{q_z^2 C_{zz}(\mathbf{r}_{\|})} e^{i\mathbf{q}_{\|}.\mathbf{r}_{\|}} . \quad (2.63)$$

The integrant in Eq. (2.63) does not tend to 0 when $\mathbf{r}_{\|}$ is increased. The integration over an infinite surface does not exist in the function sense. Indeed, $d\sigma/d\Omega$ accounts for both the coherent and incoherent contributions to the scattered power. It is possible extract the specular (coherent) and the diffuse (incoherent) components by writing the integrant in the form,

$$e^{q_z^2 C_{zz}(\mathbf{r}_{\|})} = 1 + \left(e^{q_z^2 C_{zz}(\mathbf{r}_{\|})} - 1 \right) . \quad (2.64)$$

The distributive part (or Dirac function) characterises the coherent or specular reflectivity while the regular part gives the diffuse power. Eq. (2.63) is then cast in the form,

$$\frac{d\sigma}{d\Omega} = \left(\frac{d\sigma}{d\Omega}\right)_{\text{coh}} + \left(\frac{d\sigma}{d\Omega}\right)_{\text{incoh}} \quad (2.65)$$

with

$$\begin{aligned} \left(\frac{d\sigma}{d\Omega}\right)_{\text{coh}} &= \frac{r_e^2 \rho_{el}^2 L_x L_y}{q_z^2} e^{-q_z^2\sigma^2} (\widehat{\mathbf{e}}_{\text{in}}.\widehat{\mathbf{e}}_{\text{sc}})^2 \int d\mathbf{r}_{\|} e^{i\mathbf{q}_{\|}.\mathbf{r}_{\|}} \\ &= \frac{4\pi^2 r_e^2 \rho_{el}^2 L_x L_y}{q_z^2} e^{-q_z^2\sigma^2} \delta(\mathbf{q}_{\|}) (\widehat{\mathbf{e}}_{\text{in}}.\widehat{\mathbf{e}}_{\text{sc}})^2 , \end{aligned} \quad (2.66)$$

and

$$\left(\frac{d\sigma}{d\Omega}\right)_{\text{incoh}} = \frac{r_e^2 \rho_{el}^2 L_x L_y}{q_z^2} e^{-q_z^2\sigma^2} (\widehat{\mathbf{e}}_{\text{in}}.\widehat{\mathbf{e}}_{\text{sc}})^2 \int d\mathbf{r}_{\|} \left(e^{q_z^2 C_{zz}(\mathbf{r}_{\|})} - 1 \right) e^{i\mathbf{q}_{\|}.\mathbf{r}_{\|}} . \quad (2.67)$$

The specular part is similar to that of a flat surface except that it is reduced by the roughness Debye-Waller factor $e^{-q_z^2\sigma^2}$. The diffuse scattering part may be determined numerically if one knows the functional form of the correlation function. When $q_z^2 C_{zz}(\mathbf{r}_{\|})$ is small, the exponential can be developed as $1 + q_z^2 C_{zz}(\mathbf{r}_{\|})$. In this case, the differential scattering cross-section appears to be proportional to the power spectrum of the surface $P(\mathbf{q}_{\|})$,

$$\left(\frac{d\sigma}{d\Omega}\right)_{\text{incoh}} = r_e^2 \rho_{el}^2 L_x L_y e^{-q_z^2\sigma^2} 4\pi^2 P(\mathbf{q}_{\|}) (\widehat{\mathbf{e}}_{\text{in}}.\widehat{\mathbf{e}}_{\text{sc}})^2 . \quad (2.68)$$

We see with Eqs. (2.66, 2.68) that the Born assumption permits to evaluate both the coherent and incoherent scattering cross-section of rough surfaces in a relatively simple way. This technique can be applied without additional difficulties to more complicated structures such as multilayers or inhomogeneous films. Unfortunately, in many configurations, the Born assumption proves to be too restrictive and one can miss major features of the scattering process. More accurate models such as the Distorted-wave Born approximation have been developed and are presented in Chap. 4 of this book. Yet, the expressions of the coherent and incoherent scattering cross-sections given here given by the first Born approximation provide useful insights on how the measured intensity relates to the shape (statistics) of the sample. The coherent reflectivity, Eq. (2.66), does not give direct information on the surface lateral fluctuations, except for the overall roughness σ, but it provides the electronic density of the plane substrate. Hence, reflectivity experiments are used in general to probe, along the vertical axis, the electronic density of samples that is roughly homogeneous in the (xOy) plane but varies in a deterministic way along Oz (e.g. typically multilayers). Chapter 3 of this book is devoted to this issue. On the other hand the incoherent scattering Eq. (2.68) is directly linked to the height-height correlation function of the surface. Bearing in mind the physical meaning of the power spectrum, Sect. 2.2.3, we see that measuring the diffuse intensity at increasing $q_{||}$ permits to probe the surface state at decreasing lateral scales. Hence, scattering experiments can be a powerful tool to characterise the rough sample in the lateral (Oxy) plane. This property will be developed and detailed in Chap. 4.

References

1. P. Beckmann and A. Spizzichino. *The scattering of electromagnetic waves from rough surfaces.* Pergamon Press, Oxford,UK, (1963).
2. F. G. Bass and I. M. Fuks. *Wave scattering from statistically rough surfaces.* Pergamon, New York, (1979).
3. J. A. Ogilvy. *Theory of wave scattering from random rough surfaces.* Adam Hilger, Bristol,UK, (1991).
4. G. Voronovich. *Wave scattering from rough surfaces.* Springer-Verlag, Berlin, (1994).
5. L. Mandel and E. Wolf. *Optical coherence and quantum optics.* Cambridge University Press, Cambridge USA, (1995).
6. C. A. Guérin, M. Holschneider, and M. Saillard, Waves in Random Media, **7**, 331–349, (1997).
7. B.B. Mandelbrodt, " The fractal geometry of nature", Freeman, New-York (1982).
8. M. Born and E. Wolf. *Principle of Optics.* Pergamon Press, New York, (1980).
9. D.W. Oxtoby, F. Novack, S.A. Rice, *J. Chem. Phys.* **76**, 5278 (1982).
10. S.K. Sinha, M. Tolan and A. Gibaud, Phys. Rev. B **57** , 2740 (1998)
11. M. Nieto-Vesperinas and J. C. Dainty. *Scattering in Volume and Surfaces.* Elsevier Science Publishers, B. V. North-Holland, (1990).

12. R. Petit, ed. *Electromagnetic Theory of Gratings.* Topics in Current Physics. Springer Verlag, Berlin, (1980).
13. M. Saillard and D. Maystre, J. Opt. Soc. Am. A, **7**(6), 982–990, (1990).
14. J. C. Dainty, ed. *Laser speckle and related phenomena.* Topics in Applied physics. Springer-Verlag, New York, (1975).
15. M. Saillard and D. Maystre, Journal of Optics **19**, 173–176, (1988).

3 Specular Reflectivity from Smooth and Rough Surfaces

Alain Gibaud

Laboratoire de Physique de l'Etat Condensé, UPRESA 6087 , Université du Maine Faculté des sciences, 72085 Le Mans Cedex 9, France

It is well known that light is reflected and transmitted with a change in the direction of propagation at an interface between two media which have different optical properties. The effects known as reflection and refraction are easy to observe in the visible spectrum but more difficult when x-ray radiation is used (see the introduction for a historical presentation). The major reason for this is the fact that the refractive index of matter for x-ray radiation does not differ very much from unity, so that the direction of the refracted beam does not deviate much from the incident one. The reflection of x-rays is however of great interest in surface science, since it allows the structure of the uppermost layers of a material to be probed. In this chapter, we present the general optical formalism used to calculate the reflectivity of smooth or rough surfaces and interfaces which is also valid for x-rays.[1]

3.1 The Reflected Intensity from an Ideally Flat Surface

3.1.1 Basic Concepts

It was shown in Chap. 1 that the refractive index of matter for x-ray radiation is (see Sect. 1.4.2 for more details):

$$n = 1 - \delta - i\beta. \tag{3.1}$$

The classical model of an elastically bound electron yields the following expression of δ:

$$\delta = \frac{\lambda^2}{2\pi} r_e \rho_e, \tag{3.2}$$

where r_e is the classical electron radius ($r_e = 2.810^{-5}$Å), λ is the wavelength and ρ_e is the electron density of the material. This shows that the real part

[1] The basic concepts used to determine the reflection and transmission coefficients of an electromagnetic wave at an interface were first developed by A. Fresnel [1] in his mechano-elastic theory of light.

of the refractive index mainly depends on the electron density of the material and on the wavelength. Typical values for δ are $10^{-5} - 10^{-6}$ and β is ten times smaller. A similar equation holds for neutrons where $r_e\rho_e$ has to be replaced by ρb (see Chap. 5, Eq. (5.24)).

A specific property of x-rays and neutrons is that since the refractive index is slightly less than 1, a beam impinging on a flat surface can be totally reflected. The condition to observe total external reflection is that the angle of incidence θ (defined here as the angle between the incident ray and the surface) must be less than a critical angle θ_c. This angle can be obtained by applying Snell-Descartes' law with $\cos\theta_{\mathrm{tr}} = 1$, yielding in absence of absorption:

$$\cos\theta_c = n = 1 - \delta. \tag{3.3}$$

Since δ is of the order of 10^{-5}, the critical angle for total external reflection is clearly extremely small. At small angles, $\cos\theta_c$ can be approximated as $1 - \theta_c^2/2$ and (3.3) becomes

$$\theta_c^2 = 2\delta. \tag{3.4}$$

The total external reflection of an x-ray (or neutron) beam is therefore only observed at grazing angles of incidence below about $\theta < 0.5°$. At larger angles, the reflectivity decreases very rapidly as mentioned above.

In this chapter, we will calculate the reflectivity as a function of the incident angle θ or alternatively as a function of the modulus $q = 4\pi \sin\theta/\lambda$ of the wave-vector transfer $\mathbf{q}$ (see Eq. (2.28) and Fig. 2.3 with $\psi_{\mathrm{sc}} = 0$). This means that the following ratios,

$$R(\theta) = \frac{I(\theta)}{I_0}, \tag{3.5}$$

$$R(\mathbf{q}) = \frac{I(\mathbf{q})}{I_0} \tag{3.6}$$

will be determined, where $I(\theta)$ or $I(\mathbf{q})$ is the reflected intensity (Flux of Poynting's vector through the detector area) for an angle of incidence θ (or wavevector transfer $\mathbf{q}$), and I_0 is the intensity of the incident beam. The theory of x-ray reflectivity is valid under the assumption that it is possible to consider the electron density as continuous (see Chap. 1). Under this approximation, the reflection is treated like in optics, and the reflected amplitude is obtained by writing down the boundary conditions at the interface, i.e., the continuity of the electric and magnetic fields at the interface, leading to the classical Fresnel relations.

3.1.2 Fresnel Reflectivity

The reflection and transmission coefficients can be derived by writing the conditions of continuity of the electric and magnetic fields at the interface.

The reflected intensity, which is the square of the modulus of the reflection coefficient, is the quantity measured in an experiment. Let us consider an electromagnetic **plane wave** propagating in the xOz plane of incidence, with its electric field polarised normal to this plane along the Oy direction. The interface between air and the reflecting medium which is located at $z = 0$ as shown in figure 3.1 will be assumed to be abrupt. In order to better emphasize that the same formalism applies for x-rays and visible optics we use in this section the angles defined from the surface normal as in optics, together with the grazing angles usually used in x-ray or neutron reflectivity.

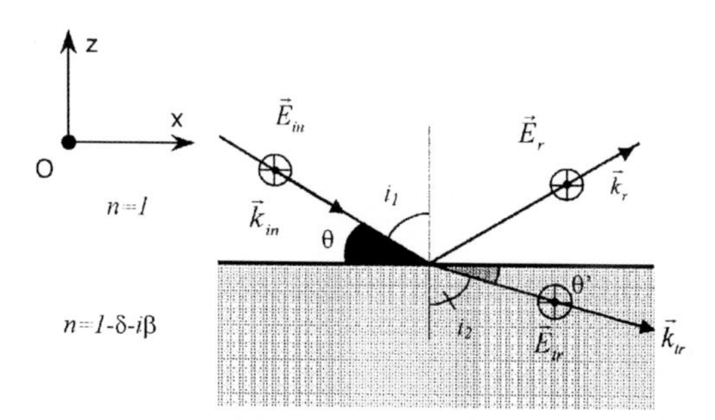

Fig. 3.1. Reflection and refraction of an incident wave polarised along y and travelling in the xOz plane of incidence

The expression for the electric field in a homogeneous medium is derived from Maxwell's equations which when combined, lead to the propagation equation of the electric field known as Helmholtz's equation (see Chap. 1, Eqs. (1.12), (1.15) for details)

$$\Delta \mathbf{E} + k_j^2 \mathbf{E} = 0, \tag{3.7}$$

where k_j is the wave-vector in medium j. The electric field which is solution of Helmoltz's equation is given for the incident (in), reflected (r) and transmitted (tr) plane waves by,

$$\mathbf{E}_j = A_j e^{i(\omega t - \mathbf{k}_j . \mathbf{r})} \hat{\mathbf{e}}_y \tag{3.8}$$

with j=in, r or tr, $k_0 = |\mathbf{k}_{\text{in}}| = |\mathbf{k}_r| = 2\pi/\lambda = |\mathbf{k}_{tr}|/n$, and $\hat{\mathbf{e}}_y$ is a unit vector along the y axis (see figure 3.1). Note that the convention of signs used in crystallography is adopted here (see Part I, Chap. 1, by F. de Bergevin for details). It is straightforward to show, that the components of the(in), (tr), and (r) wavevectors are,

$$\begin{cases} \mathbf{k}_{\text{in}} = k_0(\sin i_1 \hat{\mathbf{e}}_x - \cos i_1 \hat{\mathbf{e}}_z) \\ \mathbf{k}_{\text{r}} = k_0(\sin i_1 \hat{\mathbf{e}}_x + \cos i_1 \hat{\mathbf{e}}_z) \\ \mathbf{k}_{\text{tr}} = k_0 n(\sin i_2 \hat{\mathbf{e}}_x - \cos i_2 \hat{\mathbf{e}}_z). \end{cases} \tag{3.9}$$

The tangential component of the electric field must be continuous at the interface *($z = 0$)*. In air, the field is the sum of the incident and reflected fields. Assuming that the medium is sufficiently thick for the transmitted beam to be completely absorbed, the following relation must be fulfilled,

$$A_{\text{in}} e^{i(\omega t - k_0 \sin i_1 x)} + A_r e^{i(\omega t - k_0 \sin i_1 x)} = A_{tr} e^{i(\omega t - k_0 n \sin i_2 x)}. \tag{3.10}$$

Equation (3.10) must be valid for any value of x, so that the following condition must hold,

$$\sin i_1 = n \sin i_2 . \tag{3.11}$$

This condition is simply the well-known Snell-Descartes' second law. As a result of this, the conservation of the perpendicular component of the electric field leads to,

$$A_{\text{in}} + A_r = A_{tr} . \tag{3.12}$$

It will be assumed that the media are non-magnetic so that the tangential component of the magnetic field must also be continuous. According to the Maxwell-Faraday equation,

$$\nabla \times \mathbf{E} = -\frac{\partial \mathbf{B}}{\partial t} = -i\omega \mathbf{B} \tag{3.13}$$

the tangential component B_t is the dot product of the magnetic field with the unit vector $\hat{\mathbf{e}}_x$, i.e.,

$$B_t = \frac{(\nabla \times \mathbf{E}).\hat{\mathbf{e}}_x}{i\omega}. \tag{3.14}$$

Since the electric field is normal to the incident plane, it is polarised along the y axis and the curl of the field gives,

$$\nabla \times \mathbf{E} = \frac{\partial E_y}{\partial x} \hat{\mathbf{e}}_z - \frac{\partial E_y}{\partial z} \hat{\mathbf{e}}_x . \tag{3.15}$$

The tangential component of the magnetic field is then given by

$$B_t = -\frac{1}{i\omega}\frac{\partial E_y}{\partial z}, \tag{3.16}$$

and from equation (3.10) it is easy to show that the conservation of this quantity yields,

$$(A_{\text{in}} - A_r)\cos i_1 = nA_{tr}\cos i_2. \tag{3.17}$$

Writing the reflected amplitude $r = A_r/A_{\text{in}}$ and the transmitted one $t = A_{tr}/A_{\text{in}}$, the following relations are obtained,

$$\begin{aligned} 1 + r &= t \\ 1 - r &= nt\frac{\cos i_2}{\cos i_1}. \end{aligned} \tag{3.18}$$

Combining these two equations, the reflected amplitude coefficient in the case of a *(s)* polarisation is found to be,

$$r^{(s)} = \frac{\cos i_1 - n\cos i_2}{\cos i_1 + n\cos i_2} \tag{3.19}$$

which by the use of the Snell-Descartes' relation leads to,

$$r^{(s)} = \frac{\sin(i_2 - i_1)}{\sin(i_1 + i_2)}. \tag{3.20}$$

In the case of an electric field parallel to the plane of incidence, a similar calculation leads to,

$$r^{(p)} = \frac{\tan(i_2 - i_1)}{\tan(i_2 + i_1)}. \tag{3.21}$$

Those equations are known as the Fresnel equations [1]. It is easy to show that at small grazing angles of incidence for x-rays $r^{(p)} \approx r^{(s)} \approx r$. Only (s) polarisation (electric field polarised perpendicular to the plane of incidence) will be considered in detail below but some results will also be given for (p) polarisation).

The grazing angle of incidence θ that the incident beam makes with the reflecting surface is usually the experimental variable in a reflectivity measurement. It is therefore important to express the coefficient of reflection as a function of this angle θ and also of the refractive index n. Starting from

$$r = \frac{\cos i_1 - n\cos i_2}{\cos i_1 + n\cos i_2}, \tag{3.22}$$

and using the fact that the θ and i_1, and the θ_{tr} and i_2 are complementary angles as shown in Fig. 3.1, Eq. (3.22) becomes:

$$r = \frac{\sin\theta - n\sin\theta_{tr}}{\sin\theta + n\sin\theta_{tr}}. \tag{3.23}$$

Applying the Snell-Descartes' law,

$$\cos\theta = n\cos\theta_{tr} \tag{3.24}$$

produces the following coefficient of reflection,

$$r(\theta) = \frac{\sin\theta - \sqrt{n^2 - \cos^2\theta}}{\sin\theta + \sqrt{n^2 - \cos^2\theta}}. \tag{3.25}$$

In the case of small incident angles (for which $\cos\theta = 1 - \theta^2/2$) and for electromagnetic x-ray waves for which the refractive index (in the absence of absorption) is given by,

$$n^2 = 1 - 2\delta = 1 - \theta_c^2. \tag{3.26}$$

The general equation Eq. (3.25) becomes,

$$r(\theta) = \frac{\theta - \sqrt{\theta^2 - \theta_c^2}}{\theta + \sqrt{\theta^2 - \theta_c^2}}. \tag{3.27}$$

The reflectivity which is the square of the modulus of the reflection coefficient, is given by,

$$R(\theta) = rr^* = \left|\frac{\theta - \sqrt{\theta^2 - \theta_c^2}}{\theta + \sqrt{\theta^2 - \theta_c^2}}\right|^2. \tag{3.28}$$

Finally, if the absorption of the x-ray beam by the material is accounted for, the refractive index takes a complex value and the Fresnel reflectivity is then written,

$$R(\theta) = rr^* = \left|\frac{\theta - \sqrt{\theta^2 - \theta_c^2 - 2i\beta}}{\theta + \sqrt{\theta^2 - \theta_c^2 - 2i\beta}}\right|^2. \tag{3.29}$$

The reflectivity can equally well be given in terms of the wave-vector transfer q:

$$R(\mathbf{q}) = \left| \frac{q_z - \sqrt{q_z^2 - q_c^2 - \frac{32i\pi^2\beta}{\lambda^2}}}{q_z + \sqrt{q_z^2 - q_c^2 - \frac{32i\pi^2\beta}{\lambda^2}}} \right|^2 . \tag{3.30}$$

When the wave-vector transfer is very large compared to q_c , i.e. $q \gtrsim 3q_c$, the following asymptotic behaviour is observed:

$$R = \frac{q_c^4}{16q^4} . \tag{3.31}$$

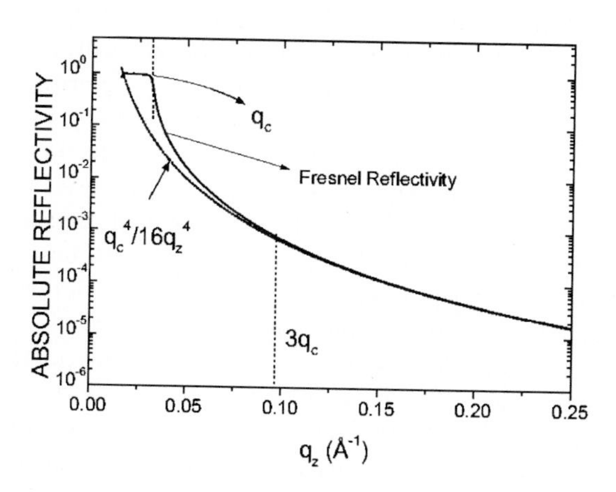

Fig. 3.2. Calculated reflectivity of a flat silicon wafer and asymptotic law

It can be seen from the Fig. 3.2, that the reflectivity curve or reflectivity profile consists of three different regimes:

- a plateau of total external reflection $R = 1$ when $q < q_c$ (see Sect. 7.1.1 for more details)
- a very steep decrease when $q = q_c$
- a $1/q^4$ power law when $q > 3q_c$.

It is worth noting that if the value of q_c is measured experimentally, this immediately yields the value of the electron density in the material (see Part I, Chap. 1 by F. de Bergevin) since,

$$q_c = 3.7510^{-2}\sqrt{\rho_e} , \tag{3.32}$$

where ρ_e is the electron density in the units $e^-/\text{Å}^3$.
Finally remembering that the reflectivity is observed under specular conditions, reference to the system of axes defined in Fig. 3.1, shows that the Fresnel reflectivity $R(\mathbf{q})$ can be written as:

$$R(\mathbf{q}) = \left| \frac{q_z - \sqrt{q_z^2 - q_c^2 - \frac{32 i \pi^2 \beta}{\lambda^2}}}{q_z + \sqrt{q_z^2 - q_c^2 - \frac{32 i \pi^2 \beta}{\lambda^2}}} \right|^2 \delta q_x \delta q_y , \tag{3.33}$$

since $q = q_z$ in equation (3.33), and the reflectivity of a flat surface is only measurable in the specular direction. Equation (3.33) completely describes the reflectivity of a homogeneous material, showing in particular that the reflectivity differs from zero only for wave-vector transfers normal to the surface of the sample.[2]

Figure 3.2 illustrates the calculated reflectivity curve for a silicon wafer in the power law regime and also in the case of a more complete dynamical calculation. The deviation from unity due to the absorption of the x-rays in the material can be seen to play a major role in determining the form of the curve in the region close to the critical edge at $q = q_c$. Equation (3.33) shows quite clearly that the calculation of a reflectivity curve requires only the electron density and the absorption of the material (for the wavelength used). Table 3.1 gives some useful data for calculating the reflectivity of various elements and compounds. A much wider data base of quantities relevant to reflectivity measurements can be found at the following web site, "http://www-cxro.lbl.gov/optical-constants/".

As a conclusion of this section we wish to stress some points concerning the validity of equation (3.33). It is important to realise that in a real experiment we never measure the theoretical reflectivity as given by Eq. (3.33) since the incident beam is not necessarily strictly monochromatic, is generally divergent, and the detector has a finite acceptance. For any instrument, the effects of the divergence of the x-ray source, of the slit settings or of the angular acceptance of the monochromator and analyser crystals used to collimate the incident and scattered beams (see Chap. 7 by J.M. Gay) must be taken into account. Those effects can be described using a 3-dimensional resolution function which is never a Dirac distribution but a 3-dimensional function having a certain width (see Chaps. 4 and 7) which precisely depends on the setup characteristics detailed above. The value of the measured reflectivity

[2] For this reason, the reflectivity of a flat surface is described as "***specular***", a term which is more normally used to describe the reflection by an ordinary mirror. It seems that Compton [2] was the first to have foreseen the possibility of totally reflecting x-rays in 1923 and that Forster [3] introduced equation (3.29). Prins [4] carried out some experiments to illustrate the predictions of this equation in 1928, using an iron mirror. He also used different anode targets to study the influence of the x-ray wavelength on the absorption. Kiessig also made similar experiments in 1931 [5] using a nickel mirror. An account of the historical development of the subject can be found in the pioneering work of L.G. Parrat [6] in 1954, and of Abeles [7]. The fundamental principles are discussed in the textbook by James [8].

Table 3.1. A few examples of useful data used in reflectivity analysis. The table contains the electron density ρ_e, the critical wave-vector q_c , the parameter δ, the absorption coefficient β, the structure of the material and its specific mass (δ and β are given at $\lambda = 1.54 \mathring{A}$). A useful formula for calculating the critical wave-vector transfer is $q_c(\text{Å}^{-1}) = 0.0375\sqrt{\rho_e\,(e^-/\text{Å}^3)}$, and conversely $\rho_e = 711 q_c^2$

Material	ρ_{el}	q_c	δ	β	Structure	ρ
	$e^-/\text{Å}^3$	Å^{-1}	10^6	10^7		kg/m^3
Si	0.7083	0.0316	7.44	1.75	cubic diamond a=5.43Å, Z=8	2330
SiO_2	0.618	0.0294	6.5	1.7		2200
Ge	1.425	0.0448	15.05	5	cubic,diamond a=5.658Å, Z=8	5320
AsGa	1.317	0.0431	13.9	4.99	cubic,diamond a=5.66Å, Z=8	5730
Glass Crown	0.728	0.0328	8.1	1.36	67.5%SiO_2,12% B_2O_3 9%,Na_2O, 9.5%K_2O,2%BaO	2520
Float Glass	0.726	0.0320	7.7	1.3		–
Nb	2.212	0.056	24.5	15.1	cubic,bcc a=3.03Å, Z=2	8580
Cu	2.271	0.0566	24.1	5.8	cubic, fcc a=3.61Å, Z=4	8930
Au	4.391	0.0787	46.5	49.2	cubic, fcc a=4.078Å	19280
Ag	2.760	0.0624	29.25	28	cubic, fcc a=4.09Å	10500
ZrO_2	1.08	0.0395	11.8			–
WO_3	1.723	0.0493	18.25	12	–	–
H_2O	0.334	0.0217	3.61	0.123	–	1000
CH_3CH_2-	0.32	0.0212			–	–
-COOH	0.53	0.0273			–	–
CCl_4	0.46	0.0254			–	–
CH_3OH	0.268	0.0194			–	–
PS-PMMA	0.377	0.0233			–	–

can be estimated through the convolution of equation (3.33) with the resolution function of the instrument. For measurements made in the incidence plane and under specular conditions, a first effect is that the convolution smears out the q_z dependence of the reflectivity. This can generally be accounted for by convolving $R(q_z)$ with a Gaussian function. Another, most important effect of the finite resolution is that beams outside the specular direction are accepted by the detector (in other words, the specular condition $\delta(q_x)\delta(q_y)$ is replaced by a function having a finite width $\Delta q_x \times \Delta q_y$). Then, if the surface to be analysed is rough, the convolution with the resolution function drastically changes the problem because part of the diffuse intensity which arises from the roughness is contained in the resolution volume. It may even happen for very rough surfaces that the diffuse intensity becomes as intense as the specular reflectivity. When this is occuring, the only way to use equation (3.33) is to subtract the diffuse part from the reflected intensity to obtain the true specular reflectivity (see Sect. 4.7 for details).

3.1.3 The Transmission Coefficient

As shown in Eq. (3.18), the amplitude of the transmission coefficient satisfies the relation, $1+r=t$. It is straightforward to show by combining Eqs. (3.18) and (3.29), that the transmitted intensity must be given by,

$$T(\theta) = tt^* = \left| \frac{2\theta}{\theta + \sqrt{\theta^2 - \theta_c^2 - 2i\beta}} \right|^2 \tag{3.34}$$

$$T(q_z) = tt^* = \left| \frac{2q_z}{q_z + \sqrt{q_z^2 - q_c^2 - \frac{32i\pi^2\beta}{\lambda^2}}} \right|^2 . \tag{3.35}$$

The transmitted intensity has a maximum at $\theta=\theta_c$ as shown in Fig. 3.3 which gives the actual variation of the transmitted intensity as a function of the incident angle θ (or q_z) in the case of silicon, germanium and copper samples irradiated with the copper K_α radiation. The transmitted intensity is nearly zero at very small angles in the regime of total reflection. It increases strongly at the critical angle and finally levels off towards a limit equal to unity at large angles of incidence. The maximum in the transmission coefficient, which is also a maximum in the field at the interface is the origin of the so-called Yoneda wings which are observed in transverse off-specular scans (see Sect. 4.3.1).

3.1.4 The Penetration Depth

The absorption of a beam in a medium depends on the complex part of the refractive index and limits the penetration of the beam inside the material. The refractive index for x-rays, defined in equation (3.1) is $n = 1-\delta-i\beta$. The amplitude of the electric field polarised along the y direction ((s) polarisation) and propagating inside the medium of refractive index n is given by,

$$E = E_0 e^{i(\omega t - k_0 n \cos\theta_{\mathrm{tr}} x + k_0 n \sin\theta_{\mathrm{tr}} z)} . \tag{3.36}$$

Since $n\cos\theta_{\mathrm{tr}} = \cos\theta$ (the Snell-Descartes'law) and $\sin\theta_{\mathrm{tr}} \approx \theta_{\mathrm{tr}}$, this equation can be written,

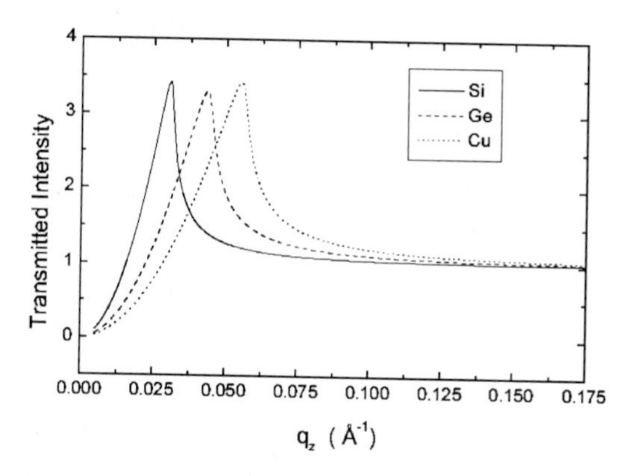

Fig. 3.3. Transmission coefficient in intensity in different materials, silicon, copper, and germanium; the maximum appears at the critical wave-vector transfer of the material

$$E = E_0 e^{+i(\omega t - k_0 \cos\theta x)} e^{ik_0 n\theta_{\mathrm{tr}} z}. \tag{3.37}$$

The absorption is governed by the real part of $e^{ik_0 n\theta_{\mathrm{tr}} z}$, with

$$n\theta_{\mathrm{tr}} = (1 - \delta - i\beta)\sqrt{\theta^2 - 2\delta - 2i\beta} = A + iB \tag{3.38}$$

The coefficients A and B can be deduced from the above equation and B is given by,

$$B(\theta) = -\frac{1}{\sqrt{2}}\sqrt{\sqrt{\left(\theta^2 - 2\delta\right)^2 + 4\beta^2} - \left(\theta^2 - 2\delta\right)}. \tag{3.39}$$

It follows that the electric field is,

$$E = E_0 e^{i(\omega t - k_0 \cos\theta x + k_0 A z)} e^{-k_0 B(\theta) z}. \tag{3.40}$$

Taking the modulus of this electric field shows that the variation of the intensity $I(z)$ with depth into the material is given by,

$$I(z) \propto EE^* = I_0 e^{-2k_0 B(\theta) z} \tag{3.41}$$

The absorption coefficient is therefore

$$\mu(\theta) = -2k_0 B(\theta) = \frac{-4\pi B(\theta)}{\lambda}, \tag{3.42}$$

and the penetration depth which is the distance for which the beam is attenuated by $1/e$ is given by,

$$z_{1/e}(\theta) = \frac{1}{\mu(\theta)} = \frac{-\lambda}{4\pi B(\theta)} = \frac{1}{2\mathcal{I}mk_{z,1}}. \tag{3.43}$$

Note that this quantity depends on the incident angle θ through the value of $B(\theta)$. In particular, in the limit $\theta \to 0$, neglecting absorption,

$$z_{1/e}(\theta_c) = \frac{\lambda}{4\pi\theta_c}. \tag{3.44}$$

In addition, the penetration depth is wavelength dependent since β depends on the wavelength. Values of β are tabulated in the International Tables of Crystallography, vol. IV [9] or they can also be found at the web site which has already been referred to, "http://www-cxro.lbl.gov/optical_constants/".

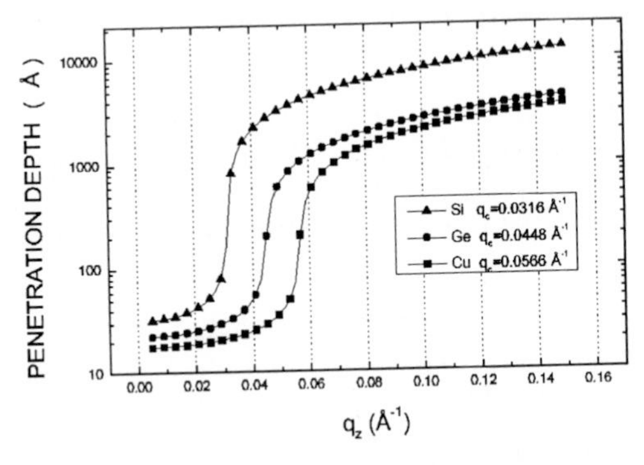

Fig. 3.4. Evolution of the penetration depth in Si, Ge and Cu irradiated with the *Kα line of a copper tube as a function of the wave vector transfer. Note that the figure is presented as a function of* $q_z = 4\pi \sin\theta/\lambda$

Figure 3.4 shows the variation of the penetration depth as a function of the incident angle in silicon, germanium and copper, for the case of CuKα radiation. The penetration depth remains small, that is below about 30$\AA$ when θ is smaller than the critical angle. This is this property which is exploited in surface diffraction, where only the first few atomic layers are analysed. The penetration depth increases steeply at the critical angle and finally slowly grows when $\theta >> \theta_c$.

3.2 X-Ray Reflectivity in Stratified Media

The simple case of a uniform substrate exhibiting a constant electron density, was considered in the previous section. This situation is of course not the most general one. For example, stratified media and multilayers are frequently encountered. Moreover, interfaces generally cannot be considered as steps, but are rough and thick. Thick interfaces may be approximated by dividing them into as many slabs of constant electron density as necessary to describe their (continuous) density profile. Again, it is not possible in this case to use the Fresnel coefficients directly to calculate the reflectivity. The calculation must be performed by applying the boundary conditions for the electric and magnetic fields at each of the interfaces between the slabs of constant electron density. The result is usually presented as the product of matrices, and multiple reflections are taken in account in the calculation known as the dynamical theory of reflection. Several excellent descriptions of this kind of calculation can be found in references [10–14].

3.2.1 The Matrix Method

Let us consider a plane wave polarised in the direction perpendicular to the plane of incidence (*(s)* polarisation) and propagating into a stratified medium. The axes are chosen so that the wave is travelling in the xOz plane as shown in Fig. 3.5.

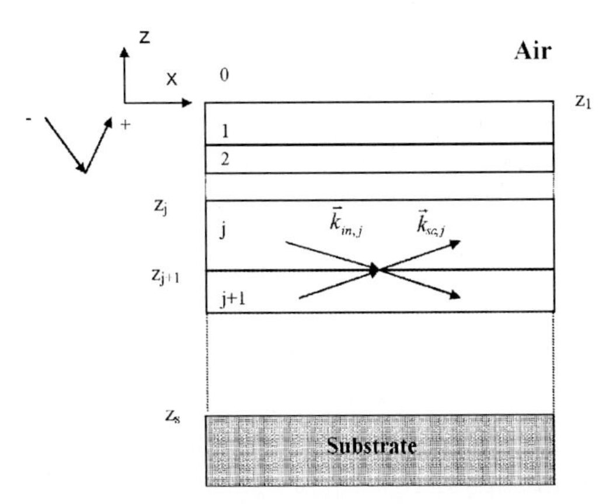

Fig. 3.5. Illustration of the plane of incidence for a stratified medium. The signs - and + label the direction of propagation of the wave; Air is labelled medium 0 and the strata are identified by $1 \leq j \leq n$ *layers in which upwards and downwards waves travel*

The air is labelled as medium 0 and the strata or layers with different electron densites are identified by $1 \leqslant j \leqslant$ n downwards. In this notation the depth Z_{j+1} marks the interface between the j and j+1 layers. The wave travelling through the material will be transmitted and reflected at each interface and the amplitudes of the upwards and downwards travelling waves will be defined as A^+ and A^- respectively. The electric field $\mathbf{E}^-$ of the downwards travelling wave in the jth stratum for example, is given by the solution of the Helmoltz's equation,

$$\mathbf{E}^- = A^- e^{+i(\omega t - k_{\mathrm{in}x,j}x - k_{\mathrm{in}z,j}z)}\hat{\mathbf{e}}_y. \tag{3.45}$$

The following notation will be adopted in the derivation,

$$\begin{aligned} k_{\mathrm{in}x,j} &= k_j \cos\theta_j \\ k_{\mathrm{in}z,j} &= -k_j \sin\theta_j = -\sqrt{k_j^2 - k_{\mathrm{in}x,j}^2}. \end{aligned} \tag{3.46}$$

Note that the value of $k_{\mathrm{in}x,j}$ is conserved at each interface since this condition is imposed by the Snell-Descartes' law of refraction. The upwards and downwards travelling waves are obviously superimposed at each interface so that at at a depth z from the surface the electric field in medium j is:

$$E_j(x,z) = (A_j^+ e^{ik_{\mathrm{in}z,j}z} + A_j^- e^{-ik_{\mathrm{in}z,j}z})e^{+i(\omega t - k_{\mathrm{in}x,j}x)}. \tag{3.47}$$

As $k_{\mathrm{in}z,j}$ takes a complex value, the magnitude of the upwards and downwards electric fields in layer j will be denoted by,

$$U(\pm k_{\mathrm{in}z,j}, z) = A_j^{\pm} e^{\pm ik_{\mathrm{in}z,j}z} \tag{3.48}$$

to simplify the notation. In addition, the quantity $k_{\mathrm{in}z,j}$ will be replaced by $k_{z,j}$. The condition of continuity of the tangential component of the electric field and the conservation of $k_{x,j}$ at the depth Z_{j+1} of the interface $j, j+1$ lead to:

$$U(k_{z,j}, Z_{j+1}) + U(-k_{z,j}, Z_{j+1}) = U(k_{z,j+1}, Z_{j+1}) + U(-k_{z,j+1}, Z_{j+1}). \tag{3.49}$$

It was shown in (3.16), that the tangential component of the magnetic field is continuous when the first derivative of the electric field is conserved. This leads to the equality below, at the *j,j+1* interface,

$$\begin{aligned} &k_{z,j}\left[U(k_{z,j}, Z_{j+1}) - U(-k_{z,j}, Z_{j+1})\right] \\ &= k_{z,j+1}\left[U(k_{z,j+1}, Z_{j+1}) - U(-k_{z,j+1}, Z_{j+1})\right]. \end{aligned} \tag{3.50}$$

The combination of these two equations can be written in a matrix form, so that the magnitudes of the electric field in media *j, j+1* at depth Z_{j+1} must satisfy

$$\begin{bmatrix} U(k_{z,j}, Z_{j+1}) \\ U(-k_{z,j}, Z_{j+1}) \end{bmatrix} = \begin{bmatrix} p_{j,j+1} & m_{j,j+1} \\ m_{j,j+1} & p_{j,j+1} \end{bmatrix} \begin{bmatrix} U(k_{z,j+1}, Z_{j+1}) \\ U(-k_{z,j+1}, Z_{j+1}) \end{bmatrix} \tag{3.51}$$

with

$$\begin{aligned} p_{j,j+1} &= \frac{k_{z,j} + k_{z,j+1}}{2k_{z,j}} , \\ m_{j,j+1} &= \frac{k_{z,j} - k_{z,j+1}}{2k_{z,j}} . \end{aligned} \tag{3.52}$$

The matrix which transforms the magnitudes of the electric field from the medium j to the medium $j+1$ will be called the refraction matrix $\mathcal{R}_{j,j+1}$. It is worth noting that $\mathcal{R}_{j,j+1}$ is not unimodular and has a determinant equal to $k_{z,j+1}/k_{z,j}$. In addition, the amplitude of the electric field within the medium j varies with depth as follows,

$$\begin{bmatrix} U(k_{z,j}, z) \\ U(-k_{z,j}, z) \end{bmatrix} = \begin{bmatrix} e^{-ik_{z,j}h} & 0 \\ 0 & e^{ik_{z,j}h} \end{bmatrix} \begin{bmatrix} U(k_{z,j}, z+h) \\ U(-k_{z,j}, z+h) \end{bmatrix} . \tag{3.53}$$

The matrix which is involved here will be denoted the translation matrix $\mathcal{T}$. The amplitude of the electric field at the surface (depth $Z_1=0$) of the layered material in Fig. 3.5 is obtained by multiplying all the refraction and the translation matrices in each layer starting from the substrate (at $z=Z_s$) as follows,

$$\begin{bmatrix} U(k_{z,0}, Z_1) \\ U(-k_{z,0}, Z_1) \end{bmatrix} = \mathcal{R}_{0,1}\mathcal{T}_1\mathcal{R}_{1,2}........\mathcal{R}_{N,s} \begin{bmatrix} U(k_{z,s}, Z_s) \\ U(-k_{z,s}, Z_s) \end{bmatrix} . \tag{3.54}$$

All the matrices involved in the above product are 2x2 matrices so that their product which is called the transfer matrix $\mathcal{M}$, is also a 2x2 matrix. We thus have,

$$\begin{aligned} \begin{bmatrix} U(k_{z,0}, Z_1) \\ U(-k_{z,0}, Z_1) \end{bmatrix} &= \mathcal{M} \begin{bmatrix} U(k_{z,s}, Z_s) \\ U(-k_{z,s}, Z_s) \end{bmatrix} \\ &= \begin{bmatrix} M_{11} & M_{12} \\ M_{21} & M_{22} \end{bmatrix} \begin{bmatrix} U(k_{z,s}, Z_s) \\ U(-k_{z,s}, Z_s) \end{bmatrix} . \end{aligned} \tag{3.55}$$

The reflection coefficient is defined as the ratio of the reflected electric field to the incident electric field at the surface of the material and is given by,

$$r = \frac{U(k_{z,0}, Z_1)}{U(-k_{z,0}, Z_1)} = \frac{M_{11}U(k_{z,s}, Z_s) + M_{12}U(-k_{z,s}, Z_s)}{M_{21}U(k_{z,s}, Z_s) + M_{22}U(-k_{z,s}, Z_s)} . \tag{3.56}$$

It is reasonable to assume that no wave will be reflected back from the substrate if the x-rays penetrate only a few microns, so that

$$U(k_{z,s}, Z_s) = 0, \tag{3.57}$$

and therefore the reflection coefficient is simply defined as

$$r = \frac{M_{12}}{M_{22}}. \tag{3.58}$$

The transmission coefficient is defined as the ratio of the transmitted electric field to the incident electric field

$$t = U(-k_{z,s}, Z_s)/U(-k_{z,0}, Z_1), \tag{3.59}$$

and is given by,

$$t = 1/M_{22}. \tag{3.60}$$

This method for the derivation of the reflection and transmission coefficients is known as the matrix technique. It is a general method which is valid for any kind of electromagnetic wave. However, it should be noted that for a plane wave of polarisation (p), the $p_{j,j+1}$ and $m_{j,j+1}$ coefficients must be modified in (3.52) by changing the wave-vector $k_{z,j}$ in medium j by $k_{z,j}/n_j^2$. One obtains:

$$\begin{aligned} p_{j,j+1}^{(p)} &= \frac{n_{j+1}^2 k_{z,j} + n_j^2 k_{z,j+1}}{2n_{j+1}^2 k_{z,j}} \\ m_{j,j+1}^{(p)} &= \frac{n_{j+1}^2 k_{z,j} - n_j^2 k_{z,j+1}}{2n_{j+1}^2 k_{z,j}}. \end{aligned} \tag{3.61}$$

Let us remark that, instead of considering the passage from $U(\pm k_{z,j}, Z_{j+1})$ to $U(\pm k_{z,j+1}, Z_{j+1})$, it is also possible to directly consider the passage from $A_j^\pm$ to $A_{j+1}^\pm$. The corresponding matrix [12] is,

$$\begin{bmatrix} A_j^+ \\ A_j^+ \end{bmatrix} = \begin{bmatrix} p_{j,j+1} e^{i(k_{z,j+1}-k_{z,j})Z_{j+1}} & m_{j,j+1} e^{-i(k_{z,j+1}+k_{z,j})Z_{j+1}} \\ m_{j,j+1} e^{i(k_{z,j+1}+k_{z,j})Z_{j+1}} & p_{j,j+1} e^{-i(k_{z,j+1}-k_{z,j})Z_{j+1}} \end{bmatrix} \begin{bmatrix} A_{j+1}^+ \\ A_{j+1}^+ \end{bmatrix}. \tag{3.62}$$

In this case, it is no longer necessary to introduce the translation matrix. A third alternative consists in defining a matrix which links the electric field and its first derivative at a depth Z_j to the same quantities at a depth Z_{j+1}. The matrix is unimodular and is defined for an *(s)* polarised wave as [11,13],

$$\begin{vmatrix} \cos \delta_{j+1} & \frac{\sin \delta_{j+1}}{k_{z,j+1}} \\ -k_{z,j+1} \sin \delta_{j+1} & \cos \delta_{j+1} \end{vmatrix} \tag{3.63}$$

with $\delta_{j+1} = k_{z,j+1} (Z_j - Z_{j+1})$. The application of this general electromagnetic formalism to the case of x-ray reflectivity is discussed in the next section.

3.2.2 The Refraction Matrix for X-Ray Radiation

As shown in the previous section (equations (3.51) and (3.52)), the refraction matrix is defined as

$$\mathcal{R}_{j,j+1} = \begin{bmatrix} p_{j,j+1} & m_{j,j+1} \\ m_{j,j+1} & p_{j,j+1} \end{bmatrix},$$

with

$$p_{j,j+1} = \frac{k_{z,j} + k_{z,j+1}}{2k_{z,j}} \qquad m_{j,j+1} = \frac{k_{z,j} - k_{z,j+1}}{2k_{z,j}}. \tag{3.64}$$

Equation (3.46) shows that $\mathrm{k}_{z,j}$ is the component of the wave-vector normal to the surface and that it is equal to,

$$k_{z,j} = -k_j \sin \theta_j = -\sqrt{k_j^2 - k_{x,j}^2}. \tag{3.65}$$

$k_{x,j}$ is conserved and is equal to $k \cos \theta$. As a result of this, the z component of k_0 in medium j is

$$k_{z,j} = -\sqrt{k_0^2 n_j^2 - k_0^2 \cos^2 \theta}, \tag{3.66}$$

where k_0 is the wave-vector in air. In the limit of small angles and substituting the expression of the refractive index for x-rays, this becomes,

$$k_{z,j} = -k_0 \sqrt{\theta^2 - 2\delta_j - 2i\beta_j}. \tag{3.67}$$

A similar expression can be obtained for $k_{z,j+1}$ so that the coefficients $p_{j,j+1}$ and $m_{j,j+1}$, and as a consequence, the refraction matrix $\mathcal{R}_{j,j+1}$ are entirely determined by the incident angle and by the value of δ and β in each layer.

3.2.3 Reflection from a Flat Homogeneous Material

For a homogeneous material, the transfer matrix between air (medium 0) and medium 1 is simply the refraction matrix, which means that $\mathcal{M} = \mathcal{R}_{0,1}$ so that the reflection coefficient r becomes

$$r = r_{0,1} = \frac{U(k_{z,0},0)}{U(-k_{z,0},0)} = \frac{M_{12}}{M_{22}} = \frac{m_{0,1}}{p_{0,1}} = \frac{k_{z,0} - k_{z,1}}{k_{z,0} + k_{z,1}}, \tag{3.68}$$

or (neglecting absorption)

$$r = \frac{-k_0\theta + k_0\sqrt{\theta^2 - 2\delta - 2i\beta}}{-k_0\theta - k_0\sqrt{\theta^2 - 2\delta - 2i\beta}} = \frac{\theta - \sqrt{\theta^2 - 2\delta - 2i\beta}}{\theta + \sqrt{\theta^2 - 2\delta - 2i\beta}}. \tag{3.69}$$

Equation (3.69) is of course identical to the one obtained by using the familiar expression for the Fresnel reflectivity (see Eq. (3.29)). Similarly, the transmission coefficient is simply given by,

$$t_{0,1} = \frac{U(-k_{z,1},0)}{U(-k_{z,0},0)} = \frac{1}{M_{22}} = \frac{1}{p_{0,1}} = \frac{2k_{z,0}}{k_{z,0} + k_{z,1}} \tag{3.70}$$

which is the same result as the one obtained earlier in equation (3.34). It should be realised that these reflection and transmission coefficients have been derived for an incident wave impinging on the surface of the material with a wave-vector k_{in}. In some cases (for example in the next chapter when treating the distorted wave Born approximation), it is important to label these coefficients to indicate which is the incident wave (Fig. 3.6). The detailed notation for the reflection and transmission coefficients will then be $r_{0,1}^{\text{in}}$ and $t_{0,1}^{\text{in}}$ when k_{in} is concerned and $r_{0,1}^{\text{sc}}$ and $t_{0,1}^{\text{sc}}$ for the wave vector k_{sc}. The explicit expressions for those coefficients are:[3]

$$t_{0,1}^{\text{in}} = \frac{2k_{\text{in}z,0}}{k_{\text{in}z,0} + k_{inz,1}} \tag{3.71}$$

$$t_{0,1}^{\text{sc}} = \frac{2k_{scz,0}}{k_{scz,0} + k_{scz,1}}. \tag{3.72}$$

[3] Let us also point out that the field in medium 1 associated with a plane-wave travelling with a wave vector $\mathbf{k}_{\text{in},1}$ is $E_1(\mathbf{k}_{\text{in},1}, \mathbf{r}) = U(\mathbf{k}_{\text{in},1}, z)e^{-k_{in,x}x} = E_0 t_{0,1}^{\text{in}} e^{-k_{\text{in},x}x} e^{-k_{\text{in}\,z,1}z}$.

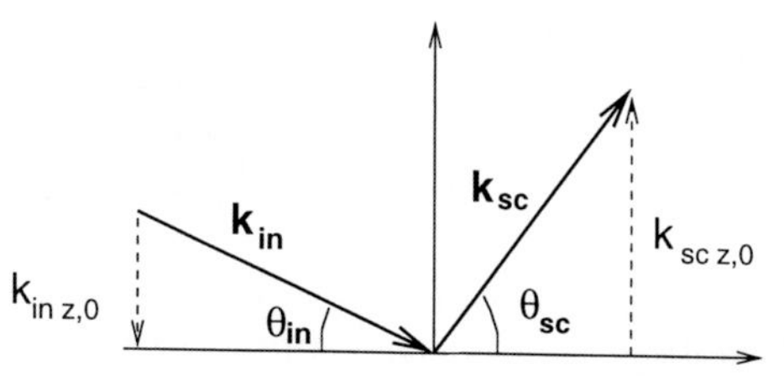

Fig. 3.6. Definition of the angles for the calculation of $t_{0,1}^{\mathrm{in}}$ and $t_{0,1}^{\mathrm{sc}}$

3.2.4 A Single Layer on a Substrate

The transfer matrix for the case of a layer of thickness $-h = Z_1 - Z_2$ (h and $k_{z,1}$ are negative) deposited on a substrate is given as

$$\mathcal{R}_{0,1}\mathcal{T}_1\mathcal{R}_{1,2} = \begin{vmatrix} p_{0,1} & m_{0,1} \\ m_{0,1} & p_{0,1} \end{vmatrix} \begin{vmatrix} e^{-ik_{z,1}h} & 0 \\ 0 & e^{+ik_{z,1}h} \end{vmatrix} \begin{vmatrix} p_{1,2} & m_{1,2} \\ m_{1,2} & p_{1,2} \end{vmatrix}, \tag{3.73}$$

and the reflection coefficient is,

$$r = \frac{M_{12}}{M_{22}} = \frac{m_{0,1}p_{1,2}e^{ik_{z,1}h} + m_{1,2}p_{0,1}e^{-ik_{z,1}h}}{m_{0,1}m_{1,2}e^{ik_{z,1}h} + p_{1,2}p_{0,1}e^{-ik_{z,1}h}}. \tag{3.74}$$

Dividing numerator and denominator by $p_{0,1}p_{1,2}$ and introducing the reflection coefficients $r_{i-1,i} = m_{i-1,i}/p_{i-1,i}$ for the two media i and $i-1$, the reflection coefficient of the electric field at the layer is then found to be,

$$r = \frac{r_{0,1} + r_{1,2}e^{-2ik_{z,1}h}}{1 + r_{0,1}r_{1,2}e^{-2ik_{z,1}h}}. \tag{3.75}$$

It is worth noting that the denominator of this expression differs from unity by a term which corresponds to multiple reflections in the material, as shown by the product of the two reflection coefficients $r_{01}r_{12}$.
It is also straightforward to determine the transmission coefficient since its value is given by $1/M_{22}$; this yields,

$$t = \frac{t_{0,1}t_{1,2}e^{-ik_{z,1}h}}{1 + r_{0,1}r_{1,2}e^{-2ik_{z,1}h}}. \tag{3.76}$$

In the case when the absorption can be neglected, the reflected intensity is therefore,

$$R = \frac{r_{0,1}^2 + r_{1,2}^2 + 2r_{0,1}r_{1,2}\cos 2k_{z,1}h}{1 + r_{0,1}^2 r_{1,2}^2 + 2r_{0,1}r_{1,2}\cos 2k_{z,1}h}. \tag{3.77}$$

The presence of the cosine terms in equation (3.77) indicates clearly that the reflectivity curve will exhibit oscillations in reciprocal space whose period will be defined by the equality

$$2k_{z,1}h \approx q_{z,1}h = 2p\pi, \tag{3.78}$$

or

$$q_{z,1} = \frac{2p\pi}{h}. \tag{3.79}$$

These oscillations are the result of the constructive interference between the waves reflected at interfaces *1* and *2*. The difference in path length which separates the two waves is

$$\delta = 2h\sin\theta_1 = p\lambda, \tag{3.80}$$

so that

$$q_{z,1} = \frac{2\pi p}{h}. \tag{3.81}$$

Figure 3.7 which shows the experimental reflectivity of a copolymer deposited onto a silicon substrate provides a good illustration of this type of interference phenomena. The experimental curve is presented in open circles and the calculated one as a solid line. The calculation is made by using the matrix technique in which we use equation (3.73) as starting point. The fact that the reflectivity is less than 1 below the critical angle is related to a surface effect. At very shallow angles, it frequently happens that the footprint of the beam is larger than the sample surface so that only part of the intensity is reflected (see Sect. 7.1.1 for details). A correction must then be applied to describe this part of the reflectivity curve. The roughness of the interfaces is also included in the calculation as discussed below.

3.2.5 Two Layers on a Substrate

The calculation of the reflectivity can also be made by the matrix technique in the case of two layers deposited on a substrate. After multiplying the five matrices of refraction and translation it is possible to express the reflection coefficient as,

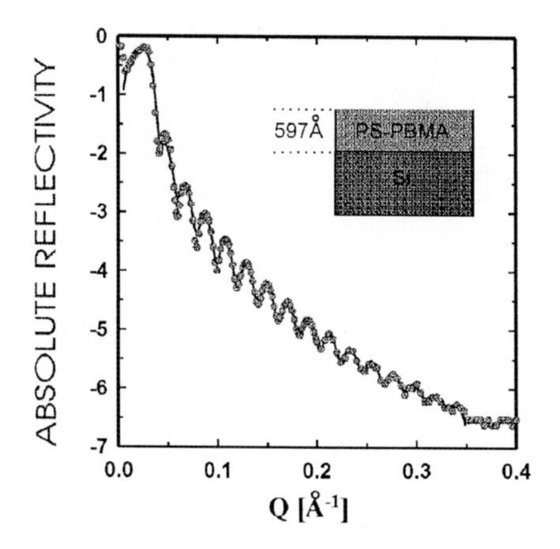

Fig. 3.7. Measured and calculated reflectivities of a thin film of a diblock copolymer PS-PBMA deposited on a silicon wafer

$$r = \frac{r_{0,1} + r_{1,2}e^{-2ik_{z,1}h_1} + r_{2,S}e^{-2i(k_{z,2}h_2+k_{z,1}h_1)} + r_{0,1}r_{1,2}r_{2,S}e^{-2ik_{z,2}h_2}}{1 + r_{0,1}r_{1,2}e^{-2ik_{z,1}h_1} + r_{1,2}r_{2,S}e^{-2ik_{z,2}h_2} + r_{2,S}r_{0,1}e^{-2i(k_{z,1}h_1+k_{z,2}h_2)}}. \tag{3.82}$$

The above expression clearly shows that for two layers on a substrate, multiple reflections at each interface appear in the matrix calculation. The phase shifts depend on the path difference calculated in each medium and therefore on the thickness of each layer, and indirectly on the angle of incidence. Examples of this kind are encountered in metallic thin films which tend to oxidise when placed in air. The reflectivity curve of an oxidised niobium thin film deposited on a sapphire substrate [15] is shown in Fig. 3.8. The upper layer obviously corresponds to the niobium oxide. The oxide layer grows as a function of time of exposure to air and reaches a maximum thickness of around *15Å* after a few hours. The reflectivity curve presented in Fig. 3.8 displays a very characteristic shape, which includes the following features,

- short wavelength oscillations which can be identified with the interferences within the (thick) niobium layer.
- a beating of the oscillations with a longer wavelength in q, which comes from the presence of the oxide on the top of the niobium layer; this leads to two interfaces at nearly the same altitude from the surface of the sapphire substrate.

There are similarities between this phenomenon and the characteristic beating of acoustic waves of similar frequency.

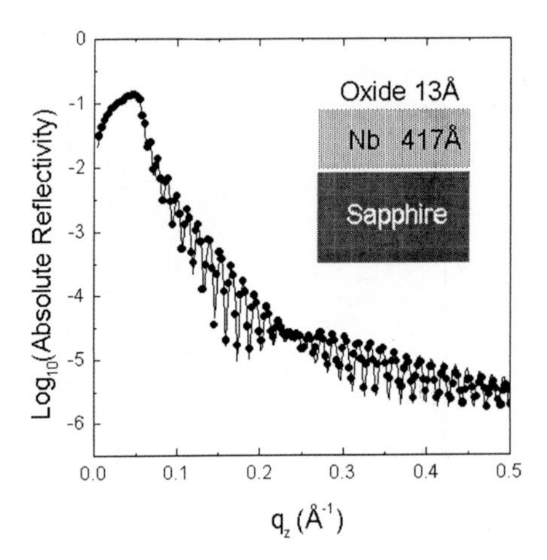

Fig. 3.8. Reflectivity of Nb thin film on sapphire showing the beating of spatial frequency between two comparable thicknesses which are the thickness of the niobium film and the thickness of the entire film (niobium and niobium oxide)

3.3 From Dynamical to Kinematical Theory

The full dynamical theory described above is exact but does not clearly show the physics of scattering because numerical calculations are necessary. Sometimes, one can be more interested in an approximated analytical expression. Different approximations can be done [13,16–18], the simplest one being the Born approximation.[4] We will start from the dynamical expression of the reflected amplitude calculated in the previous section (equation (3.77)) for a thin film of thickness h deposited on a substrate

$$r = \frac{r_{0,1} + r_{1,2}\, e^{-2ik_{z,1}h}}{1 + r_{0,1}r_{1,2}\, e^{-2ik_{z,1}h}}, \tag{3.83}$$

and degrade it to obtain approximate expressions. Here the phase shift between the reflected waves on the substrate and the layer denoted by $\varphi = -2k_{z,1}h = q_{z,1}h$ can be written as a function of either k or q. The term $r_{0,1}\, r_{1,2}\, e^{i\varphi}$ in this equation represents the effect of multiple reflections in the layer and a first step in the approximation consists in neglecting this term.

This is illustrated in Fig. 3.9 which shows a comparison between the reflectivities calculated for a diblock copolymer film on a silicon wafer with the matrix method taking into account or not the multiple reflections at the

[4] This kind of approach was first made by Rayleigh in 1912 in the context of the reflection of electromagnetic waves [16] but has since become known as the Born approximation since Born generalised it to different types of scattering processes.

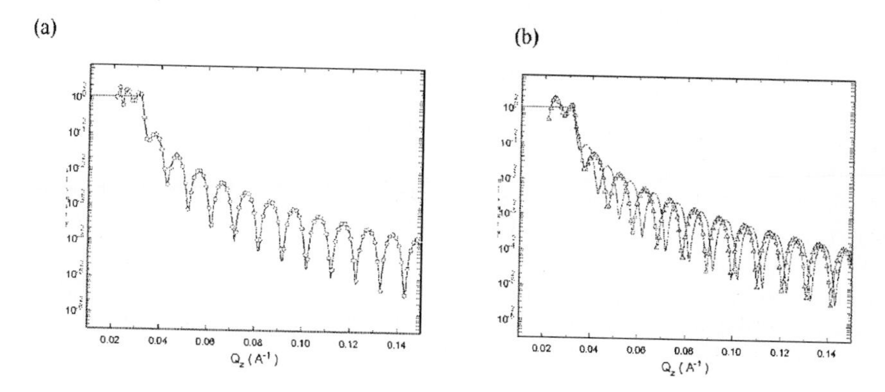

Fig. 3.9. Comparison between reflectivities calculated with the matrix technique (full line in (a) and (b)) and after neglecting the multiple reflections (squares in (a)) and in addition the refraction (triangles in (b)). Calculations are performed for a diblock copolymer ($q_c = 0.022\,\text{Å}^{-1}$) 600 Å thick on a silicon substrate

interfaces. It can be seen that the two curves are almost identical showing that this approximation is quite good. Under this approximation, the reflection coefficient r for a stratified medium composed of N layers is:

$$r = r_{0,1} + r_{1,2}e^{iq_{z,1}d_1} + r_{2,3}e^{i(q_{z,1}d_1 + q_{z,2}d_2)} + \ldots + r_{j,j+1}e^{i\sum_{k=0}^{j} q_{z,k}\,d_k} + \ldots \tag{3.84}$$

In Eq. (3.84) the ratio $r_{j,j+1}$of the amplitudes of the reflected to the incident waves at interface $j, j+1$ is

$$r_{j,j+1} = \frac{q_{z,j} - q_{z,j+1}}{q_{z,j} + q_{z,j+1}}, \tag{3.85}$$

with the wave-vector transfer in medium j:

$$q_{z,j} = (4\pi/\lambda)\sin\theta_j = \sqrt{q_z^2 - q_{c,j}^2}. \tag{3.86}$$

Finally,

$$R(q_z) = \left|\sum_{j=0}^{n} r_{j,j+1}e^{iq_z z_j}\right|^2 \quad \text{with } r_{j,j+1} = \frac{q_{z,j} - q_{z,j+1}}{q_{z,j} + q_{z,j+1}}.$$

A further approximation consists in neglecting the refraction and the absorption in the material in the phase factor in Eq. (3.84):

$$r = \sum_{j=0}^{n} r_{j,j+1} e^{iq_z \sum_{m=0}^{j} d_m} \tag{3.87}$$

In this case the approximation is more drastic and this can be seen in figure 3.9(b) showing that the region of the curve just after the critical angle is most affected, and in particular the positions of the interference fringes.
A final approximation consists in assuming that the wavevector q_z does not change significantly from one medium to the next so that the sum in the denominator of $r_{j,j+1}$ may be simplified:

$$r_{j,j+1} = \frac{q_{z,j}^2 - q_{z,j+1}^2}{(q_{z,j} + q_{z,j+1})^2} = \frac{q_{c,j+1}^2 - q_{c,j}^2}{4q_z^2} = \frac{4\pi r_e(\rho_{j+1} - \rho_j)}{q_z^2}, \tag{3.88}$$

with $q_{c,j} = \sqrt{16\pi r_e \rho_j}$ in which r_e stands for the classical radius of the electron. These approximations lead to the following expression for the reflection coefficient,

$$r = 4\pi r_e \sum_{j=1}^{n} \frac{(\rho_{j+1} - \rho_j)}{q_z^2} e^{iq_z \sum_{m=0}^{j} d_m}. \tag{3.89}$$

If the origin of the z axis is chosen to be at the upper surface (medium *0* at a depth of Z_1=*0*), then the sum over d_m in the phase factor, can be replaced by the depth Z_{j+1} of the interface *j,j+1* and the equation becomes,

$$r = 4\pi r_e \sum_{j=1}^{n} \frac{(\rho_{j+1} - \rho_j)}{q_z^2} e^{iq_z Z_{j+1}}. \tag{3.90}$$

Finally, if we consider that the material is made of an infinite number of thin layers, the sum may then be transformed into an integral over z, and the reflection coefficient r has the form,

$$r = \frac{4\pi r_e}{q_z^2} \int_{-\infty}^{+\infty} \frac{d\rho(z)}{dz} e^{iq_z z} dz. \tag{3.91}$$

A very useful, less drastic approximation is obtained by replacing $(4\pi r_e \rho_s)^2 / q_z^4$ by $R_F(q_z)$ in Eq. (3.91). Under this approximation the reflectivity can be written as [18],

$$R(q_z) = r.r^* = R_F(q_z) \left| \frac{1}{\rho_s} \int_{-\infty}^{+\infty} \frac{d\rho(z)}{dz} e^{iq_z z} dz \right|^2. \tag{3.92}$$

The above expression for $R(q_z)$ is not rigorous but it has the advantage of being easily handled in analytical calculations. In addition, if the Wiener-Kintchine theorem is applied to this result, we find

$$\frac{R(q_z)}{R_F(q_z)} = \frac{1}{\rho_s^2} TF\left[\rho'(z) \otimes \rho'(z)\right], \tag{3.93}$$

so that the data inversion gives the autocorrelation function of the first derivative of the electron density [19] or the Patterson function [20,21].

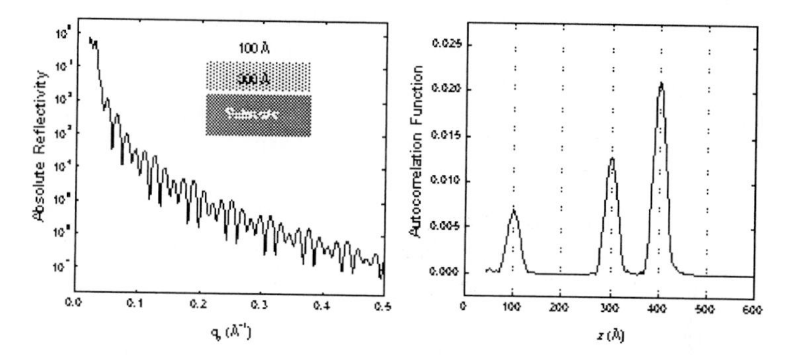

Fig. 3.10. Calculated Reflectivity of a two layers system and its Fourier transform after division of the data by the Fresnel reflectivity of the substrate. In the calculation the two layers of different electron densities are 300 and 100Å thick. The Fourier transform immediately gives the thickness of each layer without relying on any model. One can also note the an expected peak at z=400Å in the autocorrelation function

Figure 3.10, illustrates the main features of this data inversion. It is based on a calculation with a model structure [18] for a sample consisting of two layers, - a lower one of 300Å and an upper one of 100Å on a substrate. The left hand side diagram gives the calculated reflectivity curve which shows a feature similar to the "beating" effect seen in Fig. 3.8, arising here because of the similar thicknesses of the two layers. The right hand side diagram gives the auto-correlation function, which has intense peaks at the interfaces where the derivative of the electron density is maximimum. In an ideally flat sample these peaks would be delta functions, but for a real case their width depends on factors such as the roughness and degree of interdiffusion at the interfaces.

Equation (3.92) is a good starting point to introduce a last formulation for the reflected intensity. Starting from Eq. (3.92)

$$R(q_z) = rr^* = R_F(q_z) \left| \frac{1}{\rho_s} \int\limits_{-\infty}^{+\infty} \frac{d\rho(z)}{dz} e^{iq_z z} dz \right|^2 \tag{3.94}$$

and using the general relation between the Fourier transform of a function and the Fourier transform of its first derivative, we have,

$$R(q_z) = R_F(q_z) \left| \int_{-\infty}^{+\infty} \rho(z)\, e^{i\, q_z\, z}\, dz \right|^2$$
$$= R_F(q_z) \iint \rho(z)\rho(z')e^{iq_z(z-z')}dzdz'. \quad (3.95)$$

To summarise, it has been shown in this section that the kinematic theory is derived from the dynamical theory by three approximations :

(1) - no multiple reflections at the interfaces,

(2) - that the effects of refraction can be neglected and,

(3) -that the reflection coefficient at each interface is proportional to the difference of electron density.

All the expressions discussed above have been derived under the assumption of ideally flat interfaces in the samples. In such a case, the lateral position of reflecting points at the interfaces is unimportant, since all of the points are at the same depth from the surface. It is thus implicit that the intensity is localised along the specular direction. This means that the expression above can be considered as valid over the entire reciprocal space after multiplication by the delta functions δq_x and δq_y which characterise the specular character of the reflected intensity. Therefore, the last equation of (3.95) for example may as well be written as,

$$R(\mathbf{q}) = R_F(q_z) \iint \rho(z)\rho(z')e^{iq_z(z-z')}dzdz'\delta q_x \delta q_y. \quad (3.96)$$

Note that

$$R(\mathbf{q}) = \frac{(4\pi r_e)^2}{q_z^2} \iint \rho(z)\rho(z')e^{iq_z(z-z')}dzdz'\delta q_x \delta q_y. \quad (3.97)$$

is the well-known Born approximation (or kinematical) expression for x-ray scattering. It can be recovered from the integration of the scattering cross-section

$$\frac{d\sigma}{d\Omega} = r_e^2 \iint d\mathbf{r}d\mathbf{r}'\rho(\mathbf{r})\rho(\mathbf{r}')e^{i\mathbf{q}\cdot(\mathbf{r}-\mathbf{r}')} \quad (3.98)$$

as shown in Chap. 4, footnote 8.[5]

---

[5] We may notice that if applied to a flat surface this expression would lead to $R(\mathbf{q})=q_c^4\delta(q_x)\delta(q_y)/16q_z^4$

3.4 Influence of the Roughness on the Matrix Coefficients

It was shown in Chap. 2 that scattering from a rough surface/interface can be separated into two contributions, coherent and incoherent scattering. In this chapter, we are only interested in the specular intensity, i.e. the coherent intensity given by the average value of the field. We give here a simple method to take roughness into account in the reflection by a rough multilayer, using the matrix method. We rely on a more complete and rigorous treatment of the case of a single interface given in appendix 1.A to this chapter. In this appendix, it is shown that for roughnesses with in-plane characteristic lengths smaller than the extinction length $\approx 1\mu m$ for x-rays, introduced in Chap. 1,

$$r_{0,1}^{\text{rough}} = r_{0,1}^{\text{flat}} e^{-2k_{z,0}k_{z,1}\sigma_1^2}. \tag{3.99}$$

The exponential in Eq. (3.99) is known as the Croce-Névot factor [22]. We now apply the method of section 3.A.3 to the matrix method. Starting from equation (3.62)

$$\begin{bmatrix} A_j^+ \\ A_j^- \end{bmatrix} = \begin{bmatrix} p_{j,j+1}e^{i(k_{z,j+1}-k_{z,j})Z_{j+1}} & m_{j,j+1}e^{-i(k_{z,j+1}+k_{z,j})Z_{j+1}} \\ m_{j,j+1}e^{i(k_{z,j+1}+k_{z,j})Z_{j+1}} & p_{j,j+1}e^{-i(k_{z,j+1}-k_{z,j})Z_{j+1}} \end{bmatrix} \begin{bmatrix} A_{j+1}^+ \\ A_{j+1}^- \end{bmatrix} \tag{3.100}$$

which links the amplitudes of the electric field in two adjacent layers, we assume that the position of the interface Z_{j+1} between the j and $j+1$ layers fluctuates vertically as a function of the lateral position because of the interface roughness. Following a method proposed by Tolan [23], we replace the quantity Z_{j+1} by $Z_{j+1} + z_{j+1}(x,y)$ in the above matrix and we take the average value of the matrix over the whole area coherently illuminated by the incident x-ray beam (in the spirit of Sect. 3.A.3, this amounts to averaging the phase-relationship between the fields above and below the interface). This leads to (as shown in Appendix 1.A, such expressions are only valid at first order in $\langle z_j^2 \rangle$),

$$\left\langle \begin{bmatrix} A_j^+ \\ A_j^- \end{bmatrix} \right\rangle = \left\langle \begin{bmatrix} p_{j,j+1}e^{i(k_{z,j+1}-k_{z,j})Z_{j+1}}e^{i(k_{z,j+1}-k_{z,j})z_{j+1}(x,y)} & \cdots \\ m_{j,j+1}e^{i(k_{z,j+1}+k_{z,j})Z_{j+1}}e^{i(k_{z,j+1}+k_{z,j})z_{j+1}(x,y)} & \cdots \end{bmatrix} \right.$$

$$\left. \begin{bmatrix} \cdots & m_{j,j+1}e^{-i(k_{z,j+1}+k_{z,j})Z_{j+1}}e^{-i(k_{z,j+1}+k_{z,j})z_{j+1}(x,y)} \\ \cdots & p_{j,j+1}e^{-i(k_{z,j+1}-k_{z,j})Z_{j+1}}e^{-i(k_{z,j+1}-k_{z,j})z_j(x,y)} \end{bmatrix} \begin{bmatrix} A_{j+1}^+ \\ A_{j+1}^- \end{bmatrix} \right\rangle . \tag{3.101}$$

For Gaussian statistics, or at lowest order in σ_j^2, we have, assuming the independence of the different interface roughnesses:

$$\left\langle \begin{bmatrix} A_j^+ \\ A_j^- \end{bmatrix} \right\rangle = \begin{bmatrix} p_{j,j+1} e^{i(k_{z,j+1}-k_{z,j})Z_{j+1}} e^{-(k_{z,j+1}-k_{z,j})^2\sigma_{j+1}^2/2} \\ m_{j,j+1} e^{i(k_{z,j+1}+k_{z,j})Z_{j+1}} e^{-(k_{z,j+1}+k_{z,j})^2\sigma_{j+1}^2/2} \end{bmatrix}$$

$$\begin{matrix} m_{j,j+1} e^{-i(k_{z,j+1}+k_{z,j})Z_{j+1}} e^{-(k_{z,j+1}+k_{z,j})^2\sigma_{j+1}^2/2} \\ p_{j,j+1} e^{-i(k_{z,j+1}-k_{z,j+1})Z_{j+1}} e^{-(k_{z,j+1}-k_{z,j})^2\sigma_{j+1}^2/2} \end{matrix} \Bigg] \left\langle \begin{bmatrix} A_{j+1}^+ \\ A_{j+1}^- \end{bmatrix} \right\rangle . \tag{3.102}$$

The influence of the interface roughness is apparent from this result. The coefficients $m_{j,j+1}$ and $p_{j,j+1}$ are respectively reduced by the factors $e^{-(k_{z,j+1}+k_{z,j})^2\sigma_{j+1}^2/2}$ and $e^{-(k_{z,j+1}-k_{z,j})^2\sigma_{j+1}^2/2}$. It was shown in the previous section, that the ratio $m_{j,j+1}/p_{j,j+1}$ is the relevant quantity in the expression of the reflected intensity. This ratio which is the Fresnel coefficient of reflection at the altitude Z_{j+1} is therefore reduced by the amount,

$$\frac{r_{j,j+1}^{\text{rough}}}{r_{j,j+1}^{\text{flat}}} = e^{-2k_{z,j+1}k_{z,j}\sigma_{j+1}^2} = e^{-q_{z,j+1}q_{z,j}\sigma_{j+1}^2/2} \tag{3.103}$$

in the presence of interface roughness. In the particular case where the Born approximation holds ($k_{z,j} = k_{z,j+1} = (1/2)q_z$), the Fresnel coefficient is reduced by the amount

$$\frac{r_{j,j+1}^{\text{rough}}}{r_{j,j+1}^{\text{flat}}} = e^{-2k_{z,j+1}k_{z,j}\sigma_{j+1}^2} = e^{-q_z^2\sigma_{j+1}^2/2}, \tag{3.104}$$

which is the Debye-Waller factor.

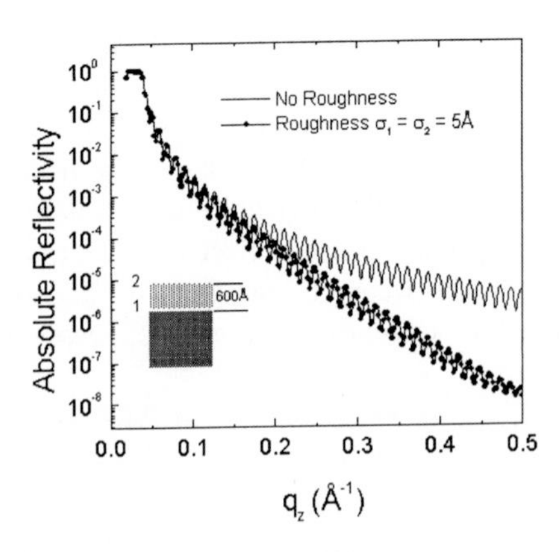

Fig. 3.11. Influence of roughness on the specular reflectivity of a 600Å thin layer deposited on a substrate

We present in Fig. 3.11 how the introduction of the roughness at the interfaces modifies the reflectivity curve. In particular, this figure shows that the reflectivity curve falls faster for rough interfaces and that the amplitude of the fringes is significantly reduced at high wave-vector transfers.

As a conlusion we have shown in this chapter that the calculation of the reflectivity can be properly handled by the matrix technique. This technique is the most widely used in the calculation of the specular reflectivity for the reason that it is simple and exact. However the main drawback of this technique is that it is only valid in specular conditions, which is an important restriction. Incoherent scattering is discussed in the next chapter, using in particular the matrix formalism described above.

3.A Appendix: The Treatment of Roughness in Specular Reflectivity

François de Bergevin, Jean Daillant, Alain Gibaud and Anne Sentenac

The aim of this appendix is to give an overview of the different methods which can be used to take roughness into account in specular reflectivity. We first present the second order Rayleigh calculation for a sinusoidal grating in order to introduce the main ideas. Then, we discuss the Distorted-wave Born approximation (DWBA) results (see Chap. 4 for a presentation of this approximation). Finally, we shortly discuss a simple method that allows one to retrieve the Debye-Waller and Croce-Névot factors which are the limiting laws, for respectively large and small in-plane correlation lengths. We consider scalar waves in all this appendix.

3.A.1 Second-Order Rayleigh Calculation for a Sinusoidal Grating

Let us consider the problem of the reflection by a rough interface (here simplified as a one dimensional sinusoidal grating of period Λ) separating two media and illuminated by a plane wave $e^{i(\omega t - k_{\mathrm{in}x}x - k_{\mathrm{in}z}z)}$. The Rayleigh method [24,25] consists in expanding the fields in both media as sets of plane waves and in writing the boundary conditions for the field and its first derivative. In order to write these boundary conditions, one has to calculate the values of the field and of its first derivative on the surface, as a series of terms like

$$a_\eta \exp -i\left(k_{\eta,x}x + k_{\eta,z}z(x)\right),$$

where η refers to both the medium (above or below the interface) and to the plane wave in the expansion (in particular, the component of its wave vector parallel to the surface describing the scattering order). One then expands:

$$a_\eta \exp -i\left(k_{\eta,x}x + k_{\eta,z}z(x)\right) \approx a_\eta \exp -i\left(k_{\eta,x}x\right)\left[1 - ik_{\eta,z}z(x) - \frac{1}{2}k_{\eta,z}^2 z^2(x) + \dots\right]. \tag{3.A1}$$

Since $z(x)$ can be expressed as a sum of two exponentials $(z_0/2)\exp(\pm 2i\pi x/\Lambda)$ (in the general case this would be a particular term in the Fourier expansion of the roughness), the expressions in the boundary conditions consist of sums of exponentials in x. For the boundary conditions to be satisfied for all x, it is necessary and sufficient that they are satisfied for each of these exponentials separately. We now have a series of equations, each corresponding to a scattering order:

$$k_{\mathrm{in}x},\ k_{\mathrm{in}x} \pm 2\pi/\Lambda, \dots$$

We define in medium 0 or 1:

$$k^{\pm 1}_{(0,1)\,z} = \sqrt{k^2_{(0,1)} - \left(k_{\text{in}x} \pm \frac{2\pi}{\Lambda}\right)^2},$$

and similarly

$$k_0 \cos\theta_{\pm 1} = k_0 \cos\theta_{\text{in}} \pm 2\pi/\Lambda.$$

A series in z_0 appears in each equation, and the system will be solved perturbatively at each order. At zeroth order we get the Fresnel coefficients. At first order in z_0 (in amplitude), we get for the intensities in the ± 1 scattering orders [26]:

$$I^{(1)}_{\pm 1} = I_0 z_0^2 k_{0,z} k^{\pm 1}_{0,z} \sqrt{R_F(\theta_{\text{in}}) R_F(\theta_{\pm 1})}, \tag{3.A2}$$

where $R_F(\theta_{\text{in}})$ and $R_F(\theta_{\pm 1})$ are the Fresnel reflection coefficients in intensity for the angles θ_{in} and $\theta_{\pm 1}$ respectively.
At second order (in amplitude) in z_0, we get in the specular:

$$I^{(2)}_0 = -I_0 z_0^2 k_{0,z} R_F(\theta_{\text{in}}) \mathcal{R}e\left(2k_{1,z} + k^{+1}_{0,z} - k^{+1}_{1,z} + k^{-1}_{0,z} - k^{-1}_{1,z}.\right) \tag{3.A3}$$

We now try to find the change in reflectivity coefficient in the limiting cases of large and small Λ values.

• *Large Λ values*

For large Λ values, the diffracted orders in both media get close to the specular and transmitted beams:

$$k^{\pm 1}_{0,z} \approx k_{0,z}, \quad k^{\pm 1}_{1,z} \approx k_{1,z}.$$

Then

$$R(\theta_{\text{in}}) \approx R_F(\theta_{\text{in}}) + I^{(2)}_0 / I_0 \approx R_F(\theta_{\text{in}})(1 - 2z_0^2 k^2_{0,z}).$$

Since one has $\langle z^2 \rangle = z_0^2/2$,

$$R(\theta_{\text{in}}) \approx R_F(\theta_{\text{in}})(1 - 4k^2_{0,z} \langle z^2 \rangle)$$

which is the first order expansion of the Debye-Waller factor in $\langle z^2 \rangle$.

• *Small Λ values*

One has:

$$k^{\pm 1}_{0,z} - k^{\pm 1}_{1,z} = \frac{k_c^2}{k^{\pm 1}_{0,z} + k^{\pm 1}_{1,z}}, \tag{3.A4}$$

where $k_c = k_0\sqrt{1-n^2}$ is the critical wave vector. For small Λ values, $k^{\pm 1}_{0,z}, k^{\pm 1}_{1,z} \gg k_c$ and therefore, using Eq. (3.A4), $k^{\pm 1}_{0,z} - k^{\pm 1}_{1,z} \ll k_c$. Since $k_{1,z}$ is never much smaller than k_c, it is the only term that survives in the sum in Eq (3.A3). Therefore,

$$R(\theta_{\text{in}}) \approx R_F(\theta_{\text{in}})(1 - 4\mathcal{R}e(k_{0,z} k_{1,z}) \langle z^2 \rangle),$$

which is the first order expansion of the Croce-Névot factor [22] in $\langle z^2 \rangle$.

3.A.2 The Treatment of Roughness in Specular Reflectivity within the DWBA

The issue of the modification of the specular intensity due to surface scattering has been considered within the distorted-wave Born approximation in particular in Refs. [27,28]. The results of Refs.[27,28] agree with the Rayleigh treatment given in the previous section. It is nevertheless interesting to note that:

Contrary to what is sometimes assumed, the specular intensity can be affected in the first-order DWBA. This is because the basis for this appromixation includes both the reflected and transmitted fields. It is therefore possible that *single* scattering events transfer energy from one field to the other (in fact, energy would be conserved at this level of approximation for the sum of the reflected and transmitted fields, see Appendix 1.A). In particular, the first order result of the DWBA Eq. (4.41) or [27] yields the Croce-Névot factor at first order in $\langle z^2 \rangle$.
Exercise: Show this.

The second order DWBA [28] shows, as did the Rayleigh calculation discussed above, that the Debye-Waller factor is obtained for large Λ values whereas the Croce-Névot factor is obtained for small Λ values.

3.A.3 Simple Derivation of the Debye-Waller and Croce-Névot Factors

The accuracy of approximated expressions for the reflectivity coefficient mainly relies on the quality of the approximations made on the local value of the electric field at the interface. Let us consider the two limiting cases of roughnesses with very small and very large in-plane characteristic length scales.
If the characteristic length scale of the roughness is much larger than the extinction length (we have a slowly varying interface height), the field can be written locally for the well-defined interface at a scale smaller than the roughness characteristic scale (this is the so-called tangent plane approximation):

$$E_j(x,z) = \left(A_j^+ e^{ik_{j,z}z} + A_j^- e^{-ik_{j,z}z}\right) e^{i\omega t - \mathbf{k}_i \mathrm{in} \| . \mathbf{r}_\|}, \tag{3.A5}$$

The field will be reflected at different heights depending on x, and the reflection coefficient is

$$r^{\mathrm{rough}} = \frac{\langle A_0^+ \rangle_x}{A_0^-},$$

where the average value is taken over the surface. Writing the boundary conditions, one obtains with the notations of Chap. 3 for a surface located at z:

$$\begin{cases} A_0^+ e^{ik_{z,0}z} + A_0^- e^{-ik_{z,0}z} & = A_1^- e^{-k_{z,1}z} \\ k_{z,0} A_0^+ e^{ik_{z,0}z} - k_{z,0} A_0^- e^{-ik_{z,0}z} & = -k_{z,1} A_1^- e^{-k_{z,1}z} . \end{cases} \tag{3.A6}$$

One obtains

$$r^{\text{rough}} = \frac{\langle A_0^+ \rangle_x}{A_0^-} = r_{0,1} \langle e^{2ik_{0,z}z} \rangle = r_{0,1} e^{-2k_{0,z} \langle z^2 \rangle},$$

which is the Debye-Waller factor as expected. We obtain this factor because the roughness characteristic length is large enough for the incident and reflected fields to have a precise phase relationship.

We now assume the characteristic length to be much smaller than the extinction length, which is on the order of $1 \mu m$. Then, the electric field is not perturbed at the roughness scale (in other words, there are no short-scale correlations between the field and the roughness). There is only an overall perturbation of the electric field which can be written to a good approximation as the combination of upwards and downwards propagating plane waves, whose amplitude will however depend on the roughness:

$$E_j(x,z) = \left(A_{j,\text{eff}}^+ e^{ik_{j,z}z} + A_{j,\text{eff}}^- e^{-ik_{j,z}z} \right) e^{i\omega t - \mathbf{k}_i \text{in} \| \cdot \mathbf{r}_\|}, \tag{3.A7}$$

where the $A_{j,\text{eff}}$ $(j = 0, 1)$ are unknown effective amplitudes for the rough interface. The reflection coefficient is defined as:

$$r^{\text{rough}} = \frac{A_{0,\text{eff}}^+}{A_{0,\text{eff}}^-}.$$

Now, we assume that the phase relationships between the field above and below the interface are only valid ***on average*** because the field "does not see" the local roughness (this is of course not a rigourous argument, the justification for Eq. (3.A8) are the calculations given in the two previous sections):

$$\begin{cases} 2k_{0,z} A_{0,\text{eff}}^+ = (k_{0,z} - k_{1,z}) A_{1,\text{eff}}^- \langle e^{-i(k_{0,z}+k_{z,1})z} \rangle \\ 2k_{0,z} A_{0,\text{eff}}^- = (k_{0,z} + k_{1,z}) A_{1,\text{eff}}^- \langle e^{i(k_{0,z}-k_{z,1})z} \rangle. \end{cases} \tag{3.A8}$$

Eq. (3.A8) can be obtained from Eq. (3.A6), or directly using the matrix method Eq. (3.62). Then

$$r^{\text{rough}} = \frac{A_{0,\text{eff}}^+}{A_{0,\text{eff}}^-} = r^{\text{flat}} e^{-2k_{z,0}k_{z,1}\langle z^2 \rangle}, \tag{3.A9}$$

one obtains the Croce-Névot factor. Note that in this case, the transition layer method would give an equally good result. Note also, that the method could be applied to the averaging of transfer matrices, as it is done in Chaps. 3 and 8.

References

1. A. Fresnel *Mémoires de l'Académie* **11**, 393 (1823).
2. A.H. Compton, *Phil. Mag,* **45**, 1121 (1923).
3. R. Forster, *Helv. Phys. Acta.*, **1**, 18 (1927).
4. H. Kiessig, *Ann. Der Physik*, **10**, 715 (1931).
5. J.A. Prins, *Z. Phys.*, **47**, 479 (1928).
6. L.G. Parrat, *Phys. Rev.* **95**, 359 (1954).
7. F. Abélès, *Ann. de Physique* **5**, 596 (1950).
8. R.W. James, The optical principles of the diffraction of x-rays, G. Bell and Sons, 9ondon, (1967).
9. International tables for x-ray crystallography, The Kynoch Press, Birmingham, 1968. vol. IV.
10. R. Petit, Ondes Electromagnétiques en radioélectricité et en optique, Ed. Masson (1989).
11. M. Born and E. Wolf, Principles of Optics, Pergamon, London, 6th Edition (1980).
12. B. Vidal and P. Vincent, *Applied Optics*, **23**, 1794 (1984).
13. J. Lekner, Theory of reflection of electromagnetic and particle waves, Martinus Nijhoff Publishers (1987).
14. T. P. Russel, *Mater. Science Rep.* **5**, 171 (1990).
15. A. Gibaud, D. McMorrow and P. P. Swadling, *J. Phys. Condens. Matter* **7**, 2645 (1995).
16. J.W.S. Rayleigh, *Proc. Roy. Soc.*, **86**, 207 (1912).
17. I. W. Hamley and J. S. Pedersen, *Appl. Cryst.* **27**, 29 (1994).
18. G. Vignaud, Thèse de l'Université du Maine (1997).
19. J. Als-Nielsen, *Z. Phys. B* **61**, 411 (1985).
20. I.M. Tidswell, B.M. Ocko, P.S. Pershan, S.R. Wassermann, G.M. Whitesides, J.D. Axe, *Phys. Rev. B* **41**, 1111 (1990).
21. G. Vignaud, A. Gibaud, G. Grübel, S. Joly, D. Ausserré, J.F. Legrand, Y. Gallot, *Physica B* **248**, 250 (1998).
22. L. Névot and P. Croce, Revue de Physique appliquée, **15**, 761 (1980).
23. M. Tolan, Rontgenstreuung an strukturierten Oberflächen Experiment &Theorie Ph.D. Thesis, Christian-Albrechts Universität, Kiel, (1993)
24. Lord Rayleigh, *Proc. R. Soc. London A* **79**, 339 (1907).
25. S.O. Rice, *Com. Pure Appl. Math.* **4**, 351 (1951).
26. D.V. Roschchupkin, M. Brunel, F. de Bergevin, A.I. Erko, *Nuclear Inst. Meth. B* **72**, 471 (1992).
27. S.K. Sinha, E.B. Sirota, S. Garroff and H.B. Stanley, *Phys. Rev. B*, **38**, 2297 (1988).
28. D.K.G. de Boer, *Phys. Rev. B* **49**, 5817 (1994).

4 Diffuse Scattering

Jean Daillant[1] and Anne Sentenac[2]

[1] Service de Physique de l'Etat Condensé, Orme des Merisiers , CEA Saclay, 91191 Gif sur Yvette Cedex, France,
[2] LOSCM/ENSPM, Université de St Jérôme, 13397 Marseille Cedex 20, France

Specular reflectivity, as described in Chap. 3 is sensitive to the average density profile along the normal (Oz) to a sample surface. Very often, one would also like to determine the statistical properties of surfaces or interfaces (i.e. the "lateral" structures in the (xOy) plane). We have seen in Chap. 2 that the scattered intensity depends on the roughness statistics of the sample (when the coherence domains are much smaller than the illuminated area). More precisely, under several simplifying assumptions, the differential scattering cross-section is related to the power spectrum of the surface. Many examples are given in the second part of this book where we shall see that, in particular because of the grazing incidence geometry, x-ray scattering experiments allow the determination of the lateral lengths of surface morphologies and of the correlations between buried interfaces over more than five orders of magnitudes from Ångtröms to tens of microns in plane (see Sect. 4.7.2).

In this chapter we present the theory of scattering by random media from an electromagnetic point of view. Starting from Maxwell equations we establish the volume integral equation giving the scattered field as the field radiated by the dipoles induced in the material (part of these results have been used without demonstration in Chap. 2). We then describe several perturbation techniques (Born approximation and Distorted-Wave Born Approximation, DWBA) that permit to obtain simple expressions for the differential scattering cross-section. The latters are then applied to scattering problems of increasing complexity: scattering by a single rough surface, surface scattering in a thin film, scattering by rough inhomogeneous multilayers. Finally, special attention is paid to the resolution function and to the determination of absolute (measured) intensities. This is necessary if one wants to draw quantitative information from an experiment.

Most of the chapter is devoted to the discussion of the so-called distorted wave Born approximation (DWBA) which presently provides the most accurate analysis of x-ray and neutron data. It is a perturbation method in which the roughness is viewed as a random perturbation of a deterministic reference state. In the simplest version of the theory presented here [1,2], the unperturbated reference (ideal) state can be a plane (in the case of the study of a rough surface) or a perfect planar multilayer (in the case of the study of rough multilayers). In more sophisticated versions, the reference state of a rough surface can be a medium with graded index whose z-dependent dielectric constant varies continuously from the air value to the material value, following the average density profile [3,4]. The electromagnetic field is calculated exactly for these reference states, hence, we expect the theory to be accurate even close to the critical angle for total external reflection. On the other hand, the radiative contributions of the permittivity fluctuations (i.e. the perturbation) are restricted to single scattering events (first order approximation) *within the reference medium*. Second-order DWBA, which accounts for doubly scattering processes, has been developed in Ref. [5] for specular reflectivity.

4.1 Differential Scattering Cross-Section for X-Rays

In this section, we first establish the propagation equation for the electric field and show that its solution can be put in the form of an integral equation using Green functions. This integral equation is the basis for the Born development (see Sect. 4.2 for the Born approximation and Sect. 4.3 for the (first order) DWBA). The definition of the distorted-wave Born approximation then amounts to the choice of an unperturbated (ideal) state for which the field in the sample and the Green functions have to be evaluated exactly. The evaluation of the Green functions is the main difficulty of the technique and we give here a simple method, based on the reciprocity theorem, to calculate them for various reference states: an infinite homogeneous medium (for the Born approximation) and a planar multilayer (for the DWBA).

The developments made in this section are valid for complicated systems like multilayers with rough interfaces and possibly density inhomogeneities. However, for simplicity the reader can refer to the case of a single rough interface separating two material media (0) and (1) depicted in Fig.4.1.

4.1.1 Propagation Equation

Using Maxwell's equations:

$$\nabla \times \mathbf{E} = -\frac{\partial \mathbf{B}}{\partial \mathbf{t}} \tag{4.1}$$

$$\nabla \times \mathbf{H} = \mathbf{j} + \frac{\partial \mathbf{D}}{\partial \mathbf{t}}, \tag{4.2}$$

one obtains the propagation equation for the electric field in the homogeneous media (0) and (1) containing no charges or currents:

$$\begin{aligned} \nabla \times \nabla \times \mathbf{E} - n^2(\mathbf{r})\frac{\omega^2}{c^2}\mathbf{E} &= -\nabla^2\mathbf{E} - n^2(\mathbf{r})\frac{\omega^2}{c^2}\mathbf{E} \\ &= \nabla \times \nabla \times \mathbf{E} - \frac{\omega^2}{c^2}\left(\mathbf{E} + \frac{\mathbf{P}}{\epsilon_0}\right) \\ &= 0, \end{aligned} \tag{4.3}$$

for waves having a $e^{i\omega t}$ time dependence. n is the refractive index; the dielectric constant and the refractive index are related by $\epsilon = n^2$; $k = n\omega/c$ is the wavevector; in the vacuum $k_0 = \omega/c = 2\pi/\lambda$ (λ is the wavelength). Note that all the possible complexity (roughness, inhomogeneities) of a sample is contained in $n^2(\mathbf{r})$.

In the case of neutrons, one has to solve Schrödinger equation which exhibits a similar structure:

$$\left(-\frac{\hbar^2}{2m}\nabla^2 + \frac{2\pi\hbar^2}{m}\sum_i b_i\rho_i\right)\psi(r) = \mathcal{E}\psi(r), \tag{4.4}$$

where b_i is the scattering length of nuclei i whose (number) density in the sample is ρ_i.

In the following we shall work out a pertubative solution for the problem of surface scattering. To do this, we decompose the index as

$$n^2(\mathbf{r}) = n^2_{\text{ref}}(\mathbf{r}) + \delta n^2(\mathbf{r}), \tag{4.5}$$

where n_{ref} is the index of a reference situation to be precised. The reference state is deterministic and simple enough for the electromagnetic field to be calculated exactly (vacuum, plane interface, planar multilayers). It represents the basis (zeroth order) of the perturbation development. Hence, it should be as close as possible to the real medium in order to minimise the influence of the perturbation. n^2_{ref} yields a specular reflection, and we will show that δn^2 yields the incoherent scattering. With Eq.(4.5), equation (4.3) is rewritten,

$$\nabla \times \nabla \times \mathbf{E}(\mathbf{r}) - n^2_{\text{ref}}(\mathbf{r})k_0^2\mathbf{E}(\mathbf{r}) = \delta n^2(\mathbf{r})k_0^2\mathbf{E}(\mathbf{r}). \tag{4.6}$$

The right-hand side of Eq. (4.6) can be considered as a ***fictitious dipole source*** $\delta\mathbf{P}(\mathbf{r}') = \epsilon_0\delta n^2(\mathbf{r}')\mathbf{E}(\mathbf{r}')$ *in the reference medium.*[1] Maxwell's equations and Eq. (4.3) being linear, the electric field can be written as $\mathbf{E} = \mathbf{E}_{\text{ref}} + \delta\mathbf{E}$ where $\mathbf{E}_{\text{ref}}$ is the field in the reference case, and $\delta\mathbf{E}$ the perturbation in the field radiated by the fictitious dipole source $\delta\mathbf{P}$.

4.1.2 Integral Equation

The field radiated at the detector by the fictitious sources $\delta\mathbf{P}$ can be calculated using Green functions. We introduce the Green tensor $\overline{\overline{\mathcal{G}}}(\mathbf{R}, \mathbf{r}')$ for the

---

[1] Note that $\mathbf{E}(\mathbf{r}')$ is the real ***unknown*** field at $\mathbf{r}'$.

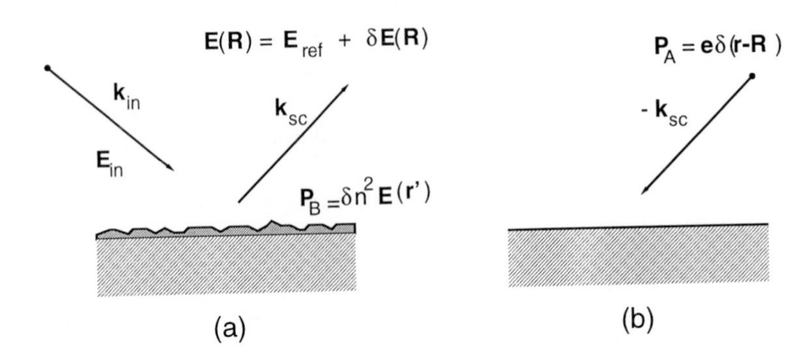

Fig. 4.1. Illustration of the reciprocity theorem in the case of a single rough surface. The rough surface (real state in (a)) is viewed as a perturbation (in gray) of the reference state (in (b)). The location of the planar interface in the reference case is arbitrary (it is here situated below the deepest incursion of the roughness, see footnote 9 on this subject). The total electric field $\mathbf{E}(\mathbf{R})$ existing in the real (rough) state is the sum of the specular field $\mathbf{E}_{\mathrm{ref}}(\mathbf{R})$ coming from the reference medium and of the scattered field $\delta\mathbf{E}(\mathbf{R})$ radiated by the dipole density $\delta\mathbf{P}(\mathbf{r}') = \epsilon_0 \delta n^2(\mathbf{r}')\mathbf{E}(\mathbf{r}')$, which is nonzero only within the grey region. In order to use the reciprocity theorem to calculate $\delta\mathbf{E}(\mathbf{R})$ we consider two distributions of sources: - A unit dipole with moment $\hat{\mathrm{e}}$ placed at $\mathbf{R}$ (detector location) creating a field distribution $\mathbf{E}_{\mathbf{A}}(\mathbf{r}') = \mathbf{E}^{\hat{\mathrm{e}}}_{\mathrm{det}}(\mathbf{R}, \mathbf{r}')$ in the reference state in (b), - A dipole density representing the perturbation brought by the roughness (gray region), $\delta\mathbf{P}(\mathbf{r}') = \epsilon_0 \delta n^2(\mathbf{r}')\mathbf{E}(\mathbf{r}')$ creating a field distribution $\mathbf{E}_{\mathbf{B}}(\mathbf{R}) = \delta\mathbf{E}(\mathbf{R})$ in (a). The reciprocity theorem yields, $\delta\mathbf{E}(\mathbf{R}).\hat{\mathrm{e}} = \int d\mathbf{r}' \delta\mathbf{P}(\mathbf{r}')\mathbf{E}^{\hat{\mathrm{e}}}_{\mathrm{det}}(\mathbf{R}, \mathbf{r}')$

propagation equation in the reference (ideal) case as the solution of [2]

$$\nabla \times \nabla \times \overline{\overline{\mathcal{G}}}(\mathbf{R}, \mathbf{r}') - n^2_{\mathrm{ref}}(\mathbf{R}) k_0^2 \overline{\overline{\mathcal{G}}}(\mathbf{R}, \mathbf{r}') = \frac{k_0^2}{\epsilon_0} \delta(\mathbf{R} - \mathbf{r}'), \tag{4.7}$$

that satisfies *out-going* wave boundary conditions, (the $\nabla \times \nabla$ operator acts on $\mathbf{R}$).[3]

[2] We have the identity $\nabla \times \nabla = \mathbf{grad}\mathrm{div} - \Delta$. In a vacuum, $\mathrm{div}\mathbf{E} = 0$ and the propagation equation reduces to a set of three Helmholtz equations $-\Delta\mathbf{E} - k_0^2\mathbf{E} = \mathbf{0}$. With this sign convention, the outgoing Green tensor $\overline{\overline{\mathcal{G}}}(r)$ reduces to $-G_-(r)$ defined in Chap. 1 for the scalar field obeying the Helmholz equation $\Delta E + k_0^2 E = 0$.

[3] The solution (in the sense of distributions) of Eq. (4.7) satisfies by construction the boundary conditions in the system (e.g. the saltus conditions at each interface if the reference state is a planar multilayer).

It is then straightforward to show (insert Eq. (4.8) in Eq. (4.6)) that,

$$\begin{aligned}
\mathbf{E}(\mathbf{R}) &= \mathbf{E}_{\mathrm{ref}}(\mathbf{R}) + \delta\mathbf{E}(\mathbf{R}) \\
&= \mathbf{E}_{\mathrm{ref}}(\mathbf{R}) + \epsilon_0 \int d\mathbf{r}\delta n^2(\mathbf{r})\overline{\overline{\mathcal{G}}}(\mathbf{R},\mathbf{r}).\mathbf{E}(\mathbf{r}) \\
&= \mathbf{E}_{\mathrm{ref}}(\mathbf{R}) + \int d\mathbf{r}\overline{\overline{\mathcal{G}}}(\mathbf{R},\mathbf{r}).\delta\mathbf{P}(\mathbf{r}) \qquad (4.8)
\end{aligned}$$

is the solution of Eq. (4.6). Eq. (4.8) is formally equivalent to Eq. (1.A4) and will be the basis for the Born (or DWBA) development. However, we first need an expression for the tensorial Green function. This can in particular be done using an elegant method due to P. Croce [6–11] based on the reciprocity theorem [12].

4.1.3 Derivation of the Green Functions Using the Reciprocity Theorem

In this paragraph we determine the Green tensor $\overline{\overline{\mathcal{G}}}(\mathbf{R},\mathbf{r}')$ introduced in Eq. (4.8) for two reference media, vacuum and a planar multilayer. For this, we use the reciprocity theorem demonstrated for example in Appendix 4.A and Ref. [12–14]. The reciprocity theorem states that, in a given reference medium, two different distributions of dipole sources $\mathbf{P_A}$ and $\mathbf{P_B}$ creating the fields $\mathbf{E_A}$ and $\mathbf{E_B}$ are linked by the relation,

$$\int d\mathbf{r}\mathbf{E_A}(\mathbf{r}).\mathbf{P_B}(\mathbf{r}) = \int d\mathbf{r}\mathbf{E_B}(\mathbf{r}).\mathbf{P_A}(\mathbf{r}). \qquad (4.9)$$

In order to calculate the perturbation in the field at the detector $\delta\mathbf{E}(\mathbf{R})$, we consider the following sources and field distributions (see Fig. 4.1),

The source with polarisation vector $\delta\mathbf{P}(\mathbf{r}') = \epsilon_0\delta n^2(\mathbf{r}')\mathbf{E}(\mathbf{r}')$ creating an unknown field $\delta\mathbf{E}(\mathbf{R})$ at the detector in the real case of the rough interface,

The unit dipole $\widehat{\mathbf{e}}\delta(\mathbf{R}-\mathbf{r}')$ located at the detector, creating a known field $\mathbf{E}_{\mathrm{det}}^{\widehat{\mathbf{e}}}(\mathbf{R},\mathbf{r}')$ at point $\mathbf{r}'$ in the roughness region (the field can be calculated exactly since the unit dipole radiates in the simple reference geometry).

We write,

$$\begin{aligned}
\mathbf{E_A} &= \mathbf{E}_{\mathrm{det}}^{\widehat{\mathbf{e}}}(\mathbf{R},\mathbf{r}') & \mathbf{P_A} &= \delta(\mathbf{R}-\mathbf{r}')\widehat{\mathbf{e}} \\
\mathbf{E_B} &= \delta\mathbf{E}(\mathbf{R}) & \mathbf{P_B} &= \epsilon_0\delta n^2\mathbf{E}(\mathbf{r}'),
\end{aligned}$$

and the reciprocity theorem Eq. (4.9) yields,

$$\int d\mathbf{r}'\epsilon_0\delta n^2(\mathbf{r}')\mathbf{E}(\mathbf{r}').\mathbf{E}_{\mathrm{det}}^{\mathbf{e}}(\mathbf{R},\mathbf{r}') = \delta\mathbf{E}(\mathbf{R}).\widehat{\mathbf{e}}. \qquad (4.10)$$

Eq. (4.10) is equivalent to,

$$\mathbf{E}(\mathbf{R}).\widehat{\mathbf{e}} = \mathbf{E}_{\mathrm{ref}}(\mathbf{R}).\widehat{\mathbf{e}} + \int d\mathbf{r}'\epsilon_0\delta n^2(\mathbf{r}')\mathbf{E}(\mathbf{r}').\mathbf{E}_{\mathrm{det}}^{\widehat{\mathbf{e}}}(\mathbf{R},\mathbf{r}'). \qquad (4.11)$$

The unit vector $\widehat{\mathbf{e}}$ being arbitrary, Eq. (4.11) is in fact a vector (and not scalar) equation (choose $\widehat{\mathbf{e}}$ equal respectively to $\widehat{x}$, $\widehat{y}$ and $\widehat{z}$ to calculate the different field components). We retrieve formally Eq. (4.8),[4]

$$\mathbf{E}(\mathbf{R}) = \mathbf{E}_{\text{ref}}(\mathbf{R}) + \epsilon_0 \int d\mathbf{r}' \delta n^2 \overline{\overline{\mathcal{G}}}(\mathbf{R},\mathbf{r}').\mathbf{E}(\mathbf{r}').$$

Comparing Eqs. (4.8) and (4.11), we see that *the Green function required to calculate the scattered field can therefore be simply calculated as the field in r' due to a unit dipole in R (detector) in the reference case.* In practice, we will directly use this property in Eq. (4.10) to calculate the scattered field. This is particularly convenient in the far-field approximation within which the dipole field is easy to calculate. Note that Eq. (4.11) is an exact relation [2] from which approximations can be made. That Eq. (4.11) is exact is verified in Appendix 4.B in the particular case of the reflection on a film.

If the polarisation (or direction) $\widehat{\mathbf{e}}_{\text{sc}}$ of the scattered field in the detector is known, the unit dipole direction $\widehat{\mathbf{e}}$ is usually taken equal to $\widehat{\mathbf{e}}_{\text{sc}}$ in order to directly obtain the scattered field amplitude,

$$\delta E(\mathbf{R}) = \int d\mathbf{r}' \epsilon_0 \delta n^2 \mathbf{E}(\mathbf{r}').\mathbf{E}_{\text{det}}^{\widehat{\mathbf{e}}_{\text{sc}}}(\mathbf{R},\mathbf{r}'). \tag{4.12}$$

In the far-field approximation, the polarisation vector $\widehat{\mathbf{e}}_{\text{sc}}$ is necessarily perpendicular to the (meaningful in far-field) sample-to-detector direction $\widehat{\mathbf{u}} = \mathbf{R}/R$. It appears convenient to introduce two main polarisation states. In polarisation (s), the field direction is normal to the scattering plane, (defined by the normal to the sample and the sample-to-detector direction (Oz,$\widehat{\mathbf{u}}$)), in polarisation (p) the field direction lies in the scattering plane.
We now explicitly calculate the Green function in a vacuum (which is the reference state in the Born approximation) and for a planar multilayer (which is the reference state for the DWBA).

4.1.4 Green Function in a Vacuum

The electric field at point $\mathbf{r}'$ created by a dipole moment $\widehat{\mathbf{e}}$ located at $\mathbf{R}$ in an homogeneous infinite medium (vacuum) can be written as (neglecting the $1/|\mathbf{R}-\mathbf{r}'|^2$ and $1/|\mathbf{R}-\mathbf{r}'|^3$ terms in the limit of large $|\mathbf{R}-\mathbf{r}'|$) [15],

$$\mathbf{E}_{\text{det}}^{\widehat{\mathbf{e}}}(\mathbf{R},\mathbf{r}') = k_0^2 (\widehat{\mathbf{u}} \times \widehat{\mathbf{e}}) \times \widehat{\mathbf{u}} \frac{e^{-ik_0|\mathbf{R}-\mathbf{r}'|}}{4\pi\epsilon_0 |\mathbf{R}-\mathbf{r}'|}, \tag{4.13}$$

where $\widehat{\mathbf{u}} = |\mathbf{R}-\mathbf{r}'|/|\mathbf{R}-\mathbf{r}'|$ is the unit vector in the direction of observation. Hereafter we assume that $R \gg r'$ so that $\widehat{\mathbf{u}} \approx \mathbf{R}/R$. Note that the dipole

[4] Note that the reciprocity theorem gives Eq. (4.11) but does not tell us that the Green tensor is reciprocal in the sense that $\overline{\overline{\mathcal{G}}}(\mathbf{R},\mathbf{r}') = \overline{\overline{\mathcal{G}}}(\mathbf{r}',\mathbf{R})$. In fact the symmetry relations on the Green tensor involve transpositions, see e.g. Ref. [14].

is here located at the detector position and that the field is observed at the sample. In the far-field approximation, one can develop

$$|\mathbf{R} - \mathbf{r}'| = |R\,\widehat{\mathbf{u}} - \mathbf{r}'| \approx R - \widehat{\mathbf{u}}.\mathbf{r}' \approx R - \mathbf{k}_{\mathrm{sc}}.\mathbf{r}'/k_0$$

(see Fig. 4.1).[5] The dipole spherical wave can therefore be developed on the tangent plane wave,

$$\mathbf{E}^{\widehat{\mathbf{e}}}_{\mathrm{det}}(\mathbf{R}, \mathbf{r}') = k_0^2(\widehat{\mathbf{u}} \times \widehat{\mathbf{e}}) \times \widehat{\mathbf{u}} \frac{e^{-ik_0R}}{4\pi\epsilon_0 R} e^{i\mathbf{k}_{\mathrm{sc}}.\mathbf{r}'} . \tag{4.14}$$

If we choose $\widehat{\mathbf{e}}$ normal to the direction of scattering, for example along the (s) and (p) polarisation directions $\widehat{\mathbf{e}}^{(s)}$ or $\widehat{\mathbf{e}}^{(p)}$, one simply has,

$$\mathbf{E}^{(s),(p)}_{\mathrm{det}}(\mathbf{R}, \mathbf{r}') = k_0^2 \frac{e^{-ik_0R}}{4\pi\epsilon_0 R} e^{i\mathbf{k}_{\mathrm{sc}}.\mathbf{r}'} \widehat{\mathbf{e}}^{(s),(p)} . \tag{4.15}$$

4.1.5 Green Function for a Stratified Medium

We now consider a planar multilayer as reference state and we want the expression of the electric field created at point $\mathbf{r}'$ by a unit dipole placed in $\mathbf{R}$. We assume that the far-field conditions are satisfied, so that the direction $\mathbf{k}_{\mathrm{sc}}$ is meaningful, and we consider the two main states of polarisation, $\widehat{\mathbf{e}} = \widehat{\mathbf{e}}^{(s)}$ and $\widehat{\mathbf{e}} = \widehat{\mathbf{e}}^{(p)}$. The point $\mathbf{r}'$ can be taken anywhere in the stratified medium, see Fig. 4.2. Using the same plane-wave limit in the general case of a stratified medium, Eq. (4.15) can be generalised for (s) or (p) polarisation for $\mathbf{r}'$ lying in layer j as,[6]

$$\mathbf{E}_{\mathrm{det}}{}^{(s),(p)}(\mathbf{R}, \mathbf{r}') = k_0^2 \frac{e^{-ik_0R}}{4\pi\epsilon_0 R} E_j^{PW\,(s),(p)}(-k_{\mathrm{scz,j}}, z')\widehat{\mathbf{e}}^{(s),(p)}_{\mathrm{sc}} e^{i\mathbf{k}_{\mathrm{sc}\parallel}.\mathbf{r}_\parallel} . \tag{4.16}$$

---

[5] The far-field conditions (or Fraunhofer diffraction) are more restricting than only $R \gg r'$. Indeed, to neglect the quadratic term in the expansion of $e^{-ik_0|\mathbf{R}-\mathbf{r}'|}$ one needs r'^2/λ to be small compared to R. Applying this approximation in Eq. (4.8) yields a condition on the whole size of the scattering object (since $\mathbf{r}'$ covers all the perturbated region). The discussion in Chap. 2 has shown that the support of the integral appearing in Eq. (4.8) can actually be restricted to the domain of coherence (induced by the incident beam and detector acceptance) of the scattering processes. In this case the far-field conditions can be written as $l_{coh}^2/\lambda \ll R$. In a typical x-ray experiment, the sample-to-detector distance is $R = 1m$, the wavelength is $\lambda = 1\mathring{A}$. The total illuminated area is a few mm but the coherence length is $l_{coh} \approx 1\mu m$, hence the far-field approximation is valid. When the coherence length is too important (very small detector acceptance) for the far-field conditions to be satisfied, we are in the frame of the Fresnel diffraction and one needs to retain the quadratic terms in the expansion of $|\mathbf{R} - \mathbf{r}'|$ [16] [17].

[6] The electric field is the solution of the inhomogeneous differential equation $\nabla \times \nabla \times \mathbf{E}^{\widehat{\mathbf{e}}}_{\mathrm{det}}(\mathbf{r}') - n^2_{\mathrm{ref}}(z')k_0^2\mathbf{E}^{\widehat{\mathbf{e}}}_{\mathrm{det}}(\mathbf{r}') = \widehat{\mathbf{e}}\delta(\mathbf{R}-\mathbf{r}')$ that satisfies *out-going* wave boundary conditions. The unit dipole at the detector position lies in medium 0 as depicted in Fig. 4.1. In the homogeneous region 0, the electric field can be written as the sum of a particular solution and a homogeneous solution. The par-

$E_j^{PW\,(s),(p)} e^{i\mathbf{k}_{\mathrm{sc}\|}\cdot\mathbf{r}_\|}$ is the field in medium j for an incident plane wave with polarisation (s) or (p) which can be computed by using standard iterative procedures [18,16]. Using the notations of chapter 3, Eqs. (3.47), (3.48), one has,

$$E_j^{PW\,(s),(p)}(k_{z,j}, z) = U_j^{(s),(p)}(k_{z,j}, Z_j)e^{-ik_{z,j}z} + U_j^{(s),(p)}(-k_{z,j}, Z_j)e^{ik_{z,j}z}, \tag{4.17}$$

where $\mathbf{r} = (\mathbf{r}_\|, z)$ with z is the z coordinate with the origin taken at $z = Z_j$, and where the superscript "PW" has been used to emphasize that E^{PW} is calculated for an ***incident plane wave***.

4.1.6 Differential Scattering Cross-Section

It is now possible to give an exact expression for the $\widehat{\mathbf{e}}$-component of the field scattered in the $\widehat{\mathbf{u}} = \mathbf{R}/R = \mathbf{k}_{\mathrm{sc}}/k_0$ direction, $\mathbf{E}_{\mathrm{sc}} = \mathbf{E} - \mathbf{E}_{\mathrm{in}}$. We first choose the vacuum as reference state. Substituting the expression (4.14) of $\mathbf{E}_{\mathrm{det}}^{\widehat{\mathbf{e}}}(\mathbf{R}, \mathbf{r}')$ in Eq. (4.8), one obtains for the component along $\widehat{\mathbf{e}}$ of the

---

ticular solution is given in Eq. (4.13) while the general homogeneous solutions are simply up-going plane waves with wavector modulus k_0. In media j with $j \neq 0 \neq s$, the electric field is solution of the homogeneous vectorial Helmholtz equation and it can be written as a sum of up-going and down-going plane waves with wavevector modulus k_j. In the substrate the general solutions are down-going plane waves with wavevector modulus k_s. To obtain the amplitudes of these plane waves we write the boundary conditions at each interface. The far-field approximation permits to simplify greatly the problem. In this case, the expression of the particular solution at $z = Z_1$ is given by Eq. (4.15). The dipole field close to the first interface can be approximated by an "incident" plane wave with wavevector $\mathbf{k}_{sc}$. Hence, the amplitudes of the other plane waves (that are the general solutions of the homogeneous Helmholtz equations) are calculated easily with the transfer matrix technique presented in Chap. 3. The problem has been reduced to the calculation of the electric field in a stratified medium illuminated by a plane wave. The meaning of superscripts (s) or (p) is always unambiguous : it indicates the direction of the radiating unit dipole in a vacuum for a given position $\mathbf{R}$ of the detector. In other words, it indicates the polarisation state of the scattered plane wave with wavevector $\mathbf{k}_{sc}$. Note that the directions of $\widehat{\mathbf{e}}_{\mathrm{sc}}^{(p)}$ and $\mathbf{k}_{sc}$ will vary from layer to layer due to refraction whereas the directions given by $\mathbf{k}_{sc\|}$ and $\widehat{\mathbf{e}}_{\mathrm{sc}}^{(s)}$ do not change.

scattered field in direction $\widehat{\mathbf{u}}$:

$$\begin{aligned}
\mathbf{E}_{\mathrm{sc}}(\mathbf{R}).\widehat{\mathrm{e}} &= \epsilon_0 \int d\mathbf{r}'\delta n^2(\mathbf{r}')\mathbf{E}^{\widehat{\mathrm{e}}}_{\mathrm{det}}(\mathbf{R},\mathbf{r}').\mathbf{E}(\mathbf{r}') \\
&= k_0^2 \frac{e^{-ik_0R}}{4\pi R} \int d\mathbf{r}'\delta n^2(\mathbf{r}')(\widehat{\mathrm{e}} - (\widehat{\mathbf{u}}.\widehat{\mathrm{e}})\widehat{\mathbf{u}}).\mathbf{E}(\mathbf{r}')e^{i\mathbf{k}_{\mathrm{sc}}.\mathbf{r}'} \\
&= \frac{e^{-ik_0R}}{4\pi R} \int d\mathbf{r}'(k^2(\mathbf{r}') - k_0^2)(\widehat{\mathrm{e}} - (\widehat{\mathbf{u}}.\widehat{\mathrm{e}})\widehat{\mathbf{u}}).\mathbf{E}(\mathbf{r}')e^{i\mathbf{k}_{\mathrm{sc}}.\mathbf{r}'} .
\end{aligned} \tag{4.18}$$

Writing Eq. (4.18) for $\widehat{\mathrm{e}}$ equal respectively to $\widehat{x}$, $\widehat{y}$ and $\widehat{z}$, one obtains:

$$\begin{aligned}
\mathbf{E}_{\mathrm{sc}}(\mathbf{R}) &= \frac{e^{-ik_0R}}{4\pi R} \int d\mathbf{r}'\delta n^2(\mathbf{r}')\mathbf{E}_\perp(\mathbf{r}')e^{i\mathbf{k}_{\mathrm{sc}}.\mathbf{r}'} \\
&= \frac{e^{-ik_0R}}{4\pi R} \int d\mathbf{r}'(k^2(\mathbf{r}') - k_0^2)\mathbf{E}_\perp(\mathbf{r}')e^{i\mathbf{k}_{\mathrm{sc}}.\mathbf{r}'} ,
\end{aligned} \tag{4.19}$$

where $\mathbf{E}_\perp = \mathbf{E} - (\widehat{\mathbf{u}}.\mathbf{E})\widehat{\mathbf{u}}$ is the component of the field normal to the direction of scattering. Note that the reference field is in this case the incident field.

To calculate the differential scattering cross-section we proceed by deriving the Poynting vector expression. In the far-field approximation,

$$\mathbf{B}_{\mathrm{sc}} = \frac{1}{c}\widehat{\mathbf{u}} \times \mathbf{E},$$

and the Poynting's vector is:

$$\mathbf{S} = \frac{|\mathbf{E}|^2}{2\mu_0 c}\widehat{\mathbf{u}}.$$

The differential scattering cross-section is obtained by calculating the flux of Poynting's vector (power radiated) per unit solid angle in direction $\mathbf{k}_{\mathrm{sc}}$ across a sphere of radius R for a unit incident flux. Using the linearity of Maxwell's equations, it can also be calculated for an incident field $\mathbf{E}_{\mathrm{in}}$ across a unit surface. One gets,

$$\frac{d\sigma}{d\Omega} = \frac{k_0^4}{16\pi^2|\mathbf{E}_{\mathrm{in}}|^2} \left| \int d\mathbf{r}'\delta n^2(\mathbf{r}')\mathbf{E}_\perp(\mathbf{r}')e^{i\mathbf{k}_{\mathrm{sc}}.\mathbf{r}'} \right|^2 . \tag{4.20}$$

This exact expression has been used in Chap. 2 to discuss the effect of an extended detector on the measured scattered intensity in relation with the statistical properties of a surface.

If one considers the issue of scattering from random media, we have seen in Chap. 2 that scattering can be separated into a coherent process and an incoherent process. The latter is the usual quantity of interest in a scattering experiment and it is given by,

$$\left(\frac{d\sigma}{d\Omega}\right)_{\text{incoh}} = \frac{k_0^4}{16\pi^2|\mathbf{E}_{\text{in}}|^2}\left\{\left|\int d\mathbf{r}'\delta n^2(\mathbf{r}')\mathbf{E}_\perp(\mathbf{r}')e^{i\mathbf{k}_{\text{sc}}.\mathbf{r}'}\right|^2 - \left|\left\langle\int d\mathbf{r}'\delta n^2(\mathbf{r}')\mathbf{E}_\perp(\mathbf{r}')e^{i\mathbf{k}_{\text{sc}}.\mathbf{r}'}\right\rangle\right|^2\right\}. \tag{4.21}$$

4.2 First Born Approximation

The first Born approximation which neglects multiple reflections can only be used far from the critical angle for total external reflection. Close to this point, the scattering cross-sections are large and the contribution to the measured intensity of at least multiple reflections cannot be neglected. The main advantage of presenting this approximation here is that it makes the structure of the scattered intensity very transparent. It has already been presented in Chap. 2 in a different context with the aim of illustrating how statistical information about surfaces or interfaces can be obtained in a scattering experiment.

4.2.1 Expression of the Differential Scattering Cross-Section

In the Born approximation, both the Green function and the electric field are evaluated in a vacuum, Eq. (4.11).

$$\mathbf{E}^{\widehat{\mathbf{e}}}_{\text{det}}(\mathbf{R}-\mathbf{r}') = k_0^2(\widehat{\mathbf{u}}\times\widehat{\mathbf{e}})\times\widehat{\mathbf{u}}\,\frac{e^{-ik_0R}}{4\pi\epsilon_0 R}e^{i\mathbf{k}_{\text{sc}}.\mathbf{r}'} \tag{4.22}$$

$$\mathbf{E}(\mathbf{r}') \approx \mathbf{E}_{\text{in}}e^{-i\mathbf{k}_{\text{in}}.\mathbf{r}'}. \tag{4.23}$$

-$\mathbf{k}_{\text{sc}}$ is the wavevector orientated from the detector to the surface which gives the dipole field of Eq. (4.10). Then, substituting into (4.11),

$$\mathbf{E}^{(s)} = \mathbf{E}_{\text{in}}e^{-i\mathbf{k}_{\text{in}}.\mathbf{r}'} + \frac{k_0^2e^{-ik_0R}}{4\pi R}(\mathbf{E}_{\text{in}}.\widehat{\mathbf{e}}_{\text{sc}})\,\widehat{\mathbf{e}}_{\text{sc}}\int d\mathbf{r}\delta n^2 e^{i\mathbf{q}.\mathbf{r}}, \tag{4.24}$$

with the wavevector transfer:

$$\mathbf{q} = \mathbf{k}_{\text{sc}} - \mathbf{k}_{\text{in}}. \tag{4.25}$$

For such a field dependence, the differential scattering cross-section (power scattered per unit solid angle, per unit incident flux) is [15]:

$$\frac{d\sigma}{d\Omega} = \frac{k_0^4}{16\pi^2}(\widehat{\mathbf{e}}_{\text{in}}.\widehat{\mathbf{e}}_{\text{sc}})^2\left|\int d\mathbf{r}\,\delta n^2 e^{i\mathbf{q}.\mathbf{r}}\right|^2. \tag{4.26}$$

Note that for small wavevector transfers, $(\widehat{\mathbf{e}}_{\text{in}}.\widehat{\mathbf{e}}_{\text{sc}}) \approx 1$.

4.2.2 Example: Scattering by a Single Rough Surface

To proceed we will first apply Eq. (4.26) to the case of a single rough surface. The more complicated case of a rough multilayer is treated in Appendix 4.3. This example of the diffuse scattering by a rough surface within the Born approximation is the simplest one can imagine and is mainly treated here to show how height-height correlation functions arise as average surface quantitites in the scattering cross-section. The scheme of the calculations will always be the same within the Born or distorted-wave Born approximations, whatever the kind of surface or interface roughness considered. We start from

$$\frac{d\sigma}{d\Omega} = \frac{k_0^4}{16\pi^2}(n^2-1)^2 (\widehat{\mathbf{e}}_{\text{in}}.\widehat{\mathbf{e}}_{\text{sc}})^2 \left| \int d\mathbf{r}_{\|} \int_{-\infty}^{z(\mathbf{r}_{\|})} dz\, e^{i\mathbf{q}.\mathbf{r}} \right|^2 . \tag{4.27}$$

The upper medium (air or vacuum) is medium 0, and the substrate (medium 1) is made slightly absorbing in order to make the integrals converge. Integrating over z first yields,

$$\frac{d\sigma}{d\Omega} = \frac{k_0^4}{16\pi^2 q_z^2}\left(n^2-1\right)^2 (\widehat{\mathbf{e}}_{\text{in}}.\widehat{\mathbf{e}}_{\text{sc}})^2 \left| \int d\mathbf{r}_{\|}\, e^{iq_z z(\mathbf{r}_{\|})} e^{i\mathbf{q}_{\|}.\mathbf{r}_{\|}} \right|^2 . \tag{4.28}$$

Equation (4.28) can be written:

$$\frac{d\sigma}{d\Omega} = \frac{k_0^4(n^2-1)^2}{16\pi^2 q_z^2} (\widehat{\mathbf{e}}_{\text{in}}.\widehat{\mathbf{e}}_{\text{sc}})^2 \int d\mathbf{r}_{\|} \int d\mathbf{r}'_{\|} e^{iq_z(z(\mathbf{r}_{\|})-z(\mathbf{r}'_{\|}))} e^{i\mathbf{q}_{\|}.\mathbf{r}_{\|}} e^{-i\mathbf{q}_{\|}.\mathbf{r}'_{\|}} . \tag{4.29}$$

Making the change of variables $\mathbf{R}_{\|} = \mathbf{r}_{\|} - \mathbf{r}'_{\|}$ and integrating over $\mathbf{R}_{\|}$:

$$\frac{d\sigma}{d\Omega} = \frac{k_0^4 A}{16\pi^2 q_z^2}\left(n^2-1\right)^2 (\widehat{\mathbf{e}}_{\text{in}}.\widehat{\mathbf{e}}_{\text{sc}})^2 \int d\mathbf{R}_{\|}\, \langle e^{iq_z(z(\mathbf{R}_{\|})-z(\mathbf{0}))}\rangle e^{i\mathbf{q}_{\|}.\mathbf{R}_{\|}} , \tag{4.30}$$

where A is the illuminated area and we have simply used the definition of the average over a surface.[7] Assuming Gaussian statictics of the height fluctuations $z(\mathbf{r}_{\|})$ (see Chap. 2), or in any case expanding the exponential to the lowest (second) order, we have:

$$\langle e^{iq_z(z(\mathbf{R}_{\|})-z(\mathbf{0}))}\rangle = e^{-\frac{1}{2}q_z^2\langle (z(\mathbf{R}_{\|})-z(\mathbf{0}))^2}. \tag{4.31}$$

We then obtain:

$$\frac{d\sigma}{d\Omega} = \frac{k_0^4 A}{16\pi^2 q_z^2}\left(n^2-1\right)^2 (\widehat{\mathbf{e}}_{\text{in}}.\widehat{\mathbf{e}}_{\text{sc}})^2 e^{-q_z^2\langle z\rangle^2} \int d\mathbf{R}_{\|} e^{q_z^2\langle z(\mathbf{R}_{\|})z(\mathbf{0})\rangle} e^{i\mathbf{q}_{\|}.\mathbf{R}_{\|}} . \tag{4.32}$$

[7] In general, this average over the surface will not be known and we will use an ensemble average as discussed in Chap. 2.

This equation also includes specular (coherent) components because it has been constructed from the general solution of an electromagnetic field in a vacuum. The diffuse intensity can be obtained by removing the specular component:[8]

$$\left(\frac{d\sigma}{d\Omega}\right)_{\mathrm{coh}} = \frac{k_0^4 A}{4 q_z^2} \left(n^2-1\right)^2 e^{-q_z^2 \langle z \rangle^2} \left(\widehat{\mathbf{e}}_{\mathrm{in}}.\widehat{\mathbf{e}}_{\mathrm{sc}}\right)^2 \delta(\mathbf{q}_{\|}), \tag{4.33}$$

where the identity for Dirac δ functions

$$\frac{1}{4\pi^2} \int d\mathbf{R}_{\|}\, e^{i\mathbf{q}_{\|}.\mathbf{R}_{\|}} = \delta(\mathbf{q}_{\|}) \tag{4.34}$$

has been used. The diffuse (incoherent) intensity is then:

$$\left(\frac{d\sigma}{d\Omega}\right)_{\mathrm{incoh}} = \frac{k_0^4 A}{16\pi^2 q_z^2} (n^2-1)^2 \left(\widehat{\mathbf{e}}_{\mathrm{in}}.\widehat{\mathbf{e}}_{\mathrm{sc}}\right)^2 \times e^{-q_z^2 \langle z^2 \rangle} \int d\mathbf{R}_{\|} \left(e^{q_z^2 \langle z(\mathbf{R}_{\|}) z(\mathbf{0}) \rangle} - 1\right) e^{i\mathbf{q}_{\|}.\mathbf{R}_{\|}}. \tag{4.35}$$

4.3 Distorted-Wave Born Approximation

We will now present an approximation with a further order of complexity. We choose as reference state the same system as the real one, but with smooth interfaces (step index profiles). The Green function and the field in Eq. (4.11) are therefore those for smooth steep interfaces and the iterative methods discussed in Chap. 3 can be used to calculate the field and the Green function. This approximation yields better results than the first Born approximation near the critical angle for total external reflection. It is currently the most popular approximation for the treatment of x-ray surface scattering data.

A first change due to the new choice of reference state is that, because refraction is taken into account, the normal component of the wavevector now depends on the local index. Using Snell-Descartes laws:

$$k_{z,i} = k_0 \sqrt{\sin^2\theta - \sin^2\theta_{ci}}, \tag{4.36}$$

[8] Integrating (4.33) over the angular acceptance of the detector $\delta\Omega = \delta S_{\mathrm{detector}}/R^2 = d\theta d\phi = (2/k_0 q_z) d\mathbf{q}_{\|}$, and normalising to the total incident flux through the area A (leading to a factor $A\sin\theta$, since contrary to the reflectivity coefficient $d\sigma/d\Omega$ is normalised to a unit incident flux), one obtains for the reflectivity coefficient R:

$$R = \frac{k_0^4 A}{q_z^4} \left(n^2-1\right)^2 e^{-q_z^2 \langle z \rangle^2} \left(\widehat{\mathbf{e}}_{\mathrm{in}}.\widehat{\mathbf{e}}_{\mathrm{sc}}\right)^2 = \left(\frac{q_c}{2q_z}\right)^4 e^{-q_z^2 \langle z \rangle^2} \left(\widehat{\mathbf{e}}_{\mathrm{in}}.\widehat{\mathbf{e}}_{\mathrm{sc}}\right)^2,$$

which shows the well-known q_z^{-4} decay. This expression also shows that within the Born approximation, the Brewster angle is 45 degrees.

where $\theta_{ci}^2 = 2(1 - n_i)$ is the critical angle for total external reflection between vacuum and medium i with $n_i = 1 - \delta_i - i\beta_i$. More precisely, the real and imaginary parts of the wavevector in medium i are:

$$\mathcal{R}e(k_{z,i}) = \frac{1}{\sqrt{2}} k_0 \sqrt{[(\theta^2 - 2\delta_i)^2 + 4\beta_i^2]^{1/2} + (\theta^2 - 2\delta_i)}, \tag{4.37}$$

$$\mathcal{I}m(k_{z,i}) = \frac{1}{\sqrt{2}} k_0 \sqrt{[(\theta^2 - 2\delta_i)^2 + 4\beta_i^2]^{1/2} - (\theta^2 - 2\delta_i)}. \tag{4.38}$$

As mentioned above, refraction also implies that the direction of the polarisation vector in (p) polarisation changes from layer to layer. To avoid the complications related to this point, unless otherwise specified, we will always limit ourselves to the case of scattering of a (s) polarised wave into (s) polarisation in the rest of this section. Then, one has $(\widehat{\mathrm{e}}_{\mathrm{in}}.\widehat{\mathrm{e}}_{\mathrm{sc}}) = \cos\psi$, in every layer. A more detailed discussion of polarisation effects will be given in section 4.4.

4.3.1 Case of a Single Rough Surface

Considering only one rough interface between media (0) and (1) and placing the reference plane above the real rough interface (Fig. 4.1),[9] we have, for

[9] The choice of the reference medium, and in particular of the location of the reference planes is important here. This question did not arise in the discussion of the first Born approximation where the reference medium is the vacuum. Three different choices are a priori possible: place the reference plane above, below, or crossing the real rough interface (Fig. 4.2). In principle, all choices are equivalent for small roughnesses owing to the continuity of the field. For larger roughnesses, using the average plane of the rough surface might be the best choice. This has been done to calculate the specular reflectivity in Ref. [1] close to the critical angle. However, this approximation is not good for larger incident angles because Fresnel eigenstates are not a good approximation of the real eigenstates of the system. In this regime, however, the first Born approximation is good far from Bragg peaks. In the present treatment, we have choosen to place the reference plane above the rough surface, hence the $t_{0,1}^{\mathrm{in}}$ and $t_{0,1}^{\mathrm{sc}}$ coefficients. This approximation is as good as that of the average plane close to the critical angle and converges to the first Born approximation at larger angles of incidence. For a multilayer, one might be worried by the phase factor corresponding to the small shift between the average plane, and the reference plane placed below the surface (this phase shift disappears in the cross-section for a single interface). On the other hand, when using the average plane, calculations become cumbersome. A reasonable solution is then to chose the average plane as reference plane, but to use the analytical continuation of the field in one of the media, for example that above the interface.

(*s*) or (*p*) polarisation:

$$\begin{aligned}\mathbf{E}_{\text{det}}^{\widehat{\mathbf{e}}_{\text{sc}}}(\mathbf{R},\mathbf{r}') &= \frac{k_0^2 e^{-ik_0R}}{4\pi\epsilon_0 R} E_1^{PW}(-k_{\text{sc}\,z,1}, z')e^{i\mathbf{k}_{\|}\cdot\mathbf{r}'_{\|}}\widehat{\mathbf{e}}_{\text{sc}} \\ &= \frac{k_0^2 e^{-ik_0R}}{4\pi\epsilon_0 R} t_{0,1}^{\text{sc}}\, e^{i\mathbf{k}_{\text{sc},1}\cdot\mathbf{r}'}\widehat{\mathbf{e}}_{\text{sc}},\end{aligned} \tag{4.39}$$

$$\begin{aligned}\mathbf{E}(\mathbf{r}') &= \mathbf{E}_{\text{in}}\, E_1^{PW}(k_{\text{in}\,z,1}, z')e^{-i\mathbf{k}_{\|}\cdot\mathbf{r}'_{\|}}\widehat{\mathbf{e}}_{\text{in}} \\ &= \mathbf{E}_{\text{in}}\, t_{0,1}^{\text{in}}\, e^{-i\mathbf{k}_{\text{in},1}\cdot\mathbf{r}'}\widehat{\mathbf{e}}_{\text{in}},\end{aligned} \tag{4.40}$$

where t^{in} and t^{sc} are the Fresnel transmission coefficients for polarisation (*s*) for respectively the angle of incidence θ_{in} and the scattering angle in the scattering plane θ_{sc}. Explicit expressions for those coefficients are given by equations (3.71) and (3.72). Putting Eqs. (4.39), (4.40) in Eq. (4.11) and following the same treatment of the integrals as in Sect. 4.2.2, we obtain a generalisation of Eq. (4.35):

$$\begin{aligned}\left(\frac{d\sigma}{d\Omega}\right)_{\text{incoh}} &= A\frac{k_0^4}{16\pi^2}(n_1^2-n_0^2)^2 \left|t_{0,1}^{\text{in}}\right|^2 \left|t_{0,1}^{\text{sc}}\right|^2 \frac{e^{-\frac{1}{2}(q_{z,1}^2+q_{z,1}^{*2})\langle z^2\rangle}}{|q_{z,1}|^2} \\ &\times \cos^2\psi \int d\mathbf{R}_{\|}\left[e^{|q_{z,1}|^2\langle z(\mathbf{R}_{\|})z(\mathbf{0})\rangle}-1\right]e^{i\mathbf{q}_{\|}\cdot\mathbf{R}_{\|}}.\end{aligned} \tag{4.41}$$

Eq. (4.41) differs from Eq. (4.35) by the additional transmission coefficients.

This expression is explicitly symmetrical in the source and detector positions as required by the reciprocity theorem.

At the critical angle for total external reflection $\theta_{\text{in}} = \theta_c$, the transmission coefficients in Eq. (4.41) have a peak value of 2. The electric field is then at its maximum value at the interface because the incident and scattered field are in phase at $z = 0$. As the dipole source equivalent to roughness $\epsilon_0\delta n^2\mathbf{E}$ is proportional to $\mathbf{E}$, there is a maximum in the scattered intensity. By using the reciprocity theorem, one can see that the Green function is also peaked near $\theta_{\text{sc}} = \theta_c$.[10] Those peaks are the so-called Yoneda peaks [19]. They can be seen on Fig. 4.4.

4.3.2 General Case of a Stratified Medium

In the general case of a stratified medium depicted in Fig. 4.2, one has in layer j for (*s*) or (*p*) polarisation:

$$\begin{aligned}\mathbf{E}_{\text{det}}^{\widehat{\mathbf{e}}_{\text{sc}}}(\mathbf{R},\mathbf{r}') &= \frac{k_0^2 e^{-ik_0R}}{4\pi\epsilon_0 R} E_j^{PW}(-k_{\text{sc}\,z,j}, z')e^{i\mathbf{k}_{\text{sc}\|}\cdot\mathbf{r}'_{\|}}\widehat{\mathbf{e}}_{\text{sc}\,j} \\ \mathbf{E}(\mathbf{r}') &\approx E_{\text{in}}\, E_j^{PW}(k_{\text{in}\,z,j}, z')e^{-i\mathbf{k}_{\text{in}\|}\cdot\mathbf{r}'_{\|}}\widehat{\mathbf{e}}_{\text{in}\,j}.\end{aligned} \tag{4.42}$$

[10] Equivalently, the peak in the Green function can be seen to arise from the angular dependence of the field emitted by a dipole placed below the interface.

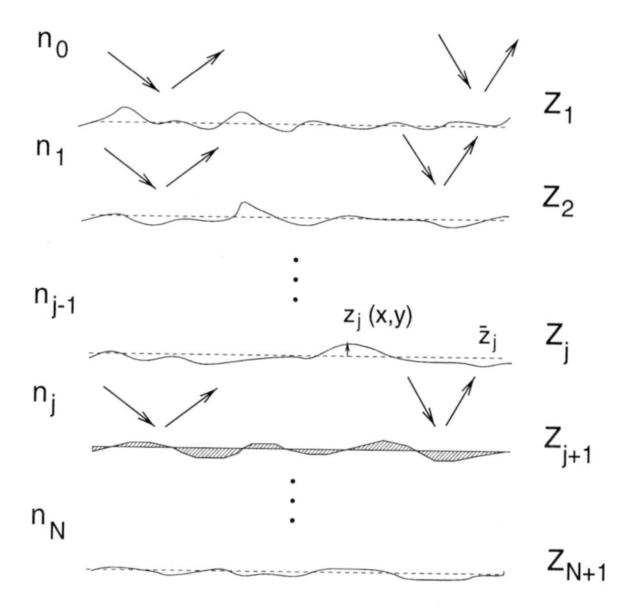

Fig. 4.2. X-ray surface scattering in a stratified rough medium. Because of multiple reflections, there are waves propagating upwards (with an amplitude $U(k_{z,j}, z)$) and downwards (with an amplitude $U(-k_{z,j}, z)$) in layer j where the total the field amplitude is $E^{PW}_{\text{in},j}$ (there is an equivalent dependence of the Green function). Multiple reflections are considered within the DWBA but not within the first Born approximation. The perturbation method consists in evaluating the field scattered by the dipolar density equivalent to the index difference $(n_{j-1} - n_j)$ between the real system where the rough interface profile is $z_j(\mathbf{r}_{\|})$ and the unperturbated system where the interface is located at Z_j, and is placed here at the average interface plane. For interface j the unperturbated and real index distributions differ in the hatched region

The DWBA method consists then in developing the E^{PW} functions defined in Eq. (4.17) in each medium as, for example in (s) polarisation, in layer j:

$$\begin{aligned} E_j^{PW\,(s)}(k_{z,j}, z) &= U^{(s)}(k_{z,j}, Z_j)e^{-ik_{z,j}z} + U^{(s)}(-k_{z,j}, Z_j)e^{+ik_{z,j}z} \\ &= \sum_{\pm} U^{(s)}(\pm k_{z,j}, Z_j)e^{\mp ik_{z,j}z}, \end{aligned} \tag{4.43}$$

where the U coefficients are the magnitudes of the upwards and downwards propagating waves which are explicitly obtained in Chap. 3 of this book, Eq. (3.48) using the "Vidal and Vincent" representation of tranfer matrices [20]. The field is then written (put Eqs. (4.42) in Eq. (4.11) and sum over all interfaces):

$$E^{(s)} = E_{\text{ref}} + E_{\text{in}} \frac{k_0^2 e^{-ik_0 R}}{4\pi\epsilon_0 R} \cos\psi \sum_{j=0}^{N} \int d\mathbf{r}_{\|}\, e^{i\mathbf{q}_{\|}\cdot\mathbf{r}_{\|}}$$

$$\int_0^{z_{j+1}(\mathbf{r}_{\|})} dz \epsilon_0 (n_{j+1}^2 - n_j^2) E_{j+1}^{PW}(k_{\text{inz,j+1}}, z) E_{j+1}^{PW}(-k_{\text{scz,j+1}}, z), \tag{4.44}$$

where it has been assumed that the reference plane is located above the interface, hence the E_{j+1}^{PW} fields; z_{j+1} is negative). Then, the generalisation of Eq. (4.41) is:

$$\left(\frac{d\sigma}{d\Omega}\right)_{\text{incoh}} = \frac{k_0^4}{16\pi^2} \sum_{j=1}^{N} \sum_{k=1}^{N} \sum_{\pm} \sum_{\pm} \sum_{\pm} \sum_{\pm}$$

$$U^{(s)}(\pm k_{\text{in }z,j}, Z_j) U^{(s)}(\pm k_{\text{sc }z,j}, Z_j) U^{(s)*}(\pm k_{\text{in }z,k}, Z_k) U^{(s)*}(\pm k_{\text{sc }z,k}, Z_k)$$

$$\tilde{Q}_{j,k}(\pm k_{\text{in }z,j} \pm k_{\text{sc }z,j}, \pm k_{\text{in }z,k} \pm k_{\text{sc }z,k}), \tag{4.45}$$

with

$$\tilde{Q}_{j,k}(q_z, q_z') = \left(n_j^2 - n_{j-1}^2\right)\left(n_k^2 - n_{k-1}^2\right)^* \cos^2\psi \int d\mathbf{r}_{\|} \int d\mathbf{r}'_{\|} e^{i\mathbf{q}_{\|}\cdot(\mathbf{r}_{\|} - \mathbf{r}'_{\|})}$$

$$\left[\left\langle \int_{z_j(\mathbf{r}_{\|})}^{0} dz \int_{z_k(\mathbf{r}'_{\|})}^{0} dz' e^{i(q_z z - q_z'^* z')} \right\rangle - \left\langle \int_{z_j(\mathbf{r}_{\|})}^{0} dz e^{iq_z z} \right\rangle \right.$$

$$\left. \left\langle \int_{z_k(\mathbf{r}'_{\|})}^{0} dz' e^{-iq_z'^* z'} \right\rangle \right], \tag{4.46}$$

where the specular (coherent) contribution, obtained as an average over the field as shown in Chap. 2, has been removed. Performing the integrations over z and z' and making the change of variables $\mathbf{r}_{\|} - \mathbf{r}'_{\|} \to \mathbf{R}_{\|}$ as previously:

$$\tilde{Q}_{j,k}(q_z, q_z') = A \frac{\left(n_j^2 - n_{j-1}^2\right)\left(n_k^2 - n_{k-1}^2\right)^*}{q_z q_z'^*} \cos^2\psi e^{-\frac{1}{2}[q_z^2 \langle z_j^2 \rangle + (q_z'^*)^2 \langle z_k^2 \rangle]}$$

$$\int d\mathbf{R}_{\|}\, e^{i\mathbf{q}_{\|}\cdot\mathbf{R}_{\|}} \left(e^{q_z q_z'^* \langle z_j(\mathbf{0}) z_k(\mathbf{R}_{\|})\rangle} - 1\right). \tag{4.47}$$

Because reflection at all interfaces is taken into account, all the possible combinations of the incident and scattered wavevectors appear in the formulae.

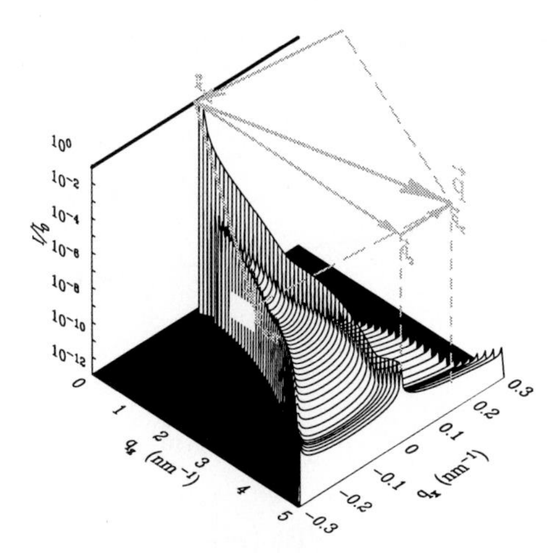

Fig. 4.3. Intensity scattered by a $2nm$ thick film on a liquid surface as a function of q_x et q_z. $I(q_z, q_x = 0)$ is is the reflectivity curve. Note that in the case represented here, the diffuse scattering is peaked in the specular direction. The roughness spectrum (Fourier transform of the height-height correlation function) is directly obtained in a q_x scan at constant q_x .The possible extension of q_x scans increases with q_z. Out of this accessible range (black surface on the figure), either the detector or the source would move below the interface. Note the Yoneda peaks near the extremety of q_x scans, where θ_{in} or θ_{sc} is equal to the very small critical angle for total external reflection

4.3.3 Particular Case of a Film

The particular case of a film has been considered in Ref. [2] where an explicit expression for the scattering cross-section as a function of the reflection and transmission coefficients has been given. In practice however, it is more convenient to write a program using Eq. (4.45) whatever the number of layers.

In a rough thin film scattering occurs at both the film surface and at the film-substrate interface. The field at the film surface is proportional to $1 + r$ where r is the *film* reflection coefficient, and the field at the substrate-film interface is proportional to the *film* transmission coefficient t. Both r and t depend on θ_{in} et θ_{sc}, Eq. (3.75), (3.76):

$$r = \frac{r_{0,1} + r_{1,2}e^{-2ik_{z,1}d}}{1 + r_{0,1}r_{1,2}e^{-2ik_{z,1}d}} \; ; \; t = \frac{t_{0,1}t_{1,2}e^{-ik_{z,1}d}}{1 + r_{0,1}r_{1,2}e^{-2ik_{z,1}d}}, \tag{4.48}$$

where $r_{i,i+1}$ $t_{i,i+1}$ are the reflection et transmission coefficients of interface $i/i+1$, d the film thickness, and $k_{z,1}$ depends on θ_{in} or θ_{sc}.
Far from an incident or exit angle close to the critical angle for total external reflection, the reflection coefficients are small and can be neglected. Then, the Born approximation is valid, and the amplitude of the field (or Green function) is 1 at the upper interface and $e^{-ik_{z,1}d}$ at the lower interface. Therefore, neglecting the polarisation factor, the scattering cross-section

will be (compare also to Eq. (4.C3) in appendix 4.C):

$$\frac{d\sigma}{d\Omega} = \frac{k_0^4}{16\pi^2 q_z^2} \left| \int d\mathbf{r}_{\|} \left[(n_1^2 - n_0^2) e^{i(k_{\mathrm{sc}z} - k_{\mathrm{in}z}) z_1} + (n_2^2 - n_1^2) e^{i(k_{\mathrm{sc}z} - k_{\mathrm{in}z})(d + z_2)} \right] e^{i\mathbf{q}_{\|} \cdot \mathbf{r}_{\|}} \right|^2 . \tag{4.49}$$

Following the usual procedure, one obtains (compare to Eq. (4.C10)):

$$\begin{aligned} \left(\frac{d\sigma}{d\Omega}\right)_{\mathrm{incoh}} &= \frac{k_0^4}{16\pi^2 q_z^2} A \\ &\left[(n_1^2 - n_0^2)^2 e^{-q_z^2 \langle z_1^2 \rangle} \int d\mathbf{R}_{\|} \left(e^{q_z^2 \langle z_1(\mathbf{0}) z_1(\mathbf{R}_{\|}) \rangle} - 1 \right) e^{i\mathbf{q}_{\|} \cdot \mathbf{R}_{\|}} \right. \\ &+ (n_2^2 - n_1^2)^2 e^{-q_z^2 \langle z_2^2 \rangle} \int d\mathbf{R}_{\|} \left(e^{q_z^2 \langle z_2(\mathbf{0}) z_2(\mathbf{R}_{\|}) \rangle} - 1 \right) e^{i\mathbf{q}_{\|} \cdot \mathbf{R}_{\|}} \\ &+ 2 (n_2^2 - n_1^2)(n_1^2 - n_0^2) e^{-\frac{1}{2} q_z^2 \langle z_2^2 \rangle - \frac{1}{2} q_z^2 \langle z_2^2 \rangle} \cos(q_z d) \\ &\left. \int d\mathbf{R}_{\|} \left(e^{q_z^2 \langle z_1(\mathbf{0}) z_2(\mathbf{R}_{\|}) \rangle} - 1 \right) e^{i\mathbf{q}_{\|} \cdot \mathbf{R}_{\|}} \right] . \end{aligned} \tag{4.50}$$

In principle, the correlation between the different interfaces can therefore be determined because the contrast of the interference pattern directly depends on this correlation and the different contributions may be separated [21].

For example, considering a single film on a substrate, the substrate roughness spectrum can be first measured without a film. Then, q_z scans at constant q_x can be performed with the film. They consist of an oscillating and a non-oscillating component. The contrast of the interference pattern yields the cross-correlation between the film-substrate and film-vacuum interfaces at the given q_x, and the non-oscillating part yields the sum of the film-substrate and film-vacuum autocorrelations. All the relevant correlation functions can therefore be determined. In a multilayer, similar constructive interference effects can occur between the beams scattered at different conformal interfaces, leading to what is called "resonant scattering" in Sect. 8.5.

In addition to the Yoneda peaks, other dynamical effects can be observed in the case of a film (or better of a multilayer, see Sect. 8.5 for a thorough discussion). Because the reflection and transmission coefficients Eq. (4.48) depend on both the incident and scattering angle if the reflection coefficients are not too small, the scattered intensity can show oscillations with a characteristic period depending on the film thickness,[11] even if both angles vary in a scan so that q_z is kept constant (Fig. 4.4 left). This dynamical effect cannot be accounted for within the first Born approximation. More generally, similar dynamical effects will occur whenever the field is modified at an interface due

[11] Note that because of the $e^{-2ik_z d}$ factor in the reflection coefficient, the periodicity associated to the dynamical effect is different from that associated to the interface cross-correlation term, allowing to distinguish between them.

to multiple reflections (see Fig. 4.4 right for an example). Again such effects are not accounted for within the first Born approximation. Many examples are given in Sect. 8.5.

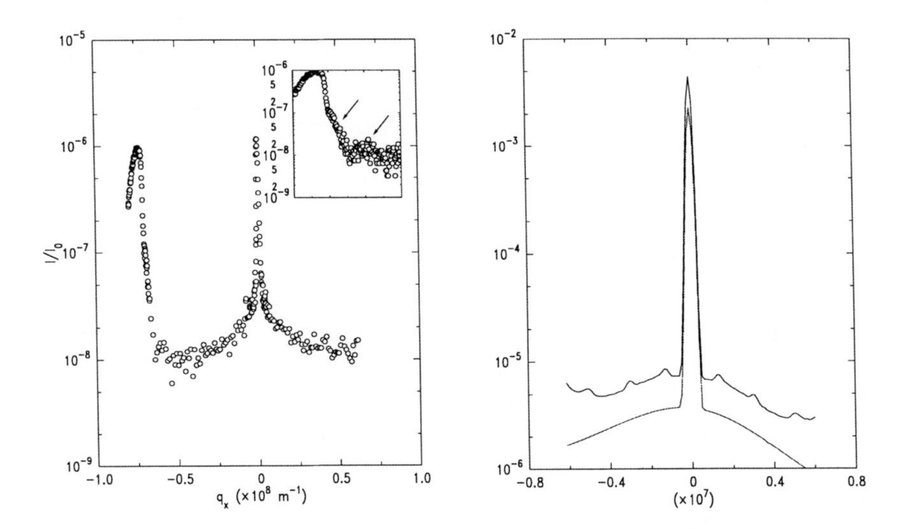

Fig. 4.4. Dynamical effects. Left: "Rocking curve" $I(q_x)$ for a polymer film (polystyrene-polymethylmetacrylate diblock copolymer) on a silicon substrate. The film is 18.9nm thick. Note in particular the Yoneda peak at $0.75 \times 10^{-8} m^{-1}$ and that its structure is related to interferences in the film (inset). In particular, the oscillations marked by arrows correspond to the dynamical effect discussed in the text. Right: "Rocking curve" $I(q_x)$ for a CdTe multilayer (20 layers). The height-height correlation function is $\langle z(0)z(x)\rangle = \sigma^2 \exp - [x/\xi]^{2\nu}$ and the interfaces are assumed to be fully correlated. The parameters used in the calculation are $\sigma = 0.25nm$, $\nu = 0.6$, $\xi = 300\mu m$. The grey curve (divided by a factor of 2 for clarity) corresponds to the first Born approximation and the black curve to the DWBA described in the text. Note the peaks in the DWBA intensity which are a dynamical effect and occur when the field is maximum at the different interfaces

4.4 Polarisation Effects

According to Fig. 2.3, the polarisation vectors are:

$$\widehat{\mathbf{e}}_{\text{in}}^{(s)} = \begin{pmatrix} 0 \\ 1 \\ 0 \end{pmatrix} \quad ; \; \widehat{\mathbf{e}}_{\text{in}}^{(p)} = \begin{pmatrix} \sin\theta_{\text{in}} \\ 0 \\ \cos\theta_{\text{in}} \end{pmatrix}$$

$$\widehat{\mathbf{e}}_{\text{sc}}^{(s)} = \begin{pmatrix} -\sin\psi \\ \cos\psi \\ 0 \end{pmatrix} \; ; \; \widehat{\mathbf{e}}_{\text{sc}}^{(p)} = \begin{pmatrix} -\sin\theta_{\text{sc}}\cos\psi \\ -\sin\theta_{\text{sc}}\sin\psi \\ \cos\theta_{\text{sc}} \end{pmatrix} . \tag{4.51}$$

At very small incident and scattering angles $\widehat{\mathbf{e}}_{\text{sc}}^{(s)} \approx \widehat{\mathbf{e}}_{\text{in}}^{(s)}$ and $\widehat{\mathbf{e}}_{\text{sc}}^{(p)} \approx \widehat{\mathbf{e}}_{\text{in}}^{(p)}$, and the E^{PW} functions for (s) and (p) polarisation are almost equal. For example, the ratio of the reflection coefficients for the perpendicular and parallel polarisations is for a single interface:

$$\frac{r^{(s)}}{r^{(p)}} \simeq 1 + 2\theta_0\theta_1 , \tag{4.52}$$

where θ_0 and θ_1 are the angles between the incident and the refracted beam with the surface. In general, at least one of the angles θ_{in} or θ_{sc} is small, and thus the difference between the (s) and (p) reflection coefficients is also small. Thus, polarisation effects are generally not very important, but they can be noticeable in some instances. Firstly, subtle effects can happen under conditions close to those responsible for the structure of the Yoneda peaks discussed above [4]. More importantly, polarisation effects must be taken into account whenever the incident or the scattering angle are larger than ≈ 10 degrees. This is the case when one tries to get information at small (atomic or molecular) lengthscales.

The treatment of polarisation effects will be different within the Born and distorted wave Born approximations. Within the Born approximation (see Sect. 4.2), the polarisation dependence is easily included in the differential scattering cross-section Eq. (4.26)

$$\frac{d\sigma}{d\Omega} = \frac{k_0^4}{16\pi^2} \left(\widehat{\mathbf{e}}_{\text{in}}.\widehat{\mathbf{e}}_{\text{sc}}\right)^2 \left| \int d\mathbf{r}\, \delta n^2 e^{i\mathbf{q}\cdot\mathbf{r}} \right|^2 ,$$

since the only dependence is through the scalar product $(\widehat{\mathbf{e}}_{\text{in}}.\widehat{\mathbf{e}}_{\text{sc}})$. There is only a simple, geometrical depolarisation corresponding to the projection of the incident polarisation on the final one. Generally, the polarisation of the scattered beam will not be known *a priori*, but we can calculate the relative scattering into (s) and (p) polarisations using Eq. (4.26). Then,

$$\left(\frac{d\sigma}{d\Omega}\right)_{\text{tot}} = \frac{d\sigma}{d\Omega}^{(s)} + \frac{d\sigma}{d\Omega}^{(p)} . \tag{4.53}$$

Within the DWBA, we must always decompose the incident and scattered field into (s) and (p) polarisations because the E^{PW} functions depend on the polarisation. Moreover, for (p) polarisation the orientation of $\widehat{\mathbf{e}}_{\text{in}}$ and $\widehat{\mathbf{e}}_{\text{sc}}$ will differ from layer to layer because of refraction. Taking these two requirements into account, the scattering cross-section can be calculated using Eq. (4.53). A simple case[12] is that of the scattering of a (s) polarised incident wave into a (s) polarised wave since a unique $\cos^2 \psi$ polarisation factor can be used in all the layers. This is the case which was considered in Sect. 4.3 for simplicity.

4.5 Scattering by Density Inhomogeneities

Only surface scattering has been considered up to this point. However, the dielectric index inhomogeneities leading to scattering can also be density fluctuations. This should always be born in mind when interpreting experiments. The scattering due to density inhomogeneities in a multilayer or at a liquid surface can be treated using a formalism similar to that used for surface scattering. The relevant correlation functions will be of the form $\langle \delta\rho(\mathbf{0}, z')\delta\rho(\mathbf{r}_\parallel, z)\rangle$. This problem was considered in the early paper of Bindell and Wainfan [22]. Again we limit the discussion to the scattering of a (s) polarised incident wave into a (s) polarised wave.

4.5.1 Density Inhomogeneities in a Multilayer

The interfaces are assumed to be perfectly smooth in this analysis. Within the DWBA, and assuming effective U functions within the layers,[13] the differential scattering cross-section will be (cf Eq. (4.45)):

$$\frac{d\sigma}{d\Omega} = \frac{k_0^4}{16\pi^2} \sum_{j=1}^{N} \sum_{k=1}^{N} \sum_{\pm} \sum_{\pm} \sum_{\pm} \sum_{\pm} U^{(s)}(\pm k_{\text{in}\,z,j}, Z_j) U^{(s)}(\pm k_{\text{sc}\,z,j}, Z_j)$$
$$U^{(s)^*}(\pm k_{\text{in}\,z,k}, Z_k) U^{(s)^*}(\pm k_{\text{sc}\,z,k}, Z_k) \tilde{B}_{j,k}(\pm k_{\text{in}z,j} \pm k_{\text{sc}z,j}, \pm k_{\text{in}z,k} \pm k_{\text{sc}z,k}), \tag{4.54}$$

[12] This is for example the case of horizontal scattering on a horizontal surface at a synchrotron source.

[13] Assuming effective U functions within the layers is only possible if the characteristic size of the inhomogeneities is much smaller than the extinction length, see appendix 3.A to Chap. 3. This might not be the case for multilayer gratings (see Sect. 8.7) or large copolymer domains [23].

where now

$$\tilde{B}_{j,k}(q_z, q_z') = \cos^2\psi \int d\mathbf{r}_\| \int d\mathbf{r}'_\| e^{i\mathbf{q}_\| \cdot (\mathbf{r}_\| - \mathbf{r}'_\|)} \int_0^{Z_{j+1}-Z_j} dz \int_0^{Z_{k+1}-Z_k} dz' \left\langle \delta n_j^2(\mathbf{r}_\|, z) \delta n_k^{2*}(\mathbf{r}'_\|, z') \right\rangle e^{i(q_z z - q_z'^* z')}. \tag{4.55}$$

Making the change of variables $\mathbf{r}_\| - \mathbf{r}'_\| \to \mathbf{R}_\|$:

$$\tilde{B}_{j,k}(q_z, q_z') = A\cos^2\psi \int d\mathbf{R}_\| e^{i\mathbf{q}_\| \cdot \mathbf{R}_\|} \int_0^{Z_{j+1}-Z_j} dz \int_0^{Z_{k+1}-Z_k} dz' \left\langle \delta n_j^2(\mathbf{0}, z) \delta n_k^{2*}(\mathbf{R}_\|, z') \right\rangle e^{i(q_z z - q_z'^* z')}. \tag{4.56}$$

In the case of a semi-infinite medium, only $U_1(-k_{\text{in},z,1}, Z_1) = t^{\text{in}}$ and $U_1(-k_{\text{sc}}, z, 1, Z_1) = t^{\text{sc}}$ are different from 0. Writing $q_{z,1} = \mathcal{R}e(q_{z,1}) + i\mathcal{I}m(q_{z,1})$, one obtains:

$$\tilde{B}_{1,1}(q_{z,1}, q_{z,1}) = A\cos^2\psi \int d\mathbf{R}_\| e^{i\mathbf{q}_\| \cdot \mathbf{R}_\|} \int_{-\infty}^0 \int_{-\infty}^0 dz dz' \, e^{i\mathcal{R}e(q_{z,1})(z-z')} e^{\mathcal{I}m(q_{z,1})(z+z')} \left\langle \delta n_1^2(\mathbf{0}, z) \delta n_1^{2*}(\mathbf{R}_\|, z') \right\rangle, \tag{4.57}$$

i.e., the bulk fluctuations are integrated over the penetration length of the beam.

Comparing Eq. (4.57) to Eq. (4.47), we note that contrary to bulk scattering, surface scattering is inversely proportional to the square of wavevector transfers. Therefore, surface scattering will generally be dominant at grazing angles whereas bulk scattering will ultimately dominate at large scattering angles (see Chap. 9, Sect. 9.3.1 for an example).

4.5.2 Density Fluctuations at a Liquid Surface

An interesting case is that of density fluctuations at a single liquid surface because an analytical calculation can be made. The liquid extends from $-\infty$ to 0 in z, and its vapor can be taken with negligible density. Eqs. (4.54),(4.57) above give:

$$\frac{d\sigma}{d\Omega} = \frac{k_0^4}{16\pi^2} A(1-n^2)^2 \left|t_{0,1}^{\text{in}}\right|^2 \left|t_{0,1}^{\text{sc}}\right|^2 \cos^2\psi \int_{-\infty}^0 dz \int_{-\infty}^0 dz' \, e^{iq_{z,1}z} e^{-iq_{z,1}^* z'} \int d\mathbf{R}_\| \frac{\langle \delta\rho(\mathbf{0}, z') \delta\rho(\mathbf{R}_\|, z) \rangle}{\rho^2} e^{i\mathbf{q}_\| \cdot \mathbf{R}_\|}, \tag{4.58}$$

where we have used $n = 1 - (\lambda^2/2\pi) r_e \rho$. Inserting the density-density correlation function for bulk liquid fluctuations

$$\langle \rho(\mathbf{r}) \rho(\mathbf{r}') \rangle = \rho^2 k_B T \, \kappa_T \delta(\mathbf{r} - \mathbf{r}'), \tag{4.59}$$

where $\kappa_T = -1/V(\partial V/\partial P)_T$ is the isothermal compressibility yields,

$$\frac{d\sigma}{d\Omega} = \frac{k_0^4}{16\pi^2} \frac{A(1-n^2)^2 \left| t^{\mathrm{in}} \right|^2 \left| t^{\mathrm{sc}} \right|^2 k_B T \kappa_T}{2\mathcal{I}m(q_{z,1})} \cos^2 \psi. \tag{4.60}$$

4.6 Further Approximations

The Distorted-wave Born approximation as presented here does not always allow an accurate enough representation of the scattered intensity close to the critical angle for total external reflection [24]. Understanding scattering at grazing angles is highly desirable because bulk scattering is minimized under such conditions. This is critical because the signal scattered by surfaces or interfaces is generally very low. Differents approaches have been attempted to improve the DWBA. Only the first one has been extensively investigated.

This approximation consists in taking into account the average interface profile in Eq. (4.11) [4]. The reference medium is now defined by the relative permittivity $\epsilon_{\mathrm{ref}}(z) = \langle \epsilon \rangle (z)$ with $\langle \epsilon \rangle (z) = 1/A \int \epsilon(\mathbf{r}_{\|}, z) d\mathbf{r}_{\|}$ in the case of a rough surface defined by an ergodic random process. (In [3] the shape of the average permittivity is approximated by an hyperbolic tangent profile to simplify the calculation of the reference Green tensor). The main interest of this new reference medium is that the reference field $\mathbf{E}_{\mathrm{ref}}$ is that of the transition layer and thus contains directly the Névot-Croce factor in the reflection coefficient. Moreover the perturbation $\delta n^2(\mathbf{r}_{\|})$ is of null average $\langle \delta n^2 \rangle = 0$ and we may expect to have minimised its value (and thus extended the validity domain of the perturbative development). This improvement has been shown to yield much better results than the classical DWBA in the optical domain where the permittivity contrasts are important [25]. In the x-ray domain its interest is more questionable since it does not lead to simple expressions for the scattering cross-section. Indeed $\delta n^2(\mathbf{r}_{\|})$ is no longer a step function and the integration along the z-axis cannot be done analytically.

Another possibility would be to directly take into account multiple surface (roughness) scattering without using the effective medium approximation. It is then necessary to iterate the fundamental equation (4.11)[4], [26–28]. This has been done up to the second order in Ref. [5] for specular reflectivity, and the corrections might be important close to the critical angle for total external reflection.

Finally there exist many approximate methods that have been developed in totally different contexts (optics, radar). In most methods, the field scattered by the rough surface is evaluated with a *surface integral equation* (given by the Huygens-Fresnel principle (or Kirchhoff integral) [29]). The integrant of the latter contains the field values and its normal derivatives at each point of the surface. The Kirchhoff approximation consists in replacing the field on the surface by the field that would exist if the surface is locally assimilated to its tangent plane. This technique, when applied to the coherent field yields the famous Debye-Waller factor on the reflection coefficient. It is a single scattering approximation (also called Physical Optics approximation). The perturbative theory (the small parameter is the rms height of the surface) has also been widely used. A possible starting point is writing the boundary conditions on the field and its derivative at the interface under the Rayleigh hypothesis. A brief survey of this method is given in Sect. 3.A.1. Note that the iteration of these methods permit to account for some multiple scattering effects, but the increasing complexity of the calculation limits their interest. It is now also possible to consider the resolution of the surface integral equation satisfied by the field without any approximation (and thus to account for all the multiple scattering). Preliminary results have been already presented in the Radar and optical domain. However in the x-ray domain those techniques have a major drawback: They only consider surface scattering (with a surface integral equation) and the generalisation to both surface and volume scattering is not straightforward. The differential method [30] which consists in solving the inhomogeneous differential equation satisfied by the Fourier component of the field (in the $\mathbf{k}_{\|}$ space) with a Runge-Kutta algorithm along the z-axis would be more adequate. It has already been used to calculate the diffraction by multilayer gratings and accounts for all multiple scattering (no approximation), but it remains difficult to use it for non-periodic (rough) surfaces because of the computing time and memory required.

4.7 The Scattered Intensity

4.7.1 Expression of the Scattered Intensity

In contrast with specular reflection where the specular condition $\delta(\mathbf{q}_{\mathbf{r}_{\|}})$ implies that resolution effects amount to a simple convolution, the scattered intensity is proportional to the resolution volume for diffuse (incoherent) scattering. In order to achieve quantitative information from an experiment, it is necessary to measure (and calculate) absolute intensities, and therefore to have a detailed knowledge of the resolution function. The differential scattering cross-section must be integrated over the detector solid angle Ω_d and the incident angle angular spread ($\Delta\theta_{\text{in}}$ in the vertical and $\Delta\theta_y$ perpendicular to the incidence plane) (Fig. 4.5). Assuming a beam cross-section $l_x \times l_y$ and a Gaussian angular distribution of the incident beam intensity:

$$I_0 e^{-\frac{\delta\theta_{\text{in}}^2}{2\Delta\theta_{\text{in}}^2} - \frac{\delta\theta_y^2}{2\Delta\theta_y^2}}$$

of width $\Delta\theta_{\text{in}}$ in the plane of incidence and $\Delta\theta_y$ normal to the plane of incidence (Fig. 4.5), the scattered intensity I_D is, using the definition of the scattering cross-section (power radiated per unit solid angle for a unit incident flux in direction $\mathbf{k}_{\text{in}}$):

$$I_D/I_0 = \frac{1}{l_x \times l_y} \int d\delta\theta_{\text{in}} d\delta\theta_y \frac{1}{2\pi\Delta\theta_{\text{in}} \times \Delta\theta_y} e^{-\frac{\delta\theta_{\text{in}}^2}{2\Delta\theta_{\text{in}}^2} - \frac{\delta\theta_y^2}{2\Delta\theta_y^2}} \int \frac{d\sigma}{d\Omega} d\Omega_d . \tag{4.61}$$

The normalisation factor $1/(l_x \times l_y)$ is required because the cross-section is defined for a unit incident flux whereas the scattered intensity is normalized to the total flux.

Since each experimental setup is different, it is impossible to give a general expression for the angular dependence of the resolution function. Different examples can be found in Chaps. 7, 8 and 9.

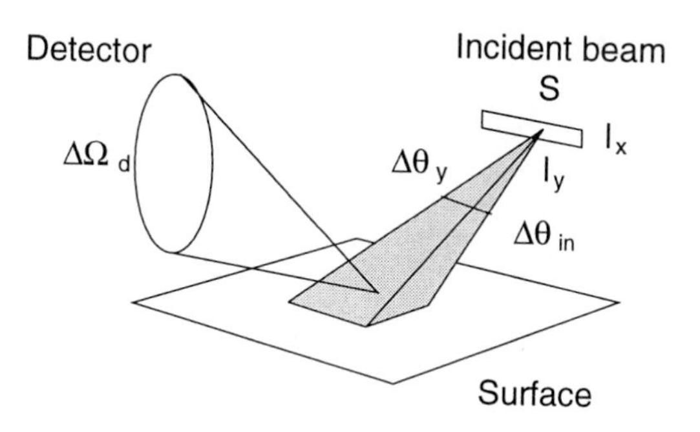

Fig. 4.5. Definition of the angles and solid angles used for calculating the scattered intensity

The cross-section is defined as a function of wave vectors, and sometimes it might be found more convenient to use wave vectors instead of angles. In fact, the use of a wave-vector resolution function is very delicate (see Ref.[31])[14] and should be avoided except if one absolutely needs an analytical expression. From a computing point of view, a numerical integration of Eq. (4.61) which very generally reduces to a multiplication with the detector solid angle is much more preferable. For this reason, we only give here a brief account of how wave-vector resolution functions can be dealt with.

[14] The major problem is that the transformation of the angular resolution function into a wave-vector resolution function leads to a function which is generally not separable in q_x and q_z (hence the projection in Fig. 4.6).

4.7.2 Wave-Vector Resolution Function

Since a rigourous transformation of the angular resolution function into a separable wave-vector transfer resolution function cannot be made generally, we use here a more simple approximation where only the resolution volume is conserved in the transformation, see Fig. 4.6. According to Fig. 4.6, close to the specular, a factor

$$\left[\frac{2}{k_0^2 q_z} \times min\left(\frac{\Delta\theta_{\rm in}}{\Delta\theta_{\rm sc}}, \frac{\Delta\theta_{\rm sc}}{\Delta\theta_{\rm in}}\right)\right]$$

must be introduced for resolution volume (and therefore intensity) conservation when performing the transformation from angular to wave-vector variables, whereas there is no factor 2 if for example $\theta_{\rm in} \ll \theta_{\rm sc}$. Considering diffuse scattering close to the specular, an approximation of Eq. (4.61) is:

$$I_D/I_0 = \frac{1}{l_x \times l_y} \frac{1}{\sqrt{2\pi}\Delta\theta_{\rm in}} \int \frac{2\Delta\theta_{\rm in}}{k_0^2 q_z \Delta\theta_{\rm sc}} \frac{d\sigma}{d\Omega} d\delta\mathbf{q}\, \mathcal{R}(\delta\mathbf{q}) \tag{4.62}$$

(in general $\Delta\theta_{\rm sc} > \Delta\theta_{\rm in}$). $\mathcal{R}(\delta\mathbf{q})$ is the resolution function in the wave-vector space.
Considering for simplicity the case of a single rough interface, we must now compute instead of Eq. (4.41), integrals of the form:[15]

$$\begin{aligned} I/I_0 = {} & \frac{k_0^3}{8\pi^2 q_z \sin\theta_{\rm in}} (n_1^2 - n_0^2)^2 \left|t_{0,1}^{\rm in}\right|^2 \left|t_{0,1}^{\rm sc}\right|^2 \frac{e^{-\frac{1}{2}(q_{z,1}^2 + q_{z,1}^{*2})\langle z^2 \rangle}}{|q_{z,1}|^2} \cos^2\psi \\ & \int d\delta\mathbf{q}_\| \int d\mathbf{R}_\| \left[e^{|q_{z,1}|^2 \langle z(\mathbf{R}_\|) z(\mathbf{0})\rangle} - 1\right] \mathcal{R}(\delta\mathbf{q}_\|) e^{i(\mathbf{q}_\| + \delta\mathbf{q}_\|).\mathbf{R}_\|}, \end{aligned} \tag{4.63}$$

where the integration over δq_z has been replaced by a factor $\sqrt{2\pi}\Delta q_z \equiv \sqrt{2\pi} k_0 \Delta\theta_{\rm sc}$ since the integrant does not significantly vary over Δq_z. The resolution function is therefore: $\mathcal{R}(\delta\mathbf{q}_\|) = e^{-1/2\,\delta q_x^2/\Delta q_x^2 - 1/2\,\delta q_y/\Delta q_y^2}$. The integration over $\delta\mathbf{q}_\|$, yields $\tilde{\mathcal{R}}$, the Fourier transform of $\mathcal{R}$:

$$\tilde{\mathcal{R}}(\mathbf{R}_\|) = 2\pi \Delta q_x \Delta q_y e^{-1/2(\Delta q_x^2 X^2 + \Delta q_y^2 Y^2)}, \tag{4.64}$$

where $\Delta q_x \sqrt{2Log2}$ and $\Delta q_y \sqrt{2Log2}$ are the half-width-at-half-maximum of the resolution function $\mathcal{R}$. According to Eq. (4.64), $1/\Delta q_x$ and $1/\Delta q_y$ represent the coherence lengths along x and y, i.e. the lengths over which correlations in the surface roughness can be observed, as discussed in Chap. 2.[16]

[15] This equation also shows that it is important to correct for the illuminated area, see Chap. 7 and in particular Sect. 7.1. In the case considered here, normalisation to the incident intensity instead of the incident flux in the differential scattering cross-section leads to the $1/\sin\theta_{\rm in}$ factor in Eq. (4.63).

[16] For a typical experiment where the resolution is mainly determined by a slit of $H \times V$ size $(h_x = 0.1mm) \times (h_y = 10mm)$ placed in front of the detector, $1m$ away

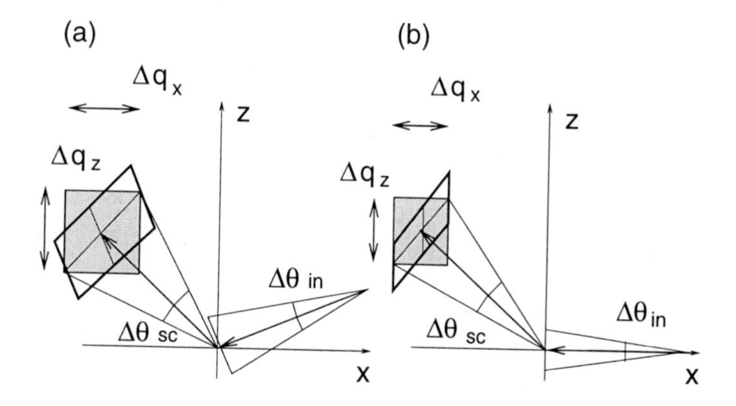

Fig. 4.6. Resolution surfaces in the plane of incidence (thick parallelograms) and their projections onto x and z in the wave-vector transfer space (grey rectangles) for scans in the plane of incidence. The parallelograms are obtained by convoluting the incident angular spread by the detector angular spread. With appropriate scaling (multiplication by k_0), all points within the parallelogram correspond to possible final wave vectors, and the wave-vector transfer can be obtained through a $-\mathbf{k}_{\text{in}}$ translation. Due to (x, z) coupling, the resolution area cannot be simply expressed as a function of Δq_x and Δq_z. In particular, because of the projection onto x or z in the definition of Δq_x and Δq_z, $\Delta q_x \times \Delta q_z \neq k_0 q_z \Delta\theta_{\text{in}} \Delta\theta_{\text{sc}}$, where $k_0 q_z$ is the Jacobian of the transformation. The approximation used here consists in using $\Delta q_x \Delta q_z$, with an appropriate factor depending on the geometry for area (i.e. intensity) conservation. (a) Reflectivity geometry. The area of the parallelogram is $1/2 k_0 q_z max(\Delta\theta_{\text{in}}/\Delta\theta_{\text{sc}}, \Delta\theta_{\text{sc}}/\Delta\theta_{\text{in}})$. (b) grazing incidence geometry. The area of the parallelogram is $k_0^2 q_z max(\Delta\theta_{\text{in}}/\Delta\theta_{\text{sc}}, \Delta\theta_{\text{sc}}/\Delta\theta_{\text{in}})$. In this last case, there is no coupling, and the transformation can be done via a straightforward Jacobian transformation. For scans limited to the plane of incidence as here, the slit openings normal to this plane only introduce an additional $k_0 \Delta_y$ in the resolution volume

The intensity scattered by a single rough interface close to the specular can finally be written:

---

from the sample, for a typical scattering angle $\theta_{\text{sc}} = 10 mrad$ and a wavelength $\lambda = 0.1 nm$, the acceptance of the detector is $\Delta\theta_y = 10^{-2} rad$ in the horizontal and $\Delta\theta_z = 10^{-4} rad$ in the vertical. The coherence lengths along x, y and z are respectively $\lambda/\theta_{\text{sc}}\Delta\theta_z = 100\mu m$, $\lambda/\Delta\theta_y = 0.01\mu m$, and $\lambda/\Delta\theta_z = 1\mu m$. This is the enhancement in the coherence along x due to the grazing incidence geometry which allows the measurement of height or density correlations over such a large range of lengthscales.

$$I/I_0 = \frac{k_0^3}{8\pi^2 q_z \sin\theta_{\text{in}}}(n_1^2 - n_0^2)^2 \left|t_{0,1}^{\text{in}}\right|^2 \left|t_{0,1}^{\text{sc}}\right|^2 \frac{e^{-\frac{1}{2}(q_{z,1}^2 + q_{z,1}^{*2})\langle z^2\rangle}}{|q_{z,1}|^2} \cos^2\psi$$
$$\times \int d\mathbf{R}_\| \left[e^{|q_{z,1}|^2\langle z(\mathbf{R}_\|)z(\mathbf{0})\rangle} - 1\right] e^{i\mathbf{q}_\|\cdot\mathbf{R}_\|}\tilde{\mathcal{R}}(\mathbf{R}_\|). \tag{4.65}$$

In the limit of small q_z one can develop the exponential in Eq. (4.65),

$$\frac{I}{I_0} = \frac{k_0^3}{8\pi^2 q_z \sin\theta_{\text{in}}}(n_1^2 - n_0^2)^2 \left|t_{0,1}^{\text{in}}\right|^2 \left|t_{0,1}^{\text{sc}}\right|^2 \cos^2\psi \left\langle z(\mathbf{q}_\|)z(-\mathbf{q}_\|)\right\rangle \otimes \mathcal{R}(\mathbf{q}_\|). \tag{4.66}$$

The scattered intensity is then simply proportional to a convolution of the roughness spectrum with the resolution function.

4.8 Reflectivity Revisited

Equation (4.66) above shows that the diffuse intensity decreases as q_z^{-2} for small q_z values, whereas it was shown in Chap. 3 that the specular (coherent) intensity decreases as q_z^{-4}. One therefore expects that diffuse scattering will eventually dominate over the specular reflectivity. Of course the wave vector at which diffuse scattering becomes dominant will depend on the experimental resolution since the diffuse intensity is proportional to the resolution volume. In fact, for reasonable experimental conditions, the corresponding wave vectors are rather small, on the order of a few nm^{-1}, and this leads to major difficulties in the treatment of reflectivity data. A "reflectivity curve" $I(q_z)$ is indeed never a pure specular reflectivity curve. Moreover, the diffuse intensity is often (but not always) peaked in the specular direction (Fig. 4.7) making the separation of the specular and diffuse components very difficult experimentally.

This is a very difficult problem since the q_z dependence of the diffuse intensity depends on the exact interface correlation function. A simple model can therefore no longer be used for the analysis of "reflectivity" curves. This is the situation found for the system of Fig. 4.8, an octadecyltrichlorosilane Langmuir film on water [32]. In this case, the surface spectrum can be calculated, and the specular and diffuse contributions to the reflectivity can be compared. The roughness spectrum (here thermally excited capillary waves) is obtained from thermodynamic considerations by Fourier decomposition of the free energy, see Chap. 9:

$$\langle z(\mathbf{q}_\|)z(-\mathbf{q}_\|)\rangle = \frac{1}{L_x \times L_y}\frac{k_B T}{\Delta\rho g + \gamma\mathbf{q}_\|^2 + \kappa\mathbf{q}_\|^4}. \tag{4.67}$$

$L_x \times L_y$ is the interfacial area, γ is the surface tension, and κ is the bending rigidity modulus. The correlation function can be obtained by Fourier transformation.

$$\langle z(\mathbf{0})z(\mathbf{r}_{\|})\rangle = k_B T/2\pi\gamma \times [K_0(r_{\|}\sqrt{\Delta\rho g/\gamma}) - K_0(r_{\|}\sqrt{\gamma/\kappa})], \qquad (4.68)$$

where K_0 is the modified Bessel of second kind of order 0. Then, for a wave vector resolution Δq_x, the intensity measured in the $\theta_{\rm sc} = \theta_{\rm in}$ direction is smaller than the reflectivity of a perfectly flat interface by a factor:

$$\pi^{-1/2}\Gamma\left[\frac{1}{2} - \frac{k_B T q_z^2}{4\pi\gamma}, \frac{1}{2}\Delta q_X^2 \frac{K}{\gamma}\right] \times \exp - \left[\frac{q_z^2 k_B T}{2\pi\gamma} ln\left(\frac{e^{\gamma_E}}{\sqrt{2}}\frac{\sqrt{\gamma/K}}{\Delta q_X}\right)\right] \qquad (4.69)$$

where Γ is the incomplete Γ function, and γ_E Euler's constant. Note that this factor is larger than $e^{-q_z^2<z^2>}$ because diffuse scattering has been taken into account in addition to specular reflectivity.

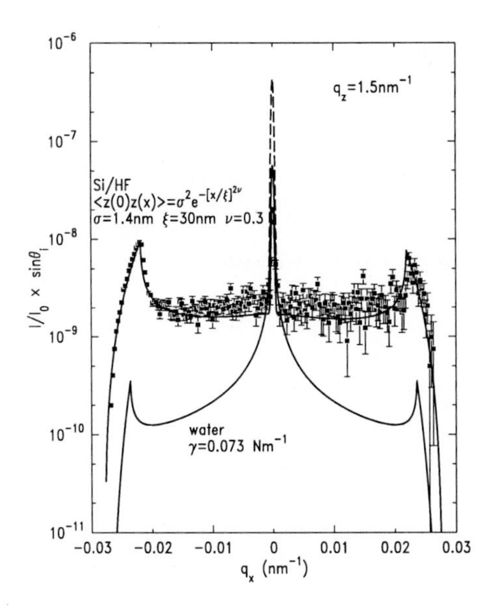

Fig. 4.7. Diffuse scattering from the water surface which is peaked in the specular direction because capillary waves of longer wavelength cost less energy (only a calculated intensity is presented here because the large background due to bulk scattering prevents from a precise measurement, see below the chapter on liquid surfaces), and a solid surface with a flat power spectrum

It can be seen on Fig. 4.8 that even for relatively small wavevectors the diffuse intensity dominates. It would not have been possible to obtain physically

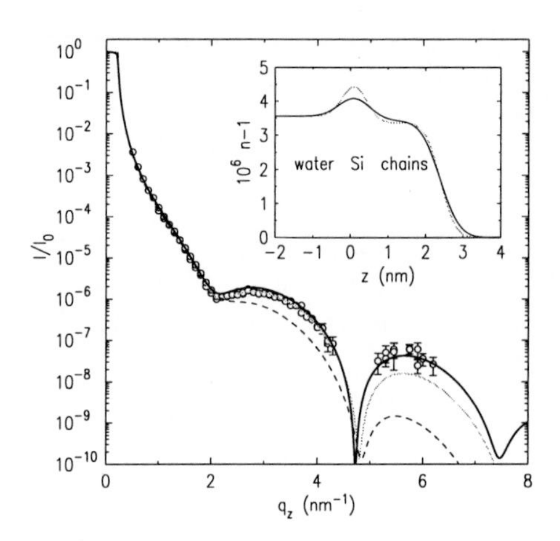

Fig. 4.8. Reflectivity of an octadecyltrichlorosilane film on water. The broken line corresponds to specular intensity. It is dominated by diffuse intensity (grey line) for wavevectors larger than $2nm^{-1}$. The black line is the total (specular + diffuse) intensity. Inset: corresponding electron densities for the complete model Eq. (4.69) (thick line) and the simple box model with error function transition layers (thin line)

reasonable parameters from the experiment without taking its contribution into account (see Fig. 4.8).

4.A Appendix: the Reciprocity Theorem

We consider a medium described by the relative permittivity $\epsilon_{\text{ref}}(\mathbf{r})$ which is assumed to be different from 1 in a localised region of space. Let two different current distribution sources $\mathbf{J}_A, \mathbf{J}_B$ (with same frequency ω) be placed in this medium. We denote by the indices A, B the fields created by these sources, separately, in the medium. They satisfy Maxwell's equations,

$$\begin{aligned} \nabla \times \mathbf{E}_{A(B)} &= -i\omega \mathbf{B}_{A(B)} \\ \nabla \times \mathbf{H}_{A(B)} &= \mathbf{J}_{A(B)} + i\omega \mathbf{D}_{A(B)}, \end{aligned} \tag{4.A1}$$

where $\mathbf{D}_{A,(B)}(\mathbf{r}) = \epsilon_0\epsilon_{\text{ref}}(\mathbf{r})\mathbf{E}_{A(B)}(\mathbf{r})$ and $\mathbf{B}_{A(B)} = \mathbf{H}_{A(B)}/\mu_0$. Substituting the Maxwell's equations in the vectorial identity,

$$\begin{aligned}\nabla.(\mathbf{E_A} \times \mathbf{H}_B - \mathbf{E}_B \times \mathbf{H}_A) = \mathbf{H}_B.\nabla \times \mathbf{E}_A - \mathbf{E}_A.\nabla \times \mathbf{H}_B \\ - \mathbf{H}_A.\nabla \times \mathbf{E}_B + \mathbf{E}_B.\nabla \times \mathbf{H}_A,\end{aligned} \tag{4.A2}$$

leads to

$$\begin{aligned}\nabla.(\mathbf{E}_A \times \mathbf{H}_B - \mathbf{E}_B \times \mathbf{H}_A) = \mathbf{E}_B.\mathbf{J}_A - \mathbf{E}_A.\mathbf{J}_B + i\omega(\mathbf{E}_B.\mathbf{D}_A - \mathbf{E_A}.\mathbf{D_B}) \\ +i\omega(\mathbf{H}_A.\mathbf{B}_B - \mathbf{H}_B.\mathbf{B}_A).\end{aligned} \tag{4.A3}$$

The last two terms on the right-hand side are zero so that we get,

$$\nabla.(\mathbf{E}_A \times \mathbf{H}_B - \mathbf{E}_B \times \mathbf{H}_A) = (\mathbf{E}_B.\mathbf{J}_A - \mathbf{E}_A.\mathbf{J}_B)\,. \tag{4.A4}$$

Integrating Eq. (4.A4) over all space gives

$$\int d^3r\nabla.(\mathbf{E}_A \times \mathbf{H}_B - \mathbf{E}_B \times \mathbf{H}_A) = \int d^3r\,(\mathbf{E}_B.\mathbf{J}_A - \mathbf{E}_A.\mathbf{J}_B)\,,$$

and using the divergence theorem

$$\int d^2r(\mathbf{E}_A \times \mathbf{H}_B - \mathbf{E}_B \times \mathbf{H}_A) = \int d^3r\,(\mathbf{E}_B.\mathbf{J}_A - \mathbf{E}_A.\mathbf{J}_B)\,. \tag{4.A5}$$

If now the current sources are limited to a finite volume, the surface of integration in Eq. (4.A5) is infinitely remote from them, and the electromagnetic field can be approximated by a plane wave with $\mathbf{E}$ and $\mathbf{H}$ orthogonal and transverse.

$$\mathbf{H} = \sqrt{\frac{\epsilon_0}{\mu_0}}\widehat{\mathbf{n}} \times \mathbf{E}.$$

It follows that,

$$\mathbf{E}_A \times \mathbf{H}_B - \mathbf{E}_B \times \mathbf{H}_A = \mathbf{0},$$

which yields

$$\int d^3r\mathbf{E}_B.\mathbf{J}_A = \int d^3r\mathbf{E}_A.\mathbf{J}_B, \tag{4.A6}$$

which is the reciprocity theorem [12–14]. Eq. (4.A6) can also be written for dipole density sources through $\mathbf{P} = (1/i\omega)\mathbf{J}$, one gets

$$\int d^3r\mathbf{E}_B.\mathbf{P}_A = \int d^3r\mathbf{E}_A.\mathbf{P}_B. \tag{4.A7}$$

4.B Appendix: Verification of the Integral Equation in the Case of the Reflection by a Thin Film on a Substrate

It has been indicated in the main text that the integral equation Eq. (4.11) obtained by applying the reciprocity theorem is an exact equation. In this appendix, we verify that this is indeed the case for a single film on a substrate. The reference situation an homogeneous medium of optical index n_0 and we want to calulate the electric field in the case where there is a film (1) of thickness d on a substrate (2). The reflection and transmission coefficients of the smooth film are respectively, Eqs. (3.75), (3.76):

$$r = \frac{r_{0,1} + r_{1,2} e^{-2ik_{z,1}d}}{1 + r_{0,1} r_{1,2} e^{-2ik_{z,1}d}}, \quad t = \frac{t_{0,1} t_{1,2} e^{-ik_{z,1}d}}{1 + r_{0,1} r_{1,2} e^{-2ik_{z,1}d}}.$$

The real case differs from the ideal one within the substrate where the refractive index difference between the real and ideal case is $(n_2^2 - n_0^2)$ and in the film where the difference is $(n_1^2 - n_0^2)$. In Eq. (4.11), we need the field in the real case, which is the transmitted field in the substrate, and is

$$E_1^{PW}(z) = \frac{t_{0,1}}{1 + r_{0,1} r_{1,2} e^{2ik_{inz,1}d}} \left[e^{-ik_{inz,1}z} + r_{1,2} e^{-ik_{inz,1}(2d-z)} \right]$$

in the film. We also need the Green function in vacuum (medium (0)),

$$\frac{k_0^2 e^{-ik_0 R}}{4\pi\epsilon_0 R} e^{i\mathbf{k}_{sc,0}\cdot\mathbf{r}}.$$

Using Eq. (4.11), the electric field can be written:

$$\begin{aligned} E = E_0 + \frac{e^{-ik_0R}}{4\pi R} \Bigg\{ \int d\mathbf{r}_{\|} e^{i\mathbf{q}_{\|}\cdot\mathbf{r}_{\|}} \Big[\ldots \\ (k_{1z}^2 - k_{0z}^2) \int_0^d e^{-ik_{scz,0}z} \frac{t_{01}(e^{-ik_{inz,1}z} + r_{12} e^{ik_{inz,1}(2d-z)})}{1 + r_{01} r_{12} e^{-2ik_{inz,1}d}} dz \\ +(k_{2z}^2 - k_{0z}^2) \int_{-\infty}^d e^{-k_{scz,0}z} t e^{-ik_{inz,2}(z-d)} \Big] \Bigg\}. \end{aligned} \quad (4.B1)$$

In Eq. (4.B1), we have used that $k_{z,j}^2 = k_j^2 - k_x^2 = n_j^2 k_0^2 - k_x^2$ implies $k_0^2(n_j^2 - n_0^2) = k_{z,j}^2 - k_{z,0}^2$. Medium (2) is considered to be slightly absorbing in order to ensure the convergence of the integration. One obtains:

$$E = E_0 - \frac{e^{-ikR}}{4\pi R} \int d\mathbf{r}_{\|} e^{i\mathbf{q}_{\|}\cdot\mathbf{r}_{\|}} (2ik_0 \sin\theta_0 r). \quad (4.B2)$$

The differential scattering cross-section is thus:

$$\frac{d\sigma}{d\Omega} = \frac{k_0^2}{4\pi^2}|r|^2 \sin^2\theta_0 4\pi^2 \mathcal{A}\delta_{\mathbf{k}_{sc\parallel},\mathbf{k}_{in\parallel}} = k_0^2 \sin^2\theta_0 |r|^2 \mathcal{A}\delta(\mathbf{q}_\parallel). \qquad (4.B3)$$

And we find for the reflection coefficient:

$$R = \frac{1}{\mathcal{A}\sin\theta_0}\int \frac{d\sigma}{d\Omega} d\Omega = |r|^2 \qquad (4.B4)$$

as expected.

4.C Appendix: Interface Roughness in a Multilayer Within the Born Approximation

In this appendix we treat the case of a rough multilayer within the Born approximation in order to show some simple properties of the scattered intensity. In the case of the rough multilayer depicted in Fig. 4.2 Eq. (4.26) gives:

$$\frac{d\sigma}{d\Omega} = \frac{k_0^4}{16\pi^2}\cos^2\psi \left| \sum_{i=1}^{N} \int d\mathbf{r}_\parallel \int_{z_i}^{z_{i+1}} dz\,(n_i^2 - 1) e^{i\mathbf{q}\cdot\mathbf{r}} \right|^2 . \qquad (4.C1)$$

The upper medium (air or vacuum) is medium 0, and the substrate (medium s) is slightly absorbing in order to make the integrals converge.

$$\frac{d\sigma}{d\Omega} = \frac{k_0^4}{16\pi^2}\cos^2\psi \left| \sum_{i=1}^{s} \int d\mathbf{r}_\parallel \frac{1}{iq_z}\left[e^{iq_z z_{i+1}} - e^{iq_z z_i}\right] n_i^2 e^{i\mathbf{q}_\parallel\cdot\mathbf{r}_\parallel} \right|^2 , \qquad (4.C2)$$

which can be written as:

$$\frac{d\sigma}{d\Omega} = \frac{k_0^4}{16\pi^2 q_z^2}\cos^2\psi \left| \sum_{i=0}^{N} \int d\mathbf{r}_\parallel \left[n_{i+1}^2 - n_i^2\right] e^{iq_z z_i} e^{i\mathbf{q}_\parallel\cdot\mathbf{r}_\parallel} \right|^2 . \qquad (4.C3)$$

Let us then define:

$$z_i = Z_i + z_i(\mathbf{r}_\parallel), \qquad (4.C4)$$

where Z_i is the height of the flat interface in the reference case. Equation (4.C3) can be written:

$$\frac{d\sigma}{d\Omega} = \frac{k_0^4}{16\pi^2 q_z^2}\cos^2\psi \times \sum_{i=0}^{N}\sum_{j=0}^{N} \int d\mathbf{r}_\parallel \int d\mathbf{r}'_\parallel \left[n_{i+1}^2 - n_i^2\right]\left[n_{j+1}^2 - n_j^2\right]$$
$$e^{iq_z(Z_i - Z_j)} e^{iq_z(z_i(\mathbf{r}_\parallel) - z_j(\mathbf{r}'_\parallel))} e^{i\mathbf{q}_\parallel\cdot(\mathbf{r}_\parallel - \mathbf{r}'_\parallel)}. \qquad (4.C5)$$

Making the change of variables $\mathbf{r}_{\|} - \mathbf{r}'_{\|} \to \mathbf{r}_{\|}$ and integrating over $\mathbf{r}'_{\|}$:

$$\frac{d\sigma}{d\Omega} = \frac{k_0^4 A}{16\pi^2 q_z^2} \cos^2\psi \times \sum_{i=0}^{N}\sum_{j=0}^{N} \int d\mathbf{r}_{\|} \left[n_{i+1}^2 - n_i^2\right]\left[n_{j+1}^2 - n_j^2\right] e^{iq_z(Z_i - Z_j)} \langle e^{iq_z(z_i(\mathbf{r}_{\|}) - z_j(\mathbf{0}))}\rangle e^{i\mathbf{q}_{\|}\cdot\mathbf{r}_{\|}}, \quad (4.C6)$$

where A is the illuminated area. Assuming Gaussian statictics of the height fluctuations $z_i(\mathbf{r}_{\|})$, or in any case expanding the exponential to the lowest (second) order, we have:

$$\langle e^{iq_z(z_i(\mathbf{r}_{\|}) - z_j(\mathbf{0}))}\rangle = e^{-\frac{1}{2}q_z^2\langle z_i(\mathbf{r}_{\|}) - z_j(\mathbf{0})\rangle^2}. \quad (4.C7)$$

We then obtain:

$$\frac{d\sigma}{d\Omega} = \frac{k_0^4 A}{16\pi^2 q_z^2} \cos^2\psi \sum_{i=0}^{N}\sum_{j=0}^{N} \left[n_{i+1}^2 - n_i^2\right]\left[n_{j+1}^2 - n_j^2\right] e^{iq_z(Z_i - Z_j)} e^{-\frac{1}{2}q_z^2\langle z_i\rangle^2 - \frac{1}{2}q_z^2\langle z_j\rangle^2} \int d\mathbf{r}_{\|}\, e^{q_z^2\langle z_i(\mathbf{r}_{\|}) z_j(\mathbf{0})\rangle^2} e^{i\mathbf{q}_{\|}\cdot\mathbf{r}_{\|}}. \quad (4.C8)$$

This equation also includes specular components because it has been constructed from the general solution of an electromagnetic field in a vacuum. The diffuse intensity can be obtained by removing the specular component:

$$\frac{d\sigma}{d\Omega} = \frac{k_0^4 A}{4 q_z^2} \sum_{i=0}^{N}\sum_{j=0}^{N} \left[n_{i+1}^2 - n_i^2\right]\left[n_{j+1}^2 - n_j^2\right] e^{iq_z(Z_i - Z_j)} e^{-\frac{1}{2}q_z^2\langle z_i\rangle^2 - \frac{1}{2}q_z^2\langle z_j\rangle^2} \delta(\mathbf{q}_{\|}). \quad (4.C9)$$

The diffuse intensity is then:

$$\left(\frac{d\sigma}{d\Omega}\right)_{\text{incoh}} = \frac{k_0^4 A}{16\pi^2 q_z^2} \sum_{i=0}^{N}\sum_{j=0}^{N} \left[n_{i+1}^2 - n_i^2\right]\left[n_{j+1}^2 - n_j^2\right] e^{iq_z(Z_i - Z_j)} e^{-\frac{1}{2}q_z^2\langle z_i\rangle^2 - \frac{1}{2}q_z^2\langle z_j\rangle^2} \cos^2\psi \int d\mathbf{r}_{\|} \left(e^{q_z^2\langle z_i(\mathbf{r}_{\|}) z_j(\mathbf{0})\rangle^2} - 1\right) e^{i\mathbf{q}_{\|}\cdot\mathbf{r}_{\|}}. \quad (4.C10)$$

For a single surface, we get Eq. (4.35). It is remarkable that equation (4.C10) has exactly the same structure as the reflectivity coefficient (Fig. 4.3),

$$\left(\frac{d\sigma}{d\Omega}\right)_{\text{coh}} = \frac{k_0^4 A}{4 q_z^2} \sum_{i=0}^{N-1}\sum_{j=0}^{N-1} \left[n_{i+1}^2 - n_i^2\right]\left[n_{j+1}^2 - n_j^2\right] e^{iq_z(Z_i - Z_j)} e^{-\frac{1}{2}q_z^2\langle z_i\rangle^2 - \frac{1}{2}q_z^2\langle z_j\rangle^2} \delta(\mathbf{q}_{\|}), \quad (4.C11)$$

each term simply being multiplied by a "transverse" coefficient:

$$\frac{1}{4\pi^2}\int d\mathbf{r}_{\parallel}\left(e^{q_z^2\langle z_i(\mathbf{r}_{\parallel})z_j(\mathbf{0})\rangle^2}-1\right)e^{i\mathbf{q}_{\parallel}\cdot\mathbf{r}_{\parallel}}.$$

4.D Appendix: Quantum Mechanical Approach of Born and Distorted-Wave Born Approximations

T. Baumbach and P. Mikulík

In this appendix we treat the formal quantum-mechanical approach to scattering by multilayers with random fluctuations. That can be interface roughness, but also porosity or density fluctuations. In particular we develop the differential scattering cross section in the kinematical approximation (first Born approximation) and in the distorted wave Born approximation in terms of the structure amplitudes of the individual layers and of their disturbances. This approach is written in a general way. In Chap. 8 it will be applied to the reflection and to diffraction under conditions of specular reflection under grazing incidence by rough multilayers and multilayered gratings. We would like to notice that we adopted here the *phase-sign* notation of this book, with plane waves e^{-ikr} and Fourier transforms e^{+iqr}, which is contrary to that used in most publications using this formalism.

4.D.1 Formal Theory

Here we develop formally the incoherent approach for the scattering by multilayers with defects independently of the specific scattering method. We make use of the (scalar) quantum mechanical scattering theory and its approximations, in particular the first order Born approximation (kinematical theory) and the distorted wave Born approximation (semi-dynamical theory).

Scattering of the incident wave $|K_0\rangle$ by the potential V produces the total wave field $|E\rangle$, described by the integral equation [33]

$$|E\rangle = |K_0\rangle + \hat{G}_0\hat{V}|E\rangle\ , \tag{4.D1}$$

where $\hat{G}_0$ is the Green function operator of the free particle. We define the *transition operator* by $\hat{T}|K_0\rangle \equiv \hat{V}|E\rangle$ and the *transition matrix* by the matrix elements $T_{0S} = \langle K_S|\hat{T}|K_0\rangle$, characterising the scattering from $|K_0\rangle$ into $|K_S\rangle$. The *differential scattering cross section* σ into an elementary solid angle $\delta\Omega$ can be expressed by the matrix elements of the transition matrix

$$d\sigma = \frac{1}{16\pi^2}\,|T_{0S}|^2\,d\Omega\ . \tag{4.D2}$$

Scattering by a randomly disturbed potential. Including a random spatial fluctuation of the scattering potential, the differential cross section averages over the statistical ensemble of all microscopic configurations

$$d\sigma = \frac{1}{16\pi^2} \left\langle |T_{0S}|^2 \right\rangle d\Omega \ . \tag{4.D3}$$

We divide $d\sigma$ into coherent and incoherent contributions

$$d\sigma = \left\{ \frac{1}{16\pi^2} | \langle T_{0S} \rangle |^2 + \frac{1}{16\pi^2} | \operatorname{Cov}(T_{0S}, T_{0S})|^2 \right\} d\Omega \equiv d\sigma_{\text{coh}} + d\sigma_{\text{incoh}} \tag{4.D4}$$

by denoting the covariance

$$\operatorname{Cov}(a,b) = \langle ab^* \rangle - \langle a \rangle \langle b \rangle^* \ . \tag{4.D5}$$

Defining the non-random part of the scattering potential by V^A (unperturbed potential) and the random (perturbed) potential by V^B, the *coherent* part of the differential cross section writes

$$d\sigma_{\text{coh}} = \frac{1}{16\pi^2} \left| T^A + \left\langle T^B \right\rangle \right|^2 d\Omega \tag{4.D6}$$

and the *incoherent* differential cross section

$$d\sigma_{\text{incoh}} = \frac{1}{16\pi^2} \operatorname{Cov}(T^B, T^B) \, d\Omega \ . \tag{4.D7}$$

If the random part V^B causes only a small disturbance to the scattering by V^A, we can calculate T^B within the distorted wave Born approximation (DWBA). It is worth noting that in contrast to the widely spread opinion it is not a small potential $V^B \ll V^A$ which defines the validity of the DWBA, but rather the *scattering by* V^B which has to be weak.

Scattering by a randomly disturbed multilayer. In a *multilayer* we represent each layer by the product of its volume polarizability $\chi_{\infty j}(\boldsymbol{r})$ and the layer size function $\Omega_j(\boldsymbol{r})$

$$\chi(\boldsymbol{r}) = \sum_{j=1}^{N} \chi_j(\boldsymbol{r}) = \sum_{j=1}^{N} \chi_{\infty j}(\boldsymbol{r}) \Omega_j(\boldsymbol{r}) \ . \tag{4.D8}$$

The optical (or scattering) potential for X-rays can be expressed by the polarizability: $V(\boldsymbol{r}) = -k_0^2 \chi(\boldsymbol{r})$. The contribution of the different layers to the scattering cross section is distinguished by considering each layer as an independent scatterer

$$\hat{V}(\boldsymbol{r}) = \sum_j v_j(\boldsymbol{r}) \ . \tag{4.D9}$$

Then Eq. (4.D4) writes

$$d\sigma = \frac{1}{16\pi^2}\left\{\left|\sum_{j=1}^{N}\langle\tau_j\rangle\right|^2 + \sum_{j=1}^{N}\sum_{k=1}^{N}\mathrm{Cov}\left(\tau_j, \tau_k\right)\right\} d\Omega \tag{4.D10}$$

with $\tau_j = \langle K_S|v_j|E\rangle$.

Separating the non-random and the random part of each layer, $v_j = v_j^A + v_j^B$, we obtain

$$d\sigma = \frac{1}{16\pi^2}\left\{\left|\sum_{j=1}^{N}\tau_j^A + \sum_{j=1}^{N}\langle\tau_j^B\rangle\right|^2 + \sum_{j=1}^{N}\sum_{k=1}^{N}\mathrm{Cov}\left(\tau_j^B, \tau_k^B\right)\right\} d\Omega\ , \tag{4.D11}$$

where the τ_j^A are the contributions of the non-perturbed layers to scattering, and τ_j^B are those of the layer disturbances. The first term is the coherent part $d\sigma_{\mathrm{coh}}$, which consists of the contribution of the ideal multilayer and of the averaged transition elements of the layer disturbances. The second, incoherent part $d\sigma_{\mathrm{incoh}}$ contains the covariance functions of all single layer transition elements.

Formally the division of $\hat{V}$ into a sum of scatterers $\sum_j v_j$ is arbitrary. The sticking point is to find a set of eigenstates, which is convenient to serve as basis for calculation of the transition elements. Finally, we remind the reader, that until now no approximation has been made.

4.D.2 Formal Kinematical Treatment by First Order Born Approximation

Within the kinematical treatment (first order Born approximation) we approximate the transition operator by the operator of the scattering potential $\hat{T}|K\rangle \approx \hat{V}|K\rangle$. The set of vacuum wave vectors $|\boldsymbol{K}\rangle = e^{-i\boldsymbol{kr}}$ provides an orthogonal basis for the calculation of the differential scattering cross section. The transition elements of the individual layers are

$$\tau_j = \langle K_S|v_j|K_0\rangle = -k_0^2 \int d\boldsymbol{r}\, \chi_j(\boldsymbol{r})\, e^{i\boldsymbol{qr}}\ , \tag{4.D12}$$

where $\boldsymbol{q} = \boldsymbol{k}_S - \boldsymbol{k}_0$. Defining the *structure factor of the layer*

$$S_j(\boldsymbol{q}) = \int_s d\boldsymbol{r}_{\|}\, F_j(q_z, \boldsymbol{r}_{\|})\, e^{i\boldsymbol{q}_{\|}\boldsymbol{r}_{\|}} \tag{4.D13}$$

with the random one-dimensional *layer form factor*

$$F_j(q_z, \boldsymbol{r}_{\|}) = \int dz\, \chi_j(\boldsymbol{r})\, e^{iq_z(z-Z_j)} \tag{4.D14}$$

the transition element becomes

$$\tau_j = -k_0^2 \, e^{iq_z Z_j} \, S_j(\boldsymbol{q}) \; . \tag{4.D15}$$

The coherent scattering cross section (4.D9) uses the statistical averages $\langle \tau_j \rangle_{\mathrm{av}}$, and so we search for the *mean layer form factor* $\langle F_j(q_z, \boldsymbol{r}_\parallel) \rangle_{\mathrm{av}}$. The incoherent differential scattering cross section contains the *covariance functions*

$$\begin{aligned} \tilde{q}_{jk} &= \mathrm{Cov}\,(S_j, S_k) \\ &= \int d\boldsymbol{r}_\parallel \int d\boldsymbol{r}_\parallel{}' \, e^{i\boldsymbol{q}_\parallel (\boldsymbol{r}_\parallel - \boldsymbol{r}_\parallel{}')} \, \mathrm{Cov}\left(F_j(q_z, \boldsymbol{r}_\parallel), F_k(q_z, \boldsymbol{r}_\parallel{}')\right) \; . \end{aligned} \tag{4.D16}$$

Substituting (4.D13) and (4.D16) into (4.D10), we obtain the kinematical differential scattering cross section of an arbitrary multilayer

$$d\sigma = \frac{k_0^4}{16\pi^2} \left\{ \left| \sum_{j=1}^{N} \langle S_j \rangle_{\mathrm{av}} \, e^{iq_z Z_j} \right|^2 + \sum_{j=1}^{N} \sum_{k=1}^{N} \tilde{q}_{jk} \, e^{iq_z (Z_j - Z_k)} \right\} d\Omega \; . \tag{4.D17}$$

4.D.3 Formal Treatment by a Distorted Wave Born Approximation

The distorted wave Born approximation takes all those effects of multiple scattering into account which are caused by the unperturbed potential V^A. It is less the method itself, but rather the right choice of V^A, which decides about the success in order to be enough transparent and sufficiently precise. We search for such a V^A which enables to explain the essential multiple scattering effects. However, it should provide the simplest possible solutions $E_{\boldsymbol{K}}^A$ used as orthonormal basis for the representation of scattering by the disturbance (perturbed potential) V^B.

Scattering by *planar multilayers with sharp interfaces* produces such simple solutions. It has been shown that rough multilayers as well as intentionally laterally patterned multilayers and gratings can be treated advantageously by starting with an ideal potential of a planar (laterally averaged) multilayer, splitting the polarizability in

$$\chi = \chi^A + \chi^B \qquad \text{with} \qquad \chi^A = \sum_{j=1}^{N} \chi_j^{A_{\mathrm{planar}}} \; . \tag{4.D18}$$

Coherent scattering by the non-perturbed multilayer generates a wave field $E_{\boldsymbol{K}}^A$, which can be decomposed into a small number of plane waves within each plane homogeneous layer, both with constant complex amplitudes and wave vectors,

$$E_{\boldsymbol{K},j}^A(\boldsymbol{r}) = \left[\sum_{n=1}^{I} E_{\boldsymbol{k}_n,j} \, e^{-i\boldsymbol{k}_{n\parallel,j} \boldsymbol{r}_\parallel} \, e^{-ik_{nz,j}(z - Z_j)} \right] \Omega_j^A(z) \; . \tag{4.D19}$$

In case of specular reflection it is $I=2$ (one transmitted and one reflected wave), for grazing incidence diffraction and strongly asymmetric X-ray diffraction $I=8$. The $E_{\boldsymbol{K}}^{A}(\boldsymbol{r})$ are used as non-perturbed states for the estimation of T^{B} (4.D6).

Within the first order DWBA one obtains

$$T^{B,\mathrm{DWBA}} = \langle E_S^{A*} | \hat{V}^B | E_0^A \rangle = -k_0^2 \int d\boldsymbol{r}\, E_S^A(\boldsymbol{r}) \chi^B(\boldsymbol{r}) E_0^A(\boldsymbol{r}) \,. \tag{4.D20}$$

Again, it is recommendable to describe the contribution of the disturbance within each plane layer separately by

$$\tau_j^B = -k_0^2 \int d\boldsymbol{r}\, E_S^A(\boldsymbol{r}) \chi_j^B E_0^A(\boldsymbol{r}) \tag{4.D21}$$

with

$$\chi_j^{B_{\mathrm{planar}}} = \chi \Omega_j^A - \chi_j^{A_{\mathrm{planar}}} \,. \tag{4.D22}$$

We define F_j^{mn} and S_j^{mn}, formally similar to the expressions (4.D13) and (4.D14), however now with respect to the *disturbance* χ_j^B and corresponding to the actual scattering vector

$$\boldsymbol{q}_j^{mn} = \boldsymbol{k}_{Sj}^m - \boldsymbol{k}_{0j}^n \tag{4.D23}$$

inside the layer

$$F_j^{mn}(q_{z,j}^{mn}, \boldsymbol{r}_{\|}) = \int dz\, \chi_j^B(\boldsymbol{r})\, e^{i q_{z,j}^{mn}(z - Z_j)} \,. \tag{4.D24}$$

Each τ_j^B consists of $I \times I$ terms

$$\tau_j^B = -k_0^2 \sum_{m=1}^{I} \sum_{n=1}^{I} E_{Sj}^m(z) S_j^{mn} E_{0j}^n(z) \,, \tag{4.D25}$$

or using the matrix formalism

$$\tau_j^B = -k_0^2 \boldsymbol{E}_{Sj}^m \hat{S}_j \boldsymbol{E}_{0j}^n \,, \tag{4.D26}$$

where the column vector $\boldsymbol{E}_{\boldsymbol{K},j}^m$ contains the amplitudes of the I plane waves of one non-perturbed state in the jth layer and $\hat{S}_j$ is the structure factor matrix of the layer disturbance, respectively. Each term in (4.D25) represents the contribution of the disturbance to the scattering from one plane wave of the initial state $E_{\boldsymbol{K}_0}^A$ in another plane wave of the final state $E_{\boldsymbol{K}_S}^A$. Each scattering process is characterised by the product of the according wave amplitudes $E_{Sj}^m E_{0j}^n$ and by the disturbance structure factors S_j^{mn}.

In order to determine the *coherent scattering cross section* we average F and S over the statistical ensemble and substitute these terms in (4.D10).

The *incoherent cross section* contains the covariance functions for each layer pair

$$\mathrm{Cov}\left(\tau_j^B, \tau_k^B\right) = k_0^4 \sum_{m,n,o,p} E_{Sj}^m E_{Sj}^n \tilde{Q}_{jk}^{mnop} E_{0k}^o E_{0k}^p \tag{4.D27}$$

with

$$\begin{aligned}\tilde{Q}_{jk}^{mnop} &= \mathrm{Cov}\left(S_j^{mn}, S_k^{op}\right) \\ &= \int d\boldsymbol{r}_\| \int d\boldsymbol{r}_\|{}' \, e^{i\boldsymbol{q}_\|(\boldsymbol{r}_\| - \boldsymbol{r}_\|{}')} \, \mathrm{Cov}\left(F_j(q_{z,j}^{mn}, \boldsymbol{r}_\|), F_k(q_{z,k}^{op}, \boldsymbol{r}_\|{}')\right) \end{aligned} \tag{4.D28}$$

Each term represents the covariance of one scattering process in layer j and a second scattering process in layer k. Adding up the contributions of all scattering processes and all layers we obtain finally

$$\begin{aligned} d\sigma = \frac{k_0^4}{16\pi^2} \Bigg\{ &\left| \sum_{j=1}^{N} \tau_j^A + \sum_{j=1}^{N} \sum_{m,n=1}^{I} E_{Sj}^m \left\langle S_j^{mn} \right\rangle E_{0j}^n \right|^2 \\ &+ \sum_{j,k=1}^{N} \sum_{m,n,o,p=1}^{I} E_{Sj}^m (E_{Sj}^n)^* \, \tilde{Q}_{jk}^{mnop} \, E_{0k}^o (E_{0k}^p)^* \Bigg\} \, d\Omega \, . \end{aligned} \tag{4.D29}$$

In *X-ray reflectivity*, each eigenstate of the unperturbed potential consists of a transmitted and reflected wave, thus $I = 2$. The four wave vector transfers $\boldsymbol{q}^{11}, \ldots, \boldsymbol{q}^{22}$, corresponding to $(\boldsymbol{k}_{\mathrm{sc}\|} - \boldsymbol{k}_{\mathrm{in}\|}, \pm k_{\mathrm{sc},z} \pm k_{\mathrm{in},z})$ in (4.46),(4.47) or (8.48), are represented in the reciprocal space in Fig. 8.40. Further, The above expressions are written explicitly for diffuse scattering in Eqs. (8.46)–(8.49) and for coherent reflectivity for deterministic (i.e. non-random) grating potential V^B in (8.72). The covariance for grazing incidence diffraction is presented by (8.62).

Simpler DWBA for multilayers. The expressions simplify enormously, if we can approximate the non-perturbed polarizability by its mean value in the multilayer, averaging vertically over the whole multilayer stack. We obtain a homogeneous "non-perturbed layer". The splitting of the potential in this way gives

$$\begin{aligned} &\chi^A(\boldsymbol{r}) = \left\langle \chi^{\mathrm{ML}}(\boldsymbol{r}) \right\rangle_{\mathrm{av}} \\ &\chi^B(\boldsymbol{r}) = \sum_{j=1}^{N} \chi_j^{B_{\mathrm{layer}}}(\boldsymbol{r}) \quad \text{with} \quad \chi_j^{B_{\mathrm{layer}}}(\boldsymbol{r}) = \left(\chi(\boldsymbol{r}) - \left\langle \chi^{\mathrm{ML}}(\boldsymbol{r}) \right\rangle_{\mathrm{av}}\right) \Omega_j^{\mathrm{id}}(\boldsymbol{r}) \, . \end{aligned} \tag{4.D30}$$

Now the non-perturbed wave field below the sample surface consists of the transmitted wave only. In consequence exclusively the *primary scattering processes*

$$\mathrm{Cov}(\tau_j^B, \tau_k^B) = K^4 \, t_S t_S^* \, \tilde{Q}_{jk}^{11} \, r_0 r_0^* \tag{4.D31}$$

and the transmission function of the sample surface are considered. Also the effect of refraction is included.

References

1. S.K. Sinha, E.B. Sirota, S. Garoff and H.B. Stanley, *Phys. Rev. B* **38**, 2297 (1988).
2. J. Daillant and O. Bélorgey, *J. Chem. Phys.* **97**, 5824 (1992).
3. I.A. Atyukov, A. Yu. Karabekov, I.V. Kozhevnikov, B.M. Alaudinov, and V.E. Asadchikov, *Physica B* **198**, 9 (1994).
4. S. Dietrich and A. Haase, Physics Reports **260**, 1 (1995).
5. D.K.G. de Boer *Phys. Rev. B* **49** 5817 (1994).
6. P. Croce, L. Névot and B. Pardo, *C.R. Acad. Sc. Paris* **274 B**, 803 (1972).
7. P. Croce, L. Névot and B. Pardo, *C.R. Acad. Sc. Paris* **274 B**, 855 (1972).
8. P. Croce and L. Névot *Revue Phys. Appl.* **11**, 113 (1976).
9. P. Croce *J. Optics (Paris)* **8**, 127 (1977).
10. L. Névot and P. Croce *Revue Phys. Appl.* **15**, 761 (1980).
11. P. Croce *J. Optics (Paris)* **14**, 213 (1983).
12. P. Lorrain and D.R. Corson "Electromagnetic Fields and Waves" W.H. Freeeman and Company (San Francisco) (1970) p.629.
13. L.D. Landau and E.M. Lifshitz, Electrodynamics of continuous media, Course of theoretical physics vol. 8, Pergamon Press, Oxford 1960, §69.
14. C.-T. Tai, Dyadic Green functions in electromagnetic theory, IEEE Press, New-York, 1994.
15. J.D. Jackson "Classical Electrodynamics" 2nd Edition Wiley (New-York) 1975.
16. M. Born and E. Wolf, "Principles of optics" 6th edition, Pergamon (London) (1980) p.51.
17. S.K. Sinha, M. Tolan. A. Gibaud, *Phys. Rev. B* **57**, 2740 (1998).
18. A. Herpin, *C. R. Acad. Sci. Paris* **225** 182 (1947).
19. Y. Yoneda, *Phys. Rev.* **131**, 2010 (1963).
20. B. Vidal et P. Vincent, *Applied Optics* **23** 1794 (1984).
21. I.M. Tidswell, T.A. Rabedeau, P.S. Pershan, S.D. Kosowsky, *Phys. Rev. Lett.* **66**, 2108 (1991).
22. J.B. Bindell and N. Waifan, *J. Appl. Phys.* **3** 503 (1970)
23. Z.-h. Cai, K. Huang, P.A. Montano, T.P. Russel, J.M. Bai, and G.W. Zajac, *J. Chem. Phys.* **98** 2376 (1993).
24. W. Weber and B. Lengeler, *Phys. Rev. B* **46**, 7953 (1992).
25. A. Sentenac and J.J. Greffet, *J. Opt. Soc. Am. A* **15**, 528 (1998).
26. G.C. Brown, V. Celli, M. Coopersmith and M. Haller *Surface Science* **129** 507 (1983)
27. G.C. Brown, V. Celli, M. Haller and A. Marvin *Surf. Sci.*, **136** 381 (1984).
28. G. Brown, V. Celli, M. Haller, A.A. Maradudin, and A. Marvin, *Phys. Rev. B*, **31** 4993 (1985).
29. M. Nieto-Vesperinas and J.C. Dainty, Scattering in volume and surfaces, Elsevier Science Publishers, B.V. North-Holland (1990)
30. R. Petit, ed, Electromagnetic theory of gratings, Topics in current physics, Springer Verlag, Berlin (1980).
31. W.H. de Jeu, J.D. Schindler, E.A.L. Mol, *J. Appl. Cryst.* **29** 511 (1996).

32. L. Bourdieu, J. Daillant, D. Chatenay, A. Braslau, and D. Colson, *Phys. Rev. Lett.* **72**, 1502 (1994).
33. A.S. Davydov, Quantum Mechanics, Pergamon Press, 1969.
1 149 (1991).

5 Neutron Reflectometry

Claude Fermon[1], Frédéric Ott[2] and Alain Menelle[2]

[1] Service de Physique de l'Etat Condensé, Orme des Merisiers, CEA Saclay, 91191 Gif sur Yvette Cedex, France,
[2] Laboratoire Léon Brillouin CEA CNRS, CEA Saclay, 91191 Gif sur Yvette Cedex, France

5.1 Introduction

Neutron reflectometry is a relatively new technique [1,2]. In the last years, it has been extensively used for solving soft matter problems like polymer mixing [3,4] or the structure of liquids at the surface [5,6] for example. The asset of neutrons for polymer studies is their small absorption compared to x-rays and the large contrast between ^{1}H and ^{2}H which allows selective "labelling" by deuteration.

In the late 80's, a new field of application of neutron reflectometry has emerged. Following the discovery of giant magnetoresistance in antiferromagnetically coupled multilayered films [7] and new magnetic phenomena in ultra-thin films, there has been an interest in the precise measurement of the magnetic moment direction in each layer of a multilayer and at the interface between layers. Owing to the large magnetic coupling between the neutron and the magnetic moment, neutron reflectometry has proved to be a powerful tool for obtaining information about these magnetic configurations and for measuring magnetic depth profiles.

In this chapter, we give an overview of the experimental and theoretical methods used for neutron reflectometry, focusing on specular reflectivity. The corresponding theory is partly derived from the previous work developed for x-rays, and we emphasize those aspects specific of neutrons.

In a first part, we will review the neutron-matter interactions. We then describe the non-magnetic scattering. In this case it is possible to introduce an optical index and give a treatment which is similar to x-ray reflectometry (Chap. 3).

In a second part, the neutron spin is introduced. In this case, optical indices cannot be used any longer and it is necessary to completely solve the Schrödinger equation. A detailed matrix formalism is presented.

We then discuss the different aspects of data processing and the problems related to the surface roughness. Two types of neutron reflectometers are described in particular: fixed-wavelength two-axis reflectometers and time-of-flight spectrometers.

The use of neutron reflectivity in the field of polymers films and of magnetic layers is then illustrated by several examples.

Notation used in this chapter

b, b_j	bound scattering length of a nucleus, mean scattering length of a layer j
b_c	bound coherent scattering length
b_i	incoherent scattering length
b_N	spin dependent scattering length
b'	real part of the scattering length
b''	imaginary part of the scattering length
$\mathcal{E}_0, \mathcal{E}_j$	energy of the neutron in the vacuum and in layer j
e	charge of the electron
d, d_j	thickness of a layer
g	Landé factor, (g =2)
$\hbar$	Planck constant
$\mathbf{I}$	nuclear spin operator
$\mathbf{k}$	wave vector
$\mathbf{M}, \mathbf{M_j}$	magnetic moment of an electron and of a layer
m	neutron mass
m_e	electron mass
n_j	refractive index of layer j
p	$p = 2.696 fm$, conversion factor of magnetisation to an effective scattering length
$\mathbf{q}$	scattering vector
$\mathbf{s}$	spin operator of the electron
σ	Pauli operator associated to the neutron spin
V_j	volume of the layer j
V(r)	interaction Hamiltonian
g_n	$g_n = -1.9132$, nuclear Landé factor of the neutron
λ, λ_0	neutron wavelength
μ_B	Bohr magneton
μ_n	nuclear magneton
ρ_j	atomic density of the layer j (atoms per cm^3)
σ_j	absorption
θ_j, φ_j	spherical angles of the magnetisation of the layer j
$\theta_{\mathrm{in}}, \theta_{\mathrm{r}}$	incident and reflected angles of the neutron beam

$\mu_B = e\hbar/(2m_e) = 9.27 \times 10^{-24}\ J.T^{-1}$,
$\mu_n = e\hbar/(2m_p) = 5.05 \times 10^{-27}\ J.T^{-1}$
$\theta(x)$ is the Heavyside function defined by:

$$\begin{cases} \theta(x) = 1 & \text{when} \quad x > 0 \\ \theta(x) = 1/2 & \text{when} \quad x = 0 \\ \theta(x) = 0 & \text{when} \quad x < 0 \end{cases}$$

We call "up" (resp. "down") the neutron polarisation parallel (resp. antiparallel) to the external applied magnetic field.
"Down-up" designates a polarised "down" incident beam and polarised "up" detected beam.
"Down-up" and "up-down" are called spin-flip processes.

5.2 Schrödinger Equation and Neutron-Matter Interactions

5.2.1 Schrödinger Equation

The neutron can be described by a wave of wavelength λ, of wave vector:

$$k_0 = \frac{2\pi}{\lambda}, \tag{5.1}$$

and of energy

$$\mathcal{E}_0 = \frac{\hbar^2 k_0^2}{2m}. \tag{5.2}$$

Its wave function verifies the Schrödinger equation (1.17):

$$\frac{\hbar^2}{2m}\frac{d^2\psi}{dr^2} + [\mathcal{E} - V(r)]\,\psi = 0, \tag{5.3}$$

where m is the neutron mass, $\mathcal{E}$ its energy and V the interaction potential. The neutron is a spin 1/2 particle so that $\psi(r)$ can be expressed on the base of the two spin states:

$$\psi_+(r)\,|+\rangle + \psi_-(r)\,|-\rangle\,. \tag{5.4}$$

When there is an external or internal magnetic field, an "up" (resp. "down") neutron designates a neutron in the eigenstate $|+\rangle$ (resp. $|-\rangle$). In the following the space dependence $(\mathbf{r})$ of the index will often be dropped.

5.2.2 Neutron-Matter Interaction

The two main interactions are the strong interaction with the nuclei and the magnetic interaction with the existing magnetic moments (nuclear and electronic). There are a large number of second order interactions which are described in [8].

Neutron Nucleus Interaction: Fermi Pseudo Potential The scattering of a neutron by a nucleus comes mainly from the strong interaction. The interaction potential is large but its extension is much smaller than the wavelength of the neutron. Hence this interaction can be considered as ponctual and isotropic. Within the Born approximation, it can be described by the Fermi pseudopotential [9]:

$$V_F(r) = b\left(\frac{2\pi\hbar^2}{m}\right)\delta(\mathbf{r}) \tag{5.5}$$

where b is the scattering length and $\mathbf{r}$ is the position of the neutron. The value of the scattering length b depends on the nucleus and on the nuclear spin of the nucleus. Formally it can be written :

$$b = b_c + \frac{1}{2} b_N \mathbf{I}.\sigma, \tag{5.6}$$

N.B.: the scattering length is generally a complex number: $b = b' + ib''$. The first term b_c is called the coherent scattering length. The second term corresponds to the strong interaction of the spin of the neutron (described by the operator $1/2\sigma$) with that of the nucleus (operator $\mathbf{I}$). The total spin $J = 1/2\sigma + I$ is a good quantum number for the neutron spin - nucleus spin interaction $1/2\sigma.\mathbf{I}$. In the manifold $\{I \pm 1/2\}$, the eigenvalues of the spin-dependent operator $\mathbf{I}.\sigma$ are I (for $J = I+1/2$) and $-(I+1)$ (for $J = I-1/2$). We name b^+ and b^- the two scattering lengths associated with these two eigenvalues, corresponding to the two states $|+\rangle$ and $|-\rangle$ of the neutron spin. The nucleus spin-dependent scattering lengths can then be written [10]:

$$\begin{cases} b^+ = b_0 + \frac{1}{2} b_n I \\ b^- = b_0 - \frac{1}{2} b_n (I+1) \end{cases}, \tag{5.7}$$

where I is the nuclear spin quantum number.
We remind that the total scattering cross section is given by (see Eq. (1.35):

$$\sigma_{\text{tot}} = 4\pi\langle |b|^2 \rangle, \tag{5.8}$$

in which the brackets designate the statistical average over the neutron and nuclear spins.

Neutron Absorption The absorption of neutrons is described by the imaginary part of the scattering length b''. The absorption cross section is given by:

$$\sigma_{\text{abs}} = (4\pi/k_0)\, b''. \tag{5.9}$$

The absorption is negligible for thin films except for some elements: Gd, Sm, B and Cd. These elements present (n,γ) nuclear resonances at thermal neutron energies which strongly increase the absorption.

Incoherent Scattering Incoherent scattering comes from the random distribution of isotopes or nuclear spin states in a material. In this case, the total scattering cross section (see equation 5.8) can be written:

$$\sigma_{\text{tot}} = 4\pi\langle|b|^2\rangle = 4\pi\left(\langle|b|\rangle^2 + \left(\langle|b|^2\rangle - \langle|b|\rangle^2\right)\right) = \sigma_{\text{coh}} + \sigma_{\text{incoh}}, \tag{5.10}$$

where σ_{coh} and σ_{incoh} are called the coherent and incoherent scattering lengths. In the presence of isotope or spin disorder, the second term in Eq. (5.10) is not zero. If for example the nucleus carries a spin (see 5.8) we have a spatial distribution b^+ and b^- of scattering lengths. In the case of an isotope distribution b_α in the material, the incoherent cross section is given by:

$$\sigma_{\text{inc,isotope}} = 4\pi \sum_{\alpha<\beta} c_\alpha c_\beta |b_\alpha - b_\beta|^2 \tag{5.11}$$

where c_α designates the fraction of isotope α in the material. Incoherent scattering appears as a q-independent background in the experiments and can be treated as an absorption plus a flat background. The incoherent scattering is particularly important for hydrogenated layers but it is small for deuterated layers. A more detailed discussion of incoherent scattering can be found in [11,12]. *Tables of the different scattering lengths (coherent, incoherent, absorption) of the different elements can be found in [12].*

Magnetic Interaction The main magnetic interaction is the dipolar interaction of the neutron spin with the magnetic field created by the unpaired electrons of the magnetic atoms. This field contains two terms, the spin part and the orbital part:

$$\mathbf{B} = \frac{\mu_0}{4\pi}\left(\nabla \times \left\{\frac{\mu_{\mathbf{e}} \times \mathbf{R}}{|\mathbf{R}|^3}\right\} - \frac{e\,\mathbf{v_e} \times \mathbf{R}}{|\mathbf{R}|^3}\right), \tag{5.12}$$

where $\mu_e = -2\mu_B\sigma$ is the magnetic moment of the electron, μ_B is the Bohr magneton, $\mathbf{v}_e$ is the speed of the electron.
The neutron magnetic moment is equal to:

$$\mu = g_n \mu_n \sigma. \tag{5.13}$$

The magnetic interaction expresses as :

$$V_M(\mathbf{r}) = -\mu.\mathbf{B} = -g_n\mu_n\sigma.\mathbf{B}. \tag{5.14}$$

Neutron reflectivity does not allow the separation of the orbital and spin contributions, it is only sensitive to the internal magnetic field.

The Zeeman Interaction It is the interaction of the neutron spin with an external magnetic field $\mathbf{B}_0$:

$$V_Z(\mathbf{r}) = -g_n\mu_n\sigma.\mathbf{B}_0. \tag{5.15}$$

5.3 Reflectivity on Non-Magnetic Systems

For non-magnetic systems we can introduce the notion of optical indices. It is an approach similar to the x-ray formalism (Chaps. 1 and 2). It can be applied to neutron reflectometry on soft matter [13] and non-magnetic systems.

We consider a neutron beam, reflected by a perfect surface with an incident angle θ. As in Chap. 3, the surface is defined by the interface between the air ($n = 1$) and a material with an optical refractive index n.

In a vacuum, the energy of the neutron is given by:

$$\mathcal{E} = \frac{\hbar^2 k_0^2}{2m} = \frac{h^2}{2m\lambda^2}. \tag{5.16}$$

Let $\mathbf{q} = \mathbf{k}_{\mathrm{r}} - \mathbf{k}_{\mathrm{in}}$ be the scattering wave vector. The projection of the scattering wave vector on the z axis (perpendicular to the surface) is given by:

$$q_z = \frac{4\pi}{\lambda} \sin\theta_{\mathrm{in}}. \tag{5.17}$$

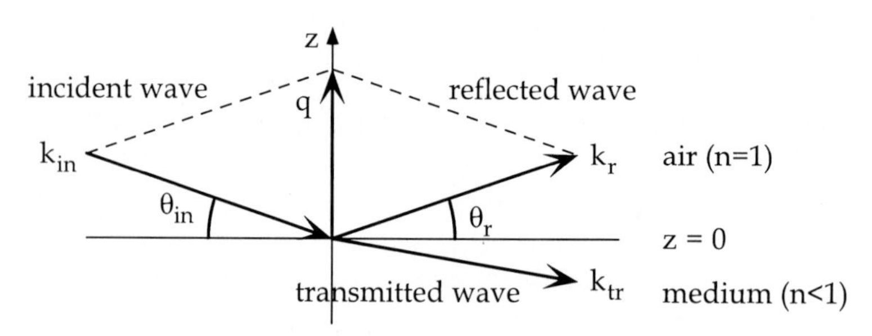

Fig. 5.1. Reflectivity on a perfect surface

5.3.1 Neutron Optical Indices

The neutron indices are very different from the x-ray indices and we will determine their expression from the Schrödinger equation. We suppose that the interaction potential $V(\mathbf{r})$ in the medium is independent of the in-plane coordinates x and y. The mean potential V in the medium is given by the integration of the Fermi pseudo-potential:

$$V = \frac{1}{v} \int_v V(\mathbf{r}) d^3\mathbf{r} = \frac{2\pi\hbar^2}{m} b\rho, \tag{5.18}$$

where ρ is the number of atoms per unit volume.

In the absence of any magnetic field, the Schrödinger equation can be written:

$$\frac{\hbar^2}{2m} \frac{d^2\psi}{dr^2} + [\mathcal{E} - V]\,\psi = 0. \tag{5.19}$$

The equation (5.19) can be written in the form of a Helmholtz propagation equation similar to the electromagnetic case:

$$\frac{d^2\psi}{dr^2} + k^2\psi = 0, \tag{5.20}$$

with

$$k^2 = \frac{2m}{\hbar^2}\left[\mathcal{E} - V\right]. \tag{5.21}$$

We define the optical index as:

$$n^2 = \frac{k^2}{k_0^2} \tag{5.22}$$

The optical index n can be written:

$$n^2 = 1 - \frac{V}{\mathcal{E}} = 1 - \frac{\lambda^2}{\pi}\rho b. \tag{5.23}$$

It is in most cases smaller than one except for materials with a negative scattering length (e.g. Ti and Mn). The quantity $1-n$ is of the order of 10^{-5} and thus n can be written:

$$n \approx 1 - \frac{\lambda^2}{2\pi}\rho b. \tag{5.24}$$

5.3.2 Critical Angle for Total External Reflection

At the interface between two media, the Snell's law applies:

$$\cos\theta_{\text{in}} = n\cos\theta_{\text{tr}}. \tag{5.25}$$

Since we have shown that the index is smaller than one, for angles $\theta \leq \theta_c$, there is a total reflection of the incident wave like in the case of x-ray reflection. The critical angle θ_c is given by the condition $\theta_{\text{tr}} = 0$, i.e.:

$$\cos\theta_c = n. \tag{5.26}$$

Since θ_c is very small it is possible to use a Taylor expansion. Using (5.24) and (5.26) the expression of θ_c is given by:

$$\theta_c = \sqrt{\frac{\rho b}{\pi}}\lambda. \tag{5.27}$$

The corresponding critical wave vector is:

$$q_c = \frac{4\pi sin\theta_c}{\lambda} = 4\sqrt{\pi\rho b}. \tag{5.28}$$

5.3.3 Determination of Scattering Lengths and Optical Indices

In the case of pure materials, the knowledge of b and ρ fully characterises the material. In the case of crystalline solids of the type $A_x B_y$ for example, one can consider unit cells. The volume of the unit cell allows one to calculate a density ρ of unit cells per unit volume. The average scattering length b_{av} in the unit cell is simply given by $b_{av} = (xb_A + yb_B)/(x+y)$. The value ρb_{av} can be used to calculate the index of the material. The case of liquids and polymers is more complex since it is usually more difficult to define a "unit cell". Thus, the best method is to calibrate the index of each polymer or liquid that one wants to study. The Table 5.1 gives the scattering length, optical index $\delta = 1 - n$ and critical wave vector q_c (at 0.4 nm) for various elements and compounds.

Table 5.1. Scattering length, atomic density, optical index $\delta = 1 - n$ (at 0.4 nm), and critical wave-vector of some common materials. More exhaustive data can be found at *"www.neutron.anl.gov"*

Material	b_n (fm)	ρ $(10^{28} m^{-3})$	ρb $(10^{13} m^{-2})$	δ (10^{-6})	q_c (nm^{-1})
H (hydrogen)	-3.73				
D (deuterium)	6.67				
C (graphite)	6.64	11.3	75	19.1	0.19
C (diamond)	6.64	17.6	117	29.8	0.24
O	5.80				
Si	4.15	5.00	20.8	5.28	0.10
Ti	-3.44	5.66	-19.5	-5.0	-
Fe	9.45	8.50	80.3	20.45	0.20
Co	3.63	8.97	32.6	8.29	0.13
Ni	10.3	9.14	94.1	24.0	0.22
Cu	7.72	8.45	65.2	16.6	0.18
Ag	5.92	5.85	34.6	8.82	0.13
Au	7.63	5.90	45	11.5	0.15
H_2O	-1.68	3.35	-5.63	-1.43	-
D_2O	19.1	3.34	63.8	16.2	0.18
SiO_2	15.8	2.51	39.7	10.1	0.14
GaAs	13.9	2.21	30.7	7.82	0.12
Al_2O_3 (sapphire)	24.3	2.34	56.9	14.5	0.17
pyrex			42	10.7	0.14
polystyrene	23.2	0.61	14.2	3.6	0.084
polystyrene (deuterated)	106.5	0.61	65	16.5	0.18

5.3.4 Reflection on a Homogeneous Medium

As shown by equation (5.19) the problem of the reflection of a neutron beam on a non-magnetic medium can be treated exactly in the same way as the reflection of x-rays. Since the potential V is only z dependent, the Schrödinger equation (5.19) reduces to the 1 dimensional equation:

$$\frac{\hbar^2}{2m}\frac{d^2\psi_z}{dz^2} + [\mathcal{E}_z - V_z]\,\psi_z = 0, \tag{5.29}$$

with a wave function of the form $\psi = e^{i(k_{\text{in}\,x}x + k_{\text{in}\,y}y)}\psi_z$.
In the medium, the general solution is given by:

$$\psi_z = Ae^{ik_{\text{tr}\,z}z} + Be^{-ik_{\text{tr}\,z}z}. \tag{5.30}$$

The transmitted wave vector can be related to the incident wave vector using (5.21):

$$k^2 = \frac{2m}{\hbar^2}[\mathcal{E} - V] = k_{\text{in}}^2 - 4\pi\rho b. \tag{5.31}$$

At the interface we have to apply the continuity condition on ψ and $\nabla\psi$. In a way similar to the x-ray case, it is then possible to show that the parallel components of the incident and reflected waves are continuous [9]. The continuity of the parallel components allows us to write:

$$k_{\text{tr}\,z}^2 = k_{\text{in}\,z}^2 - 4\pi b\rho. \tag{5.32}$$

Considering equations (5.20) and (5.32), the problems of neutron and x-ray reflectivity are formally the same. It is possible to use the same formalism as the one developed in Chap. 3 for x-ray reflectivity.

In particular, it is possible to use the classical Fresnel formulae. The reflected and transmitted amplitudes are given by:

$$r = \frac{\sin\theta_{\text{in}} - n\sin\theta_{\text{tr}}}{\sin\theta_{\text{in}} + n\sin\theta_{\text{tr}}} \tag{5.33}$$

$$t = \frac{2\sin\theta_i}{\sin\theta_i + n\sin\theta_{\text{tr}}}. \tag{5.34}$$

In terms of scattering wave vector, the reflected intensity is given by:

$$R = \left|\frac{k_{0z} - k_{\text{tr}z}}{k_{0z} + k_{\text{tr}z}}\right|^2. \tag{5.35}$$

The Fig. 5.2 shows a typical curve calculated for a perfect surface.

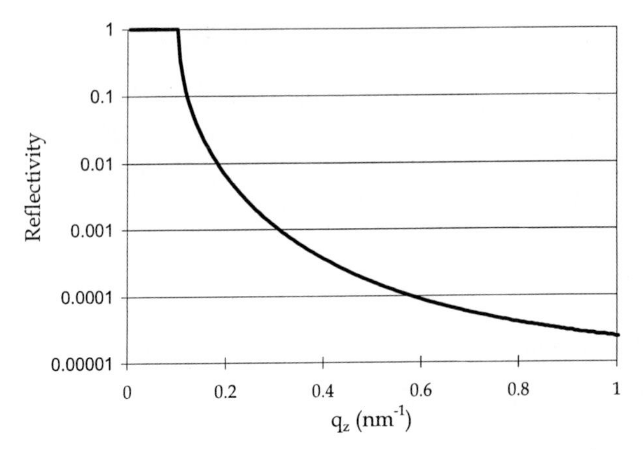

Fig. 5.2. Reflected intensity as a function of q_z for a silicon substrate (at $\lambda = 0.4nm$)

5.4 Neutron Reflectivity on Magnetic Systems

If the system is magnetic or if there is an external magnetic field on the sample, we need to take into account the spin of the neutron. It is not possible to use optical indices and it is always necessary to completely solve the Schrödinger equation [14–16]. In the case of homogeneous, infinite magnetic layers, the problem can be solved using a formalism very similar to the non-magnetic case developed in the previous part.

5.4.1 Interaction of the Neutron with an Infinite Homogeneous Layer

We consider a magnetic layer of thickness d, the neutron interacts with the different unpaired electrons. We perform a direct integration on the layer in order to obtain the potential V for the Schrödinger equation.

The Magnetic Interaction A first approach is to assume that the neutron is sensitive to the internal magnetic field in the magnetic layer. The interaction potential is then written:

$$-g_n \mu_n \sigma . \left[\mu_0 \left(1 - D\right) \mathbf{M} + \mathbf{B_0}\right] , \tag{5.36}$$

where $\mathbf{M}$ is the magnetisation of the layer, D is the demagnetising factor and $\mathbf{B}_0$ is the external magnetic field. In the case of an infinite magnetic thin film, $(1 - D)\,\mathbf{M}$ is equal to the in-plane component of the magnetisation $\mathbf{M}_{\parallel}$. It is possible to demonstrate this result but the calculations are somewhat lengthy. This is developed below for the interested reader but it can be skipped at the first reading.

The magnetic interaction can be written :

$$-g_n\mu_n\sigma.\mathbf{B} = -g_n\mu_n\sigma.\left(\nabla\times\left\{\frac{\mu_e\times\mathbf{R}}{|\mathbf{R}|^3}\right\} - \frac{e}{c}\frac{\mathbf{v_e}\times\mathbf{R}}{|\mathbf{R}|^3}\right), \tag{5.37}$$

or

$$-g_n\mu_n\left\{\sigma.\nabla\times\left(\frac{\mu_e\times\mathbf{R}}{|R|^3}\right) - \frac{e}{2m_ec}\left(\mathbf{p}_e.\frac{\sigma\times\mathbf{R}}{|\mathbf{R}|^3} + \frac{\sigma\times\mathbf{R}}{|\mathbf{R}|^3}.\mathbf{p}_e\right)\right\}, \tag{5.38}$$

with

$$\mathbf{p}_e = -i\hbar\nabla_e. \tag{5.39}$$

If we first consider only the spin dependent part of the interaction, we can write:

$$\begin{aligned}\nabla\times\left(\frac{\mu_e\times\mathbf{r}}{r^3}\right) &= -\nabla\times\left(\mu_e\times\nabla\left(\frac{1}{\mathbf{r}}\right)\right)\\ &= \frac{1}{2\pi^2}\int\frac{1}{q^2}(\mathbf{q}\times(\mu_e\times\mathbf{q}))\exp(i\mathbf{q}.\mathbf{r})d\mathbf{q}\end{aligned} \tag{5.40}$$

Integration on a Homogeneous Layer We suppose a constant atomic density ρ. We replace $\mathbf{r}$ by $\mathbf{r}+\mathbf{r}_0$. where $\mathbf{r}_0$ is the distance between the neutron and the center of the layer. $\mathbf{r}$ is the distance between the center and the volume $d\mathbf{r}$ in the layer. The spin dependent part of the interaction is:

$$2g_n\mu_n\mu_B\sigma\frac{1}{2\pi^2}\int\frac{1}{q^2}\int_V\rho(r)(\mathbf{q}\times\overline{s}(\mathbf{r})\times\mathbf{q})\exp(i\mathbf{q}.\mathbf{r}_0)\exp(i\mathbf{q}.\mathbf{r})d\mathbf{r}d\mathbf{q}. \tag{5.41}$$

where $\rho(\mathbf{r})$ is the density and $\overline{s}(\mathbf{r}) = \overline{s}$ is the mean value of the spin magnetisation in the volume $d\mathbf{r}$. The two first integrations over x and y give Dirac distributions:

$$4g_n\mu_n\mu_B\rho\sigma\int\frac{1}{q_z^2}\int_{-L/2}^{L/2}dr_z(\mathbf{q}_z\times\hat{s}\times\mathbf{q}_z)\exp(iq_z.r_{0z})\exp(iq_zr_z)dq_z. \tag{5.42}$$

The third integration gives:

$$8\pi g_n\mu_n\mu_B\rho\sigma.\overline{s}_{\|}\left[\theta(r_{0z}+L/2)-\theta(r_{0z}-L/2)\right]. \tag{5.43}$$

We can do the same calculation on the orbital part and we obtain :

$$\frac{2\pi\hbar^2}{m}p\sigma.\mathbf{M}_{\|}\rho\left[\theta(r_{0z}+L/2)-\theta(r_{0z}-L/2)\right], \tag{5.44}$$

with $p = 2.696fm$. $\mathbf{M}_{\|}$ is given in μ_B per atom and represents the in-plane component of the magnetisation and not necessarily the magnetisation perpendicular to the wave vector.

Conclusion From equation (5.44), we can deduce two very important points: it is only possible to measure the in-plane magnetisation and the magnetic interaction is zero out of the layer. These two properties are essential, the first is the main limitation to the use of neutrons for the study of magnetic thin films, the second is the justification of solving the Schrödinger equation in each layer, independently of the others. Thus the formalism developed for non-magnetic systems can be adapted to the magnetic case, however with some complications.

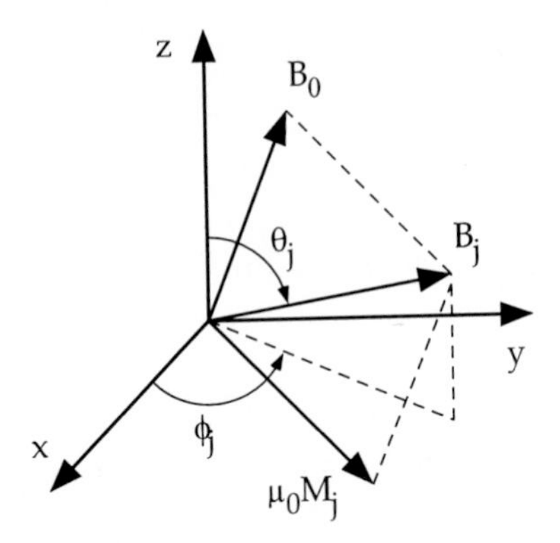

Fig. 5.3. Effective field $\mathbf{B}_{\mathrm{eff,j}}$ in the layer, sum of the external field $\mathbf{B}_0$ and of the magnetisation of the layer $\mathbf{M}_\mathrm{j}$. Definition of the spherical coordinates ϕ_j and θ_j

5.4.2 Solution of the Schrödinger Equation

The interaction potential for a layer j is given by :

$$V_j = \frac{2\pi\hbar^2}{m}\rho_j b_j - \frac{2\pi\hbar^2}{m}\rho_j p\, \sigma.\mathbf{M}_{j\parallel} - g_n\mu_n\, \sigma.\mathbf{B}_0. \tag{5.45}$$

We introduce an effective field $\mathbf{B}_{\mathrm{eff}}$ defined by:

$$\mathbf{B}_{\mathrm{eff}} = \mathbf{B}_0 + \mu_0 \mathbf{M}_{\parallel}. \tag{5.46}$$

If we introduce the spherical angles θ and ϕ to describe the effective field:

$$\begin{cases} B_{\mathrm{eff},jx} = B_{0x} + \mu_0 M_{jx} = B_{\mathrm{eff},j}\sin(\theta_\alpha)\cos(\varphi_\alpha) \\ B_{\mathrm{eff},jy} = B_{0y} + \mu_0 M_{jy} = B_{\mathrm{eff},j}\sin(\theta_\alpha)\sin(\varphi_\alpha) \\ B_{\mathrm{eff},jz} = B_{0z} = B_{\mathrm{eff},j}\cos(\theta_j) \end{cases}, \tag{5.47}$$

the interaction potential V can then be written in the compact form:

$$V_j = \frac{2\pi\hbar^2}{m}\rho_j b_j - g_n\mu_n\sigma.\mathbf{B}_{\mathrm{eff},\alpha}. \tag{5.48}$$

The Schrödinger equation:

$$-\frac{\hbar^2}{2m}\Delta\psi + V(r)\psi = \mathcal{E}\psi \tag{5.49}$$

is a vectorial equation in the basis of the two spin states $|+\rangle$ and $|-\rangle$. We have to solve the Schrödinger equation (5.49) with a wave function expressed with its two spinors components ψ_+ and ψ_-. It can be written explicitly as:

$$\begin{aligned} &-\frac{\hbar^2}{2m}\Delta\begin{pmatrix}\psi_+\\ \psi_-\end{pmatrix} + \left(\frac{2\pi\hbar^2}{m}\rho_j b_j\right)\begin{pmatrix}\psi_+\\ \psi_-\end{pmatrix} \\ &- g_n\mu_n\left(B_{\mathrm{eff},x}\sigma_x + B_{\mathrm{eff},y}\sigma_y + B_{\mathrm{eff},z}\sigma_z\right)\begin{pmatrix}\psi_+\\ \psi_-\end{pmatrix} = \mathcal{E}\begin{pmatrix}\psi_+\\ \psi_-\end{pmatrix}, \end{aligned} \tag{5.50}$$

where the Pauli spin operators σ are given by:

$$\sigma_x = \begin{pmatrix} 0 & 1 \\ 1 & 0 \end{pmatrix} \quad \sigma_y = \begin{pmatrix} 0 & -i \\ i & 0 \end{pmatrix} \quad \sigma_z = \begin{pmatrix} 1 & 0 \\ 0 & -1 \end{pmatrix}.$$

We obtain the two coupled equations involving the two spinor components ψ_+ and ψ_-:

$$\begin{aligned}
&\left(-\frac{\hbar^2}{2m}\nabla^2 + \frac{2\pi\hbar^2}{m} b_j\rho_j - g_n\mu_n B_{\mathrm{eff},z}\right)\psi_+(r) \\
&+ \left(-g_n\mu_n B_{\mathrm{eff},x} + i g_n\mu_n B_{\mathrm{eff},y}\right)\psi_-(r) = \mathcal{E}\psi_+(r) \\
&\left(-\frac{\hbar^2}{2m}\nabla^2 + \frac{2\pi\hbar^2}{m} b_j\rho_j + g_n\mu_n B_{\mathrm{eff},z}\right)\psi_-(r) \\
&+ \left(-g_n\mu_n B_{\mathrm{eff},x} - i g_n\mu_n B_{\mathrm{eff},y}\right)\psi_+(r) = \mathcal{E}\psi_-(r).
\end{aligned} \tag{5.51}$$

5.4.3 General Solution

We search solutions of the form : $\psi_+(\mathbf{r}) = a_+ \exp(i\mathbf{k}.\mathbf{r})$ and $\psi_-(\mathbf{r}) = a_- \exp(i\mathbf{k}.\mathbf{r})$ The possible values for $\mathbf{k}$ are given by the possibility of finding non zero solutions of the previous system (i.e. zero determinant condition). These conditions give four possible values for $\mathbf{k}$:

$$k_j^{\pm 2} = \frac{2m}{\hbar^2}\mathcal{E} - 4\pi\rho_j b_j \pm \frac{2m g_n\mu_n}{\hbar^2}|B_j|. \tag{5.52}$$

A general solution of the form:

$$\begin{aligned}
&\exp(ik_{\parallel,j} r_\parallel)\left(a\exp(ik^+_{z,j}z) + b\exp(-ik^+_{z,j}z)\right. \\
&\qquad \left.+ c\exp(ik^-_{z,j}z) + d\exp(-ik^-_{z,j}z)\right)
\end{aligned} \tag{5.53}$$

is not valid when there is an external magnetic field because $k^+_{\parallel,0} \neq k^-_{\parallel,0}$. At this point there are two ways of solving the problem. The first way is to solve the problem for each eigenstate $|+\rangle$ and $|-\rangle$. The second way consists in taking the general solution which is expressed as:

$$\begin{aligned}
&\exp(ik^{++}_{\parallel,j} r_\parallel)\left(a\exp(ik^{++}_{z,j}z) + b\exp(-ik^{++}_{z,j}z)\right) \\
&+ \exp(ik^{--}_{\parallel,j} r_\parallel)(c\exp(ik^{--}_{z,j}z) + d\exp(-ik^{--}_{z,j}z)) \\
&+ \exp(ik^{+-}_{\parallel,j} r_\parallel)\left(e\exp(ik^{+-}_{z,j}z) + f\exp(-ik^{+-}_{z,j}z)\right) \\
&+ \exp(ik^{-+}_{\parallel,j} r_\parallel).(g\exp(ik^{-+}_{z,j}z) + h\exp(-ik^{-+}_{z,j}z)),
\end{aligned} \tag{5.54}$$

with

$$\begin{cases} k_j^{++2} = k_j^{+2}, & k_{\parallel,j}^{++2} = k_{\parallel,0}^{+2} \\ k_j^{+-2} = k_j^{+2}, & k_{\parallel,j}^{+-2} = k_{\parallel,0}^{-2} \\ k_j^{-+2} = k_j^{-2}, & k_{\parallel,j}^{-+2} = k_{\parallel,0}^{+2} \\ k_j^{--2} = k_j^{-2}, & k_{\parallel,j}^{--2} = k_{\parallel,0}^{-2} \end{cases} . \tag{5.55}$$

The solution of equation (5.4.2) is the solution (5.54) rotated by the angles of the quantisation axis.

$$\begin{aligned} &\psi_j^+(r) \\ = \quad & \exp(ik_{\parallel,j}^{++} r_\parallel)(a_j^{++} \exp(i\,k_{z,j}^{++} z) + b_\alpha^{++} \exp(-i\,k_{z,\alpha z}^{++} z))\ \cos(\theta_\alpha/2) \\ + \quad & \exp(ik_{\parallel,j}^{+-} r_\parallel)(a_j^{+-} \exp(i\,k_{z,j}^{+-} z) + b_j^{+-} \exp(-i\,k_{z,jz}^{+-} z))\ \cos(\theta_j/2) \\ - \quad & \exp(ik_{\parallel,j}^{-+} r_\parallel)(a_j^{-+} \exp(i\,k_{z,j}^{-+} z) + b_j^{-+} \exp(-i\,k_{z,jz}^{-+} z))\ e^{-i\varphi_j} \sin(\theta_\alpha/2) \\ - \quad & \exp(ik_{\parallel,j}^{--} r_\parallel)(a_j^{--} \exp(i\,k_{z,j}^{--} z) + b_j^{-+} \exp(-i\,k_{z,jz}^{--} z))\ e^{-i\varphi_j} \sin(\theta_\alpha/2), \end{aligned} \tag{5.56}$$

and

$$\begin{aligned} &\psi_j^-(r) \\ = \quad & \exp(ik_{\parallel,j}^{++} r_\parallel)(a_j^{++} \exp(i\,k_{z,j}^{++} z) + b_j^{++} \exp(-i\,k_{z,jz}^{++} z))\ e^{i\varphi_j} \sin(\theta_j/2) \\ + \quad & \exp(ik_{\parallel,j}^{+-} r_\parallel)(a_j^{+-} \exp(i\,k_{z,j}^{+-} z) + b_j^{+-} \exp(-i\,k_{z,jz}^{+-} z))\ e^{i\varphi_j} \sin(\theta_j/2) \\ + \quad & \exp(ik_{\parallel,j}^{-+} r_\parallel)(a_j^{-+} \exp(i\,k_{z,j}^{-+} z) + b_j^{-+} \exp(-i\,k_{z,jz}^{-+} z))\ \cos(\theta_j/2) \\ + \quad & \exp(ik_{\parallel,j}^{--} r_\parallel)(a_j^{--} \exp(i\,k_{z,j}^{--} z) + b_j^{-+} \exp(-i\,k_{z,jz}^{--} z))\ \cos(\theta_j/2). \end{aligned} \tag{5.57}$$

5.4.4 Continuity Conditions and Matrices

The eight constants $a_j^{\pm}$ et $b_j^{\pm}$ are fixed by the continuity of ψ and $\nabla\psi$ at the interface. This gives exactly 8 equations. The reflection matrix $\mathcal{M}$ is then a 8×8 matrix but with 2 non zero 4×4 blocks (there are no cross terms between the group with "$k_{\parallel,0}^{+}$" components and the group with "$k_{\parallel,0}^{-}$" components in their wave vector. It is possible to split the problem in two calculations using 4×4 matrices. The continuity relations can be written:

$$\mathcal{D}_j(\mathbf{r}_j) \begin{pmatrix} a_j^{++} \\ b_j^{++} \\ a_j^{-+} \\ b_j^{-+} \\ a_j^{--} \\ b_j^{--} \\ a_j^{+-} \\ b_j^{+-} \end{pmatrix} = \mathcal{D}_{j+1}(\mathbf{r}_j) \begin{pmatrix} a_{j+1}^{++} \\ b_{j+1}^{++} \\ a_{j+1}^{-+} \\ b_{j+1}^{-+} \\ a_{j+1}^{--} \\ b_{j+1}^{--} \\ a_{j+1}^{+-} \\ b_{j+1}^{+-} \end{pmatrix}, \tag{5.58}$$

where the 8×8 matrix $\mathcal{D}_j(\mathbf{r}_j)$ is written :

$$\mathcal{D}_j(\mathbf{r}_j) = \begin{pmatrix} \mathcal{DA}_j & 0 \\ 0 & \mathcal{DB}_j \end{pmatrix}. \tag{5.59}$$

We give here the explicit expression of the two matrices $\mathcal{DA}$ and $\mathcal{DB}$ (we omit the index j and we write ($\theta' = \theta_j/2$)):

$$\mathcal{DA}_j = \left(\begin{array}{ll} e^{ik_\parallel^{++}r} \cos(\theta') e^{ik_z^{++}z} & e^{ik_\parallel^{++}r} \cos(\theta') e^{-ik_z^{++}z} \\ k_z^{++} e^{ik_\parallel^{++}r} \cos(\theta') e^{ik_z^{++}z} & -k_z^{++} e^{ik_\parallel^{++}r} \cos(\theta') e^{-ik_z^{++}z} \\ e^{ik_\parallel^{++}r} e^{i\varphi} \sin(\theta') e^{ik_z^{++}z} & e^{ik_\parallel^{++}r} e^{i\varphi} \sin(\theta') e^{-ik_z^{++}z} \\ k_z^{++} e^{ik_\parallel^{++}r} e^{i\varphi} \sin(\theta') e^{ik_z^{++}z} & -k_z^{++} e^{ik_\parallel^{++}r} e^{i\varphi} \sin(\theta') e^{-ik_z^{++}z} \end{array} \right. \tag{5.60}$$

$$\left. \begin{array}{ll} -e^{ik_\parallel^{-+}r} e^{-i\varphi} \sin(\theta') e^{ik_z^{-+}z} & -e^{ik_\parallel^{-+}r} e^{-i\varphi} \sin(\theta) e^{-ik_z^{-+}z} \\ -k_z^{-+} e^{ik_\parallel^{-+}r} e^{-i\varphi} \sin(\theta) e^{ik_z^{-+}z} & k_z^{-+} e^{ik_\parallel^{-+}r} e^{-i\varphi} \sin(\theta) e^{-ik_z^{-+}z} \\ e^{ik_\parallel^{-+}r} \cos(\theta') e^{ik_z^{-+}z} & e^{ik_\parallel^{-+}r} \cos(\theta') e^{-ik_z^{-+}z} \\ -k_z^{-+} e^{ik_\parallel^{-+}r} \cos(\theta') e^{ik_z^{-+}z} & k_z^{-+} e^{ik_\parallel^{-+}r} \cos(\theta') e^{-ik_z^{-+}z} \end{array} \right)$$

$$\mathcal{DB}_j = \left(\begin{array}{ll} -e^{ik_\parallel^{--}r} e^{-i\varphi} \sin(\theta') e^{ik_z^{--}z} & -e^{ik_\parallel^{--}r} e^{-i\varphi} \sin(\theta') e^{-ik_z^{--}z} \\ -k_z^{--} e^{ik_\parallel^{--}r} e^{-i\varphi} \sin(\theta') e^{ik_z^{--}z} & k_z^{--} e^{ik_\parallel^{--}r} e^{-i\varphi} \sin(\theta') e^{-ik_z^{--}z} \\ e^{ik_\parallel^{--}r} \cos(\theta') e^{ik_z^{--}z} & e^{ik_\parallel^{--}r} \cos(\theta') e^{-ik_z^{--}z} \\ k_z^{--} e^{ik_\parallel^{--}r} \cos(\theta') e^{ik_z^{--}z} & -k_z^{--} e^{ik_\parallel^{--}r} \cos(\theta') e^{-ik_z^{--}z} \end{array} \right. \tag{5.61}$$

$$\left. \begin{array}{ll} e^{ik_\parallel^{+-}r} \cos(\theta') e^{ik_z^{+-}z} & e^{ik_\parallel^{+-}r} \cos(\theta') e^{-ik_z^{+-}z} \\ k_z^{+-} e^{ik_\parallel^{+-}r} \cos(\theta') e^{ik_z^{+-}z} & -k_z^{+-} e^{ik_\parallel^{+-}r} \cos(\theta') e^{-ik_z^{+-}z} \\ e^{ik_\parallel^{+-}r} e^{i\varphi} \sin(\theta') e^{ik_z^{+-}z} & e^{ik_\parallel^{+-}r} e^{i\varphi} \sin(\theta') e^{-ik_z^{+-}z} \\ k_z^{+-} e^{ik_\parallel^{+-}r} e^{i\varphi} \sin(\theta') e^{ik_z^{+-}z} & -k_z^{+-} e^{ik_\parallel^{+-}r} e^{i\varphi} \sin(\theta') e^{-ik_z^{+-}z} \end{array} \right).$$

The reflection matrix $\mathcal{M}$ is defined by:

$$\mathcal{M} = \prod_{j=0}^{j=N} \mathcal{D}_j^{-1}(\mathbf{r}_j)\mathcal{D}_{j+1}(\mathbf{r}_j) = \begin{pmatrix} \mathcal{MA} & 0 \\ 0 & \mathcal{MB} \end{pmatrix}, \tag{5.62}$$

where

$$\mathcal{MA} = \prod_{j=0}^{N} \mathcal{DA}_j^{-1} \mathcal{DA}_{j+1} \quad \text{and} \quad \mathcal{MB} = \prod_{j=0}^{N} \mathcal{DB}_j^{-1} \mathcal{DB}_{j+1}. \tag{5.63}$$

We have the relation:

$$\begin{pmatrix} a_0^{++} \\ b_0^{++} \\ a_0^{-+} \\ b_0^{-+} \\ a_0^{--} \\ b_0^{--} \\ a_0^{+-} \\ b_0^{+-} \end{pmatrix} = \mathcal{M} \begin{pmatrix} a_s^{++} \\ b_s^{++} \\ a_s^{-+} \\ b_s^{-+} \\ a_s^{--} \\ b_s^{--} \\ a_s^{+-} \\ b_s^{+-} \end{pmatrix} . \tag{5.64}$$

In the case of incident "up" neutrons, equation (5.64) gives:

$$\begin{pmatrix} 1 \\ r_0^{++} \\ 0 \\ r_0^{-+} \\ 0 \\ 0 \\ 0 \\ 0 \end{pmatrix} = \mathcal{M} \begin{pmatrix} t_s^{++} \\ 0 \\ t_s^{-+} \\ 0 \\ 0 \\ 0 \\ 0 \\ 0 \end{pmatrix} . \tag{5.65}$$

For "down" neutrons we have :

$$\begin{pmatrix} 0 \\ 0 \\ 0 \\ 0 \\ 1 \\ r_0^{--} \\ 0 \\ r_0^{+-} \end{pmatrix} = \mathcal{M} \begin{pmatrix} 0 \\ 0 \\ 0 \\ 0 \\ t_s^{--} \\ 0 \\ t_s^{+-} \\ 0 \end{pmatrix} . \tag{5.66}$$

Let r_0^{++} , r_0^{-+} be the reflectivity amplitudes for a neutron "up" (resp. "down"), reflected "up" (resp. "down"). The corresponding transmission coefficients are given by t_s^{++} , t_s^{-+}. We deduce:

$$\begin{cases} r_0^{++} = \dfrac{MA_{21}MA_{33} - MA_{23}MA_{31}}{MA_{11}MA_{33} - MA_{13}MA_{31}} , \\ r_0^{-+} = \dfrac{MA_{41}MA_{33} - MA_{43}MA_{31}}{MA_{11}MA_{33} - MA_{13}MA_{31}} \end{cases} \tag{5.67}$$

and

$$\begin{cases} t_s^{++} = \dfrac{MA_{33}}{MA_{11}MA_{33} - MA_{13}MA_{31}} . \\ t_s^{-+} = \dfrac{-MA_{31}}{MA_{11}MA_{33} - MA_{13}MA_{31}} \end{cases} \tag{5.68}$$

We find similar relations for the 4 other coefficients. The reflected intensities are given by:

$$R^{++} \propto \left|r^{++}\right|^2 , \tag{5.69}$$

and

$$R^{-+} \propto \left|r^{-+}\right|^2 . \tag{5.70}$$

In the case of small external magnetic field, we have $R^{+-} \approx R^{-+}$.

5.4.5 Reflection on a Magnetic Dioptre

Let q_z be the (Oz) component of the scattering vector $\mathbf{q} = \mathbf{k}_r - \mathbf{k}_{in}$. We will consider the case of a reflection on a magnetic substrate.

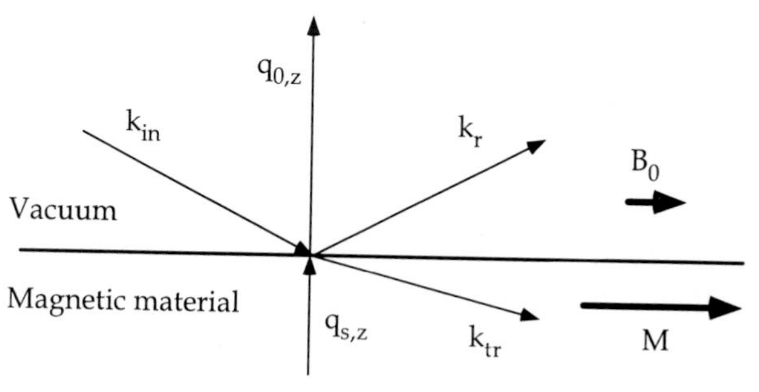

Fig. 5.4. Neutron beam incident on a magnetic substrate of magnetisation **M** in an applied field $\mathbf{B}_0$

To simplify the problem, we assume that the applied magnetic field $\mathbf{B}_0$ is small so that $q_{0z}^{+} \approx q_{0z}^{-} = q_{0z}$. The component of the $\mathbf{q}$ vector in the magnetic medium is given by (see 5.32):[1]

$$q_{sz}^{\pm} \approx \sqrt{q_{0z}^2 - 16\pi\rho(b_n \pm b_m)} . \tag{5.71}$$

We will assume that the external field $\mathbf{B}_0$ and the magnetisation **M** lie in the layer plane. This corresponds to $\theta = 90°$ (see Fig. 5.3). Let ϕ be the angle between $\mathbf{B}_0$ and **M**. The expressions of the reflection coefficients deduced from the expressions of the $\mathcal{M}$ matrices are given by:

$$r^{++} = \frac{\cos^2\frac{\phi}{2}\left(q_{0z} - q_{sz}^{+}\right)\left(q_{0z} + q_{sz}^{-}\right) + \sin^2\frac{\phi}{2}\left(q_{0z} - q_{sz}^{-}\right)\left(q_{0z} + q_{sz}^{+}\right)}{\cos^2\frac{\phi}{2}\left(q_{0z} + q_{sz}^{+}\right)\left(q_{0z} + q_{sz}^{-}\right) + \sin^2\frac{\phi}{2}\left(q_{0z} + q_{sz}^{-}\right)\left(q_{0z} + q_{sz}^{+}\right)} \tag{5.72}$$

[1] we remind that "$q_z = 2k_z$", the scattering wave vector in the substrate is equal to twice the projection of the incident wave vector on the (Oz) axis.

$$r^{+-} = \frac{2\, q_{0z} \cos\frac{\phi}{2}\, \sin\frac{\phi}{2}\, \left(q_{sz}^{+} - q_{sz}^{-}\right)}{\cos^2\frac{\phi}{2}\, \left(q_{0z} + q_{sz}^{+}\right)\left(q_{0z} + q_{sz}^{-}\right) + \sin^2\frac{\phi}{2}\, \left(q_{0z} + q_{sz}^{-}\right)\left(q_{0z} + q_{sz}^{+}\right)} . \tag{5.73}$$

The measured intensities are given by:

$$R^{++} = \left|r^{++}\right|^2 \quad et \quad R^{+-} = \left|r^{+-}\right|^2 . \tag{5.74}$$

Case of a Non-Magnetic Substrate In this case, corresponding to a zero magnetisation ($b_m = 0$), the scattering vectors q_{sz}^{+} and q_{sz}^{-} are equal (eq. 5.71). The reflection coefficients simplify and can be written in the form of classical Fresnel coefficients:

$$r^{++} = \frac{q_{0z} - q_{sz}}{q_{0z} + q_{sz}} \quad and \quad r^{+-} = 0. \tag{5.75}$$

The reflected intensity is given by:

$$R^{++} = \left|\frac{q_{0z} - q_{sz}}{q_{0z} + q_{sz}}\right|^2 = \left|\frac{q_{0z} - \sqrt{q_{0z}^2 - q_c^2}}{q_{0z} + \sqrt{q_{0z}^2 - q_c^2}}\right|^2 , \tag{5.76}$$

where the critical wave vector q_c is equal to $\sqrt{16\pi\rho b_n}$. When $q_{0z} < q_c$, q_{sz} is a pure imaginary number and one finds a reflected intensity equal to 1. When q_{0z} is very large, one can show that the intensity decreases as $1/q_{0z}^4$.

Case of a Magnetic Substrate in a Magnetic Field $\mathbf{B_0}$ Aligned with the Magnetisation M ($\phi = 0$) In this simple case, the expressions of the reflection coefficients simplify and can be written as:

$$r^{++} = \frac{q_{0z} - q_{sz}^{+}}{q_{0z} + q_{sz}^{+}}, \quad r^{--} = \frac{q_{0z} - q_{sz}^{-}}{q_{0z} + q_{sz}^{-}} \quad and \quad r^{+-} = 0. \tag{5.77}$$

These expressions still correspond to Fresnel reflectivities. The only modification introduced by the magnetism is a difference in the critical angle. The critical angles for the reflectivity curves "up-up" and "down-down" are given by:

$$q_c^{\pm} \approx \sqrt{16\pi\rho\,(b_n \pm b_m)}. \tag{5.78}$$

The spin-flip signal ($R^{\pm}$ and $R^{\mp}$) is zero.
N.B.: the coefficients r^{++} and r^{--} can be deduced one from the other by a 180° ϕ rotation.
The non spin-flip signals are plotted in solid lines on Fig. 5.5. We find classical shapes for the reflectivity curves, with a total reflectivity plateau followed by a sharp decrease. The main difference between the "up-up" and "down-down" curves is the extension of the total reflectivity plateau.

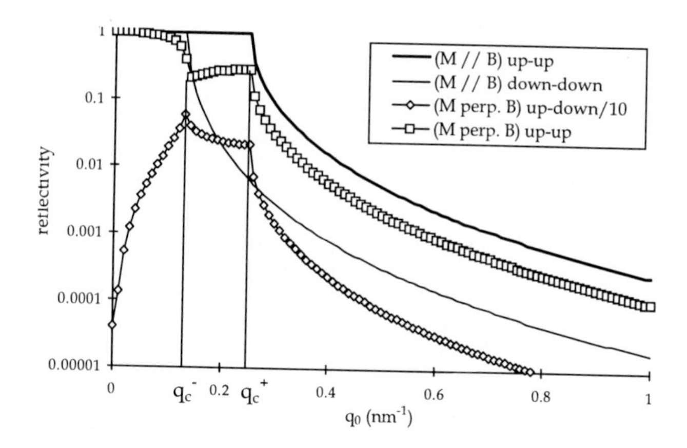

Fig. 5.5. Reflectivity curves in the case of a magnetisation parallel and perpendicular to the magnetic field $\mathbf{B}_0$. When the magnetisation is parallel to the field, the non spin flip curves "up-up" and "down-down" are distinct (solid lines), the spin-flip signal is zero. When the magnetisation is perpendicular to the field $\mathbf{B}_0$, the spin-flip curves superimpose (squares), a very large spin-flip signal appears (losanges) (the spin-flip signal has been divided by a factor 10 for clarity

Case of a Magnetic Field Perpendicular to the Substrate Magnetisation ($\phi = 90°$) In this case the reflection coefficients become:

$$r = \frac{\left(q_{0z} - q_{sz}^{+}\right)\left(q_{0z} + q_{sz}^{-}\right) + \left(q_{0z} - q_{sz}^{-}\right)\left(q_{0z} + q_{sz}^{+}\right)}{2\left(q_{0z} + q_{sz}^{+}\right)\left(q_{0z} + q_{sz}^{-}\right)} = \frac{1}{2}\left(r^{++} + r^{--}\right) \tag{5.79}$$

$$r^{+-} = \frac{q_{0z}\left(q_{sz}^{+} - q_{sz}^{-}\right)}{2\left(q_{0z} + q_{sz}^{+}\right)\left(q_{0z} + q_{sz}^{-}\right)} . \tag{5.80}$$

The reflected intensities are given by:

$$R^{++} = R^{--} = |r|^2 = \frac{1}{4}\left|r^{++}\right|^2 + \frac{1}{4}\left|r^{--}\right|^2 + \frac{1}{2}Re\left(r^{++} \times r^{--}\right) . \tag{5.81}$$

One can notice that the up-up and down-down intensities are the sum of three terms. The first two correspond to the intensities of the non spin-flip signals in the case of a magnetisation aligned with the external magnetic field ; they are weighted by a 1/4 coefficient. These terms introduce two discontinuities at the positions q_c^+ et q_c^- in the reflectivity curve (see figure 5.5, white square curve). To these two terms, an " interference " term adds $\frac{1}{2}Re\left(r^{++} \times r^{--}\right)$

whose analytical expression is not simple. Its variations are plotted on Fig. 5.6. For $q_{0z} = 0$, this term is equal to 1/2 and the intensity is totally reflected. Its value decreases as soon as q_{0z} increases and becomes negative around q_c^-. It becomes positive again around q_c^+, then decreases very quickly. However, this contribution does not modify qualitatively the form of the non spin-flip curve except that there is no total reflectivity plateau.

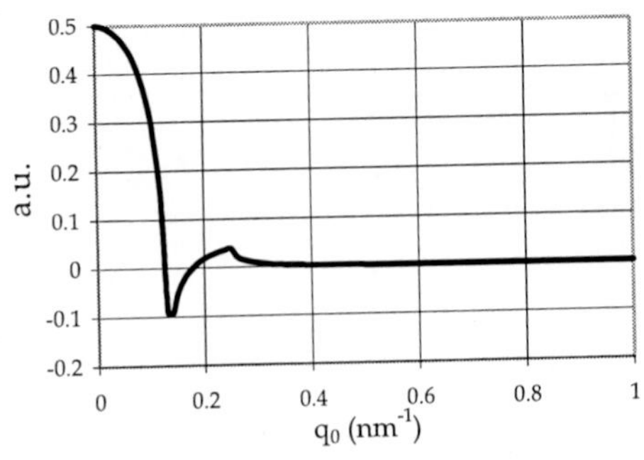

Fig. 5.6. Contribution of the interference term $Re(r^{++} \times r^{--})$ between the r^{++} and r^{--} amplitudes in the non spin-flip intensities for a reflection on a substrate whose magnetisation is perpendicular to the applied magnetic field $\mathbf{B}_0$

The spin-flip intensity is given by:

$$R^{+-} = \left|r^{+-}\right|^2 = \frac{1}{4}\left|\frac{q_{0z}\left(q_{sz}^+ - q_{sz}^-\right)}{\left(q_{0z} + q_{sz}^+\right)\left(q_{0z} + q_{sz}^-\right)}\right|^2 . \tag{5.82}$$

The characteristic form of the spin-flip signal (see figure 5.4) is given by the term $\left|q_{sz}^+ - q_{sz}^-\right|^2$. The variations of this term are plotted on the figure 5.7 (thick lines). Two successive regime changes appear at the points q_c^- and q_c^+. They correspond to the points where q_{sz}^- and q_{sz}^+ successively change from pure imaginary to real values. This signal is slightly modulated by the factor q_{0z} which gives a linear increase. The factor $1/\left|\left(q_{0z} + q_{sz}^+\right)\left(q_{0z} + q_{sz}^-\right)\right|^2$ gives a very fast decrease at large q_z. Its variations are plotted on Fig. 5.7 (thin lines). In the case where the magnetisation is not fully perpendicular to the applied magnetic field, the three terms in the R^{++} intensity are weighted by $\cos^4\frac{\phi}{2}$, $\sin^4\frac{\phi}{2}$ and $2\cos^2\frac{\phi}{2}\sin^2\frac{\phi}{2}$ factors, ϕ being the angle between the field and the magnetisation. In the case of a magnetic layer deposited on a non-magnetic substrate, the above considerations are not qualitatively modified. The main difference is that Kiessig fringes appear after the plateau of total reflection.

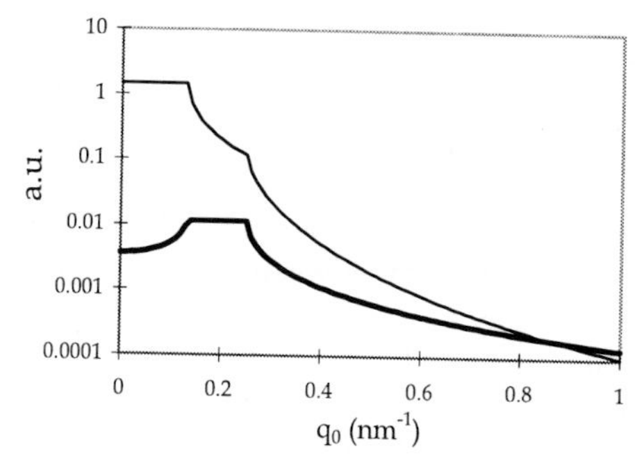

Fig. 5.7. Variations of two factors of the spin-flip intensity: (bold lines) factor $\left|q_{sz}^{+} - q_{sz}^{-}\right|^2$; (thin lines) factor $1/\left|\left(q_{0z} + q_{sz}^{+}\right)\left(q_{0z} + q_{sz}^{-}\right)\right|^2$

5.5 Non Perfect Layers, Practical Problems and Experimental Limits

5.5.1 Interface Roughness

Most of the studied systems show imperfect interfaces depending on the deposition process of the layer. We will consider three roughness scales: interface roughness, atomic interdiffusion and homogeneity of the layer thickness. Let ξ represent the characteristic lateral length-scale for the roughness. A perfect knowledge of the surface would correspond to the knowledge of $z(x, y)$ for all in-plane lengthscales. The treatment of the roughness is very similar to that described in Chaps. 2 and 3. According to the resolution of a typical neutron reflectivity experiment, one can (somewhat arbitrarily) distinguish three typical types of roughness which have different origins.

- *Interdiffusion of the species between two successive layers.* This happens during the deposition of a top layer which is miscible with the bottom material. This process is strongly temperature dependent. It corresponds to a typical lengthscale of $\xi < 0.5 \mu m$.
- *A roughness induced by rough edges on the substrate or by grains in the case of two successive layers.* This roughness usually occurs during thin film growth. It is also the type of roughness which is difficult to take into account in models. It corresponds to $1 \mu m < \xi < 100 \mu m$.
- *Flatness of the sample.* Depending on the deposition process, the atomic flux may have an angular dependence which can lead to an uneven thickness over the sample surface. It corresponds to $\xi > 100 \mu m$.

These three roughness scales can be modelled in three different ways to account for their effects on the measured reflectivity curves. They induce very different effects on the experimental signals. One has to keep in mind the following limitation: if the lateral fluctuations are not small compared to the layers thicknesses the following treatments are inadequate.

Thickness Inhomogeneity of the Sample Thickness variation in a thin film sample (usually between the middle and the sample edges) is a "large" lateral scale problem (a few mm). The experimental measured curve can be treated as the superposition of reflectivity curves calculated for the thicknesses spectrum weighted by the corresponding area. The resulting effect is a blurring of the coherent oscillations for large $\mathbf{q}$.
N.B.: Since the Kiessig fringes period is inversely proportionnal to the wavelength of the incident beam, a thickness fluctuation (which reflects in the Kiessig oscillations period) can be taken into account as an incident wavelength spread $\delta\lambda$. Figure 5.8 (thin line, $\delta\lambda = 10\%$) illustrates the effect of a wavelength spread; it also corresponds to what would be observed for a 10% sample thickness fluctuation.

Roughness and Interdiffusion Specular reflectivity cannot distinguish between these two type of roughness. The measurement of the coherent scattering length density ρb probes a large planar scale compared to the size of the roughness: for a given z depth, one measures a mean value of ρb averaged over a large surface.

First solution: Névot-Croce factors If one assumes a flat distribution of x, the two types of interface can be treated by a single model where the step function is replaced by the following error function:

$$erf\left(\frac{z - z_j}{\sigma_j}\right) = \frac{2}{\sqrt{\pi}} \int_0^{(z-z_j)/\sigma_j} e^{-t^2} dt. \tag{5.83}$$

This curves shows an inflexion point at z_j. The value σ_j is given by the inverse of the curve slope at z_j. The thickness is given by $2\sigma_j$. The effect of a smooth interface surface described by 5.83 is to multiply the reflectivity $\mathcal{R}$ of a perfect flat interface by a Debye-Waller (or better Névot-Croce, see Chap. 3, Appendix 3.A) factor [17]:

$$\exp(-2k_{z,j}k_{z,j+1}\sigma_j^2) \tag{5.84}$$

In the case of a stack of multilayers each having a specific roughness, the Névot-Croce factor is applied to each transfer matrix. Unfortunately, this cannot be applied in the magnetic case, the formalism preventing an easy calculation of the reflectivity R at each interface. However one can introduce a global factor and then apply this factor to each diagonal elements at each interface. In practice this works quite well except in the case of rather strong magnetic roughness like domains. The main effect of this factor is to decrease the reflectivity at high $\mathbf{q}$.

Second solution: discretisation This technique is efficient to model atomic interdiffusion. The interface is replaced by a finite number of discrete layers describing the concentration index. Either an error function or a linear function profile can be used. For real systems of thin solid films, one layer with an average ρb usually works well.

Intermediate Roughness In the case of the intermediate roughness, the previous methods are not completely satisfactory. Actually, this type of roughness not only decreases the specular reflectivity but also creates a non specular diffuse background which can modify the results. In this case, the diffuse scattering should be measured and the specular reflectivity should be corrected accordingly. This treatment is quite complex and will not be detailed here.

Magnetic Roughness This problem is very complex. A typical example where magnetic roughness appears is the case of a demagnetised sample. In this case each domain has an effective scattering length very different from its neighbour. This appears for neutrons as a giant roughness. There is no simple theoretical way of taking this into account yet. Non specular magnetic measurements can in some cases give the average size of the magnetic domains.

5.5.2 Angular Resolution

The different expressions given above are valid for a perfect incident beam. For the fit of experimental data, it is important to have a good knowledge of the beam divergence and homogeneity. The beam angular divergence and wavelength dispersion must be taken into account in the simulations. The divergence of the incident beam, $\delta\theta$, is usually determined by two slits if the beam is smaller than the effective width of the sample seen by the neutron beam, or by the first slit and the sample itself if the sample is small enough to be totally illuminated by the neutron beam. Usually, $\delta\theta$ is fixed during the experiment. We have then to convolute the calculated reflectivity with a function which is the experimental shape of the beam divergence. However a square function gives in most cases a good approximation of that function. In the case of curved samples, $\delta\theta$ can be slightly adjusted during the treatment. $\delta\theta$ has two effects: a decrease of the amplitude of the oscillations and a rounding of the discontinuity at the critical angle. Figure 5.8 gives an example of this effect. Wavelength dispersion is strongly dependent on the monochromator or on the time resolution in the case of time of flight spectrometers (see below). The effect of that dispersion is different from an angular divergence : the oscillations disappear at high angles (see Fig. 5.8). We remind that if a sample has a non homogeneous thickness, the effect is very similar (see above). A wavelength dispersion can be used to model thickness variations over the sample surface.

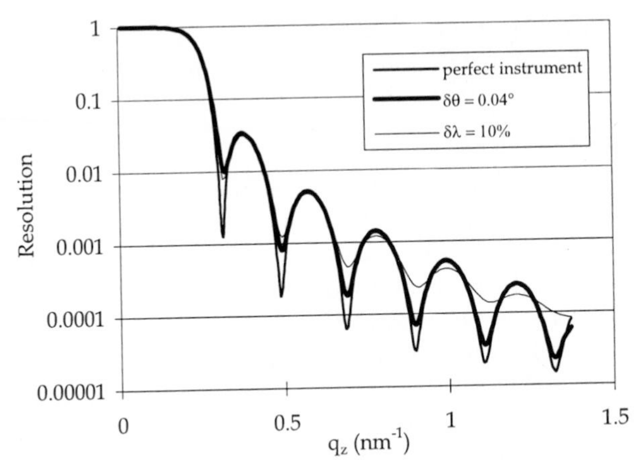

Fig. 5.8. Effect of $\delta\theta$ and $\delta\lambda$. Comparison between a perfect instrument, an instrumental $\delta\theta$, and a $\delta\lambda$ for a measurement on a single 30 nm thick layer on a substrate

5.5.3 Analysis of Experimental Data

Reflectivity curves cannot be directly inverted. For a non magnetic system, it is even possible to build a family of profiles which give the same reflectivity curve. This is due to the fact that we measure only the intensity and loose the phase of the reflectivity [18]. For magnetic systems, the problem of the signal phase is less critical. However, the main source of uncertainty on the result is in general due to the lack of intensity at high angles. The analysis of experimental data is done by adjusting the different parameters involved in the problem until a good fit is obtained. In the case of magnetic systems, we usually know rather well the composition of the different layers. We have then to adjust the roughness, the thicknesses and the magnetic moments magnitude. It is in general very useful to have some external information like x-rays reflectometry and magnetic hysteresis curves.

5.6 The Spectrometers

5.6.1 Introduction

The spectrometers can be divided in two different groups: time of flight reflectometers like EROS at the Laboratoire Léon Brillouin (LLB), CRISP and SURF at ISIS, and monochromatic reflectometers like PADA at LLB and ADAM at the Institut Laue Langevin (ILL). Time of flight spectrometers are necessary for reflectometry studies on liquids.

5.6.2 Time of Flight Reflectometers

The time of flight technique consists in sending a pulsed white beam on the sample. Since the speed of the neutron varies as the inverse of the wavelength, the latter is directly related to the time taken by the neutron to travel from the pulsed source to the detector (over the distance L) by :

$$\lambda = \frac{h}{mL}t. \tag{5.85}$$

This relation is also written as :

$$\lambda(nm) = \frac{t(\mu s)}{2527L(m)}. \tag{5.86}$$

On a spallation source, the neutron beam is "naturally" pulsed and the time of flight technique is used. On a reactor, pulsed neutrons are produced by a chopper.

For a reflectivity measurement, the angle is fixed and the reflectivity curve is obtained by measuring the reflectivity signal for each wavelength of the available spectrum, each wavelength corresponding to a different scattering wave-vector magnitude. Sometimes, it is necessary to use several angles because the q_z range is not large enough. An example of time of flight spectrometer is presented on Fig. 5.9.

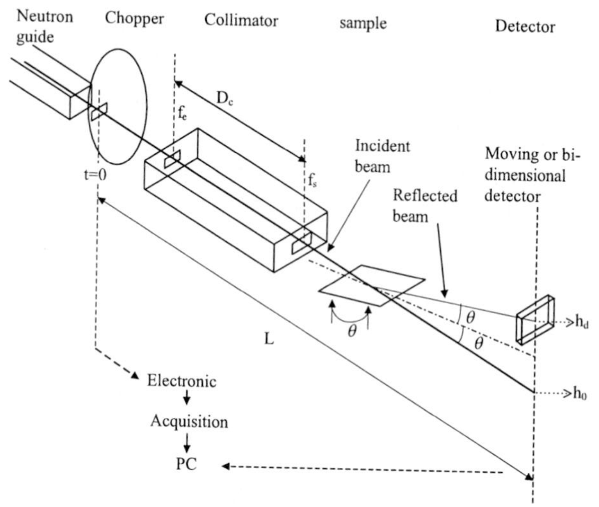

Fig. 5.9. Description of the time of flight reflectometer EROS at the LLB [19]

5.6.3 Monochromatic Reflectometers

Monochromatic reflectometers are basically two axes spectrometers. The wavelength is fixed (0.4 nm for PADA) and the reflectivity curve is obtained by changing the incident angle θ. In this case, the sample is usually vertical. On this type of reflectometers it is easy to put a polariser and an analyser in order to select the spin states of the incident and reflected neutrons.The flippers can be of Mezei type (2 orthogonal coils) [20]. They allow to flip the neutron spin state from "up" to "down". An example of two-axis spectrometer is presented on Fig. 5.10.

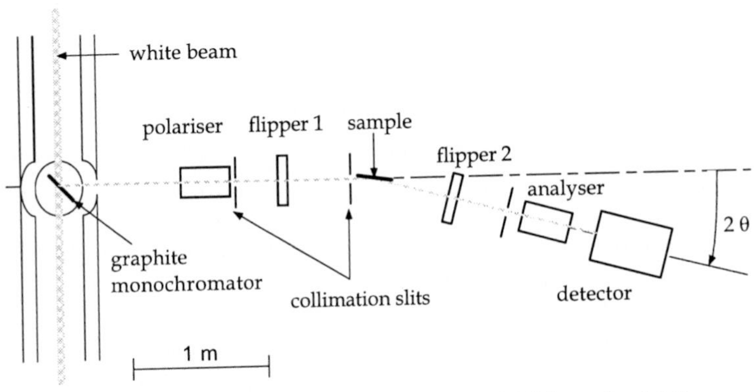

Fig. 5.10. The two axis reflectometer PADA at the LLB

5.7 Polymer Examples

Reflectometry is widely used for soft matter studies due to the large penetration depth of the neutron beam and the large contrast between deuterated ($b_D = 6.67 fm$) and protonated ($b_H = -3.7 fm$) systems [21,22]. We shall illustrate here the use of selective deuteration to study the polymer interdiffusion. By spin coating, it is possible to deposit polymers layers on glass or silicon with a roughness below 1 nm. It is then possible to deposit a second layer on the first one. If one of the layers is deuterated, it is possible to study the interdiffusion as a function of time and annealing temperature. The diffusion will appear as a smearing of the interface between the two layers and thus a decrease of the Kiessig fringes amplitude.

Interdiffusion Between Diblock Copolymer Layers Diblock copolymers are made of two chains A and B linked together $(A-B)$. These systems present a large variety of interesting properties. For example, if A and B are not miscible, they can form self-organised multilayers of a fixed thickness parallel to the surface where the solution is deposited. The observed structure is of the type $(substrate; A-B; B-A; A-B; B-A...)$. We have studied diblock copolymers of the type (polystyrene - polybutylmetacrylate: PS-PBMA). The initial system consisted in a layer of partially deuterated

PS-PBMA copolymer deposited on a trilayer of totally hydrogenated copolymers. The reflectivity of this system is shown on Fig. 5.11. The numerical fit shows a large index at the top of the system corresponding to the deuterated copolymer. The system has then been annealed for 12 hours at 400K and then remeasured (Fig.5.11). On this reflectivity curve, one can observe a clear "Bragg" peak at the position $q = 0.1\,nm^{-1}$. This indicates the diffusion of the deuterated polymer to the inner layers. Since the diblock copolymers are ordered in multilayers, a periodic variation of the index appears (see insert on Fig. 5.11). Many other examples will be found in Chap. 10 which is entirely devoted to the discussion of polymer studies.

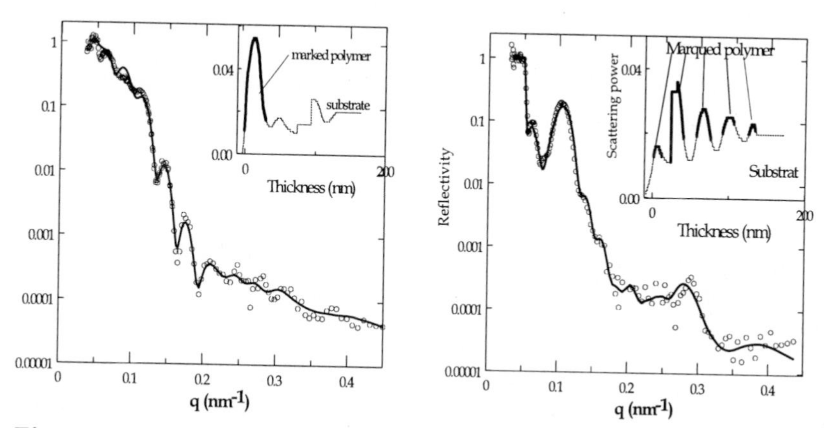

Fig. 5.11. Left: Reflectivity of a quadrilayer consisting in a partially deuterated PS-PBMA copolymer layer deposited on a trilayer of totally hydrogenated polymer (measured on the EROS reflectometer at the Laboratoire Léon Brillouin.) Right: Reflectivity of the quadrilayer after annealing for 1 hour at $115°C$

5.8 Examples on Magnetic Systems

In this part, we shall give some examples in order to highlight the information that can be obtained by polarised neutron reflectometry. All the experiments shown here have been performed on the reflectometer PADA.

5.8.1 Absolute Measurement of a Magnetic Moment

Neutron reflectometry can be used to measure absolute moments (in μ_B per atom). The obtained value is independent of the layer thickness and of the surface of the sample. As an example, Fig. 5.12 shows a curve obtained on a NiFe single layer. By fitting the curves, we obtain the ratio between Ni and Fe with an error of about 2% and the absolute moment with a precision of about $0.02\mu_B$. That measurement took only 15 minutes on a 1 cm^2 sample.

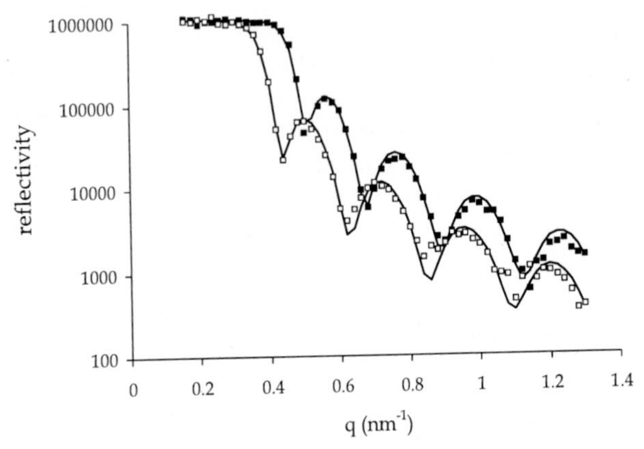

Fig. 5.12. reflectivity curve on a 25 nm thick NiFe layer. Black squares are the R^{++} intensity and white squares R^{--} intensity

5.8.2 Bragg Peaks of Multilayers

Periodic Multilayers In the case of periodic multilayers, we can observe Bragg peaks corresponding to the period of the multilayer. In the case of antiferromagnetic coupling or variable angle coupling, it is possible to obtain directly a mean angle between the different magnetic layers. With polarised neutrons, it is possible to measure very rapidly a precise value of the average moments. If high order Bragg peaks are observed, a good estimate of the chemical and magnetic interface can be obtained. In the literature, there is a large amount of results on magnetic multilayers [18,23,24].

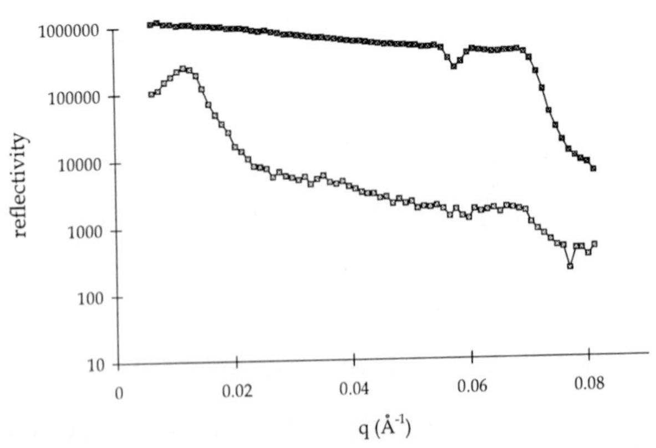

Fig. 5.13. Example of a polarizing mirror. The two curves correspond respectively to the "up" (black squares) and "down" (white squares) state of incident neutron spin

Supermirrors [25] For technical purposes it is interesting to build systems exhibiting an articially large optical index. One can can build such a structure by stacking periodic multilayers with an almost continuous variation of the period. In such a system, if the periodicity range is well choosen, a large number of Bragg peaks follow the total reflectivity plateau. Since the periodicity of the multilayer is varying continuously, all these Bragg peaks add constructively. Using this technique it is possible to enhance the length of the total reflection plateau by a factor 3 to 4. Such mirrors are now widely used for neutron guides and for polarisation devices. Figure 5.13 gives an example of a polarising mirror.

5.8.3 Measurement of the In-Plane and Out-Of-Plane Rotation of Moments. Measurement of the Moment Variation in a Single Layer

This kind of measurement is perhaps the most important information given by polarised neutron reflectivity (PNR) for magnetic thin films. We shall give here two examples of determination of in-depth magnetic profiles.

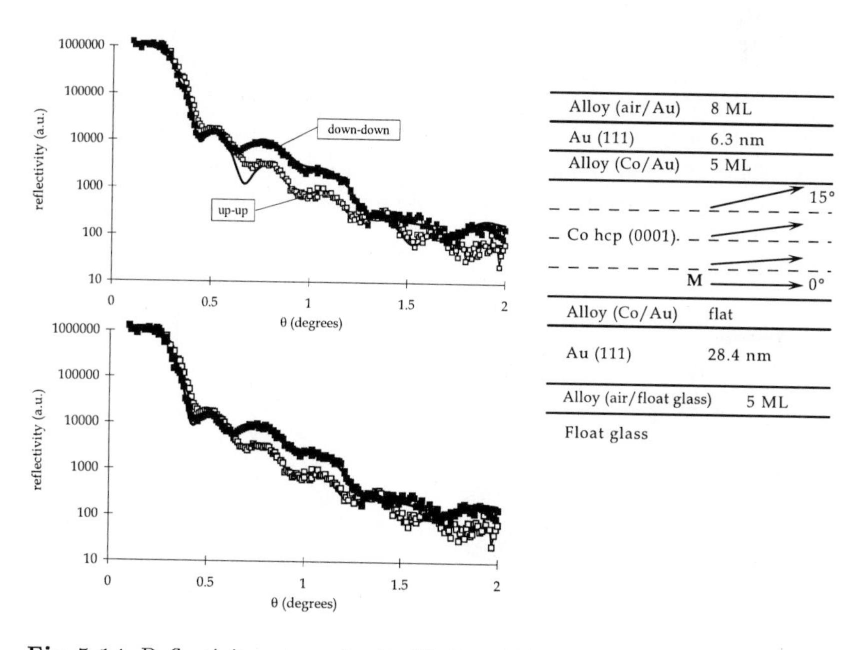

Fig. 5.14. Reflectivity curve of a Au/Co(3nm)/Au trilayer system. Empty squares: up-up reflectivity; filled squares: down-down reflectivity; lines: best fits. Top left: fit assuming a uniformily aligned magnetisation through the layer. Bottom left: fit with a model allowing magnetisation rotation; this model provides a better fit than the model using a uniform magnetisation. Right: Thicknesses and moment directions giving the best fit parameters

Out-Of-Plane Rotation of the Moments in Au/Co/Au [26]. In the case of very thin cobalt layers, thinner than 2.5 nm, the magnetisation is perpendicular to the surface layer. We have studied a 3 nm thick Co layer sandwiched between two Au layers. Magneto-optic Kerr effect measurements have shown that the moment is mainly in-plane but with a small out of plane contribution. In order to understand the magnetic behaviour of such a layer, we have fitted the PNR curves with two different models: for the first one, we have considered that the whole layer was uniformly aligned. In the second one, we have allowed a rotation in the layer. The Fig. 5.14 gives the reflectivity curves and the corresponding model.

Rotation in Strained Nickel Layers In single magnetic thin films, the microstructure can be such that the magnetoelastic (ME) properties vary throughout the sample: a gradient of the ME coefficient $B(z)$ can appear, related to surface relaxation effects. It can be written in a form similar to surface anisotropy constants [27]:

$$B(z) = B_{bulk} + \frac{B_{surf}}{(z - z_0)} \tag{5.87}$$

where z is the depth in the thin film and z_0 an adjustable parameter. When a mechanical strain is applied on a magnetic thin film, the magnetisation tends to rotate either along or perpendicular to the applied strain. A ME coefficient gradient will then lead to a gradient of magnetisation rotation through the thin film. This has been measured on single nickel layers as illustrated on Fig. 5.15. The numerical fit shows that the average rotation under a 0.03% deformation is 30° but there is a 15° gradient of rotation between the surfaces and the bulk of the material [28].

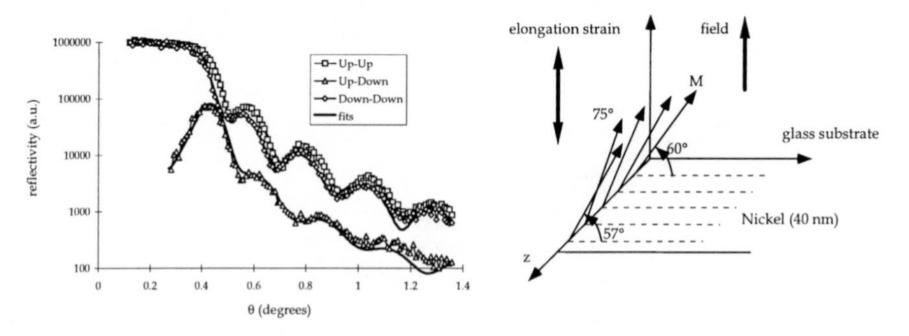

Fig. 5.15. Left: Reflectivity curves on a strained nickel layer (thickness 40 nm) for each state of neutron spin. The deformation applied to the substrate is 0.03 %. Right: Diagram showing the magnetisation rotation gradient in the strained 40 nm Ni layer deduced from the neutron fit

5.8.4 Selective Hysteresis Loops

A complete set of reflectivity curves (R^{++}, R^{--} and R^{+-}) takes about 12 hours to be performed. If we want to follow the magnetisation of different layers as a function of field or temperature, the total experiment would be far too long compared to the time usually allocated on a neutron reflectometer (typically one week per year). So the idea is the following [29]: from the fits of the reflectivity curves in the saturated state we know the different thicknesses and magnetic states of the multilayer system. We are then able to calculate the reflectivity curves for different module and orientation of the magnetic layers. It is then not necessary to perform complete reflectivity curves for each value of the magnetic field, but we can measure only the reflectivity for $(n + 1)$ well chosen θ values where n is the number of different magnetic layers. Comparison of the experimental values obtained and calculations using the parameters obtained from the saturated state allows us to rebuild the magnetic evolution as a function of the applied external field.

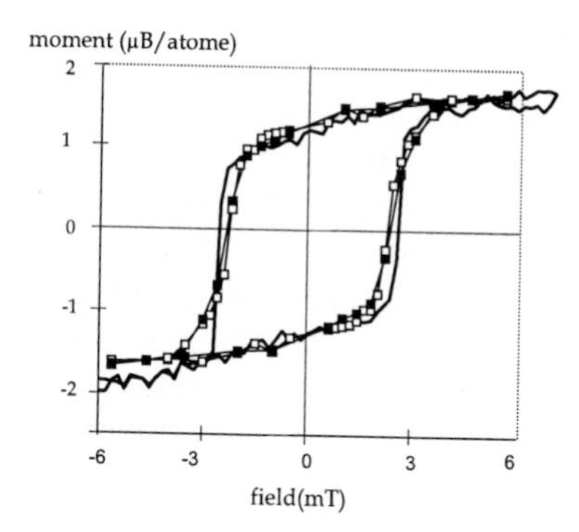

Fig. 5.16. Hysteresis loops measured for a single cobalt layer : by MOKE (continuous curve) and from the reflectivity measurements at only one angle (0.3° white squares, 0.4° black squares)

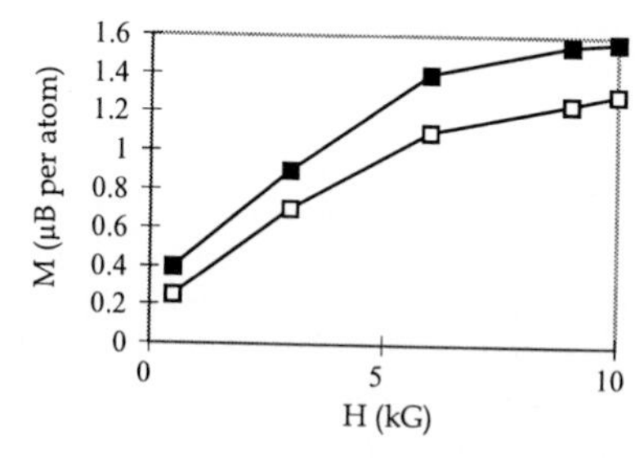

Fig. 5.17. Magnetisation of each magnetic layer of a /Pt(98Å) /Co(7.6Å) /Pt(33Å) /Co(3.8Å) /Pt(32Å) thin film. The white squares correspond to the thinnest layer

Figure 5.16 gives the example of an hysteresis loop obtained for a single magnetic layer. Figure 5.17 corresponds to a more complicated system: substrate /Pt(98Å) /Co(7.6Å) /Pt(33Å) /Co(3.8Å) /Pt(32Å). Such a technique has given the moment of each layer separately. The sum of the two magnetisations agrees well with the magnetisation given by conventional measurements and the saturation value of each layer corresponds to the values measured on other samples with just one layer.

5.9 Conclusion on Neutron Reflectometry

This chapter has given an overview of the neutron reflectometry as a tool for the investigation of surfaces. We have presented a matrix formalism which makes it possible to describe the specular reflectivity on non-magnetic and magnetic systems. Neutron reflectivity is especially suited for polymer and magnetic thin film systems. This has been illustrated with a few typical examples. We have not given here examples of non-specular and surface diffraction experiments. This kind of experiment has suffered until now from the lack of intensity on the neutron spectrometers. Moreover, the formalism necessary to analyse the experiments in the case of magnetic surface diffraction is still being developed. The neutron has a good energy for inelastic scattering on condensed matter but we have not spoken here on this aspect of reflectometry which is rather new. A beautiful example of inelastic scattering is the measurement of the Zeeman energy [30,31]. The problem of phase determination in neutron reflectometry is also an active field of research [32–34]. If not only the intensity but also the phase of the reflectivity could be measured a direct inversion of the reflectivity profile would be possible.

References

1. G.P. Felcher, R.O. Hilleke, R.K. Crawford, J. Haumann, R. Kleb and G. Ostrowski, *Rev. Sci. Instr.* **58**, 609 (1987).
2. C. F. Majkrzak, J. W. Cable, J. Kwo, M. Hong, D. B. McWhan, Y. Yafet and J. Waszcak, *Phys. Rev. Lett.* **56**, 2700 (1986).
3. J. Penfold, R.K. Thomas, *J. Phys. Condens. Matter* **2**, 1369-1412 (1990).
4. T.P. Russel, *Mat. Sci. Rep.* **5**, 171-271 (1990); T.P. Russel, *Physica B* **221**, 267-283 (1996).
5. L.T. Lee, D. Langevin, B. Farnoux, *Phys. Rev. Lett.* **67**, 2678-81 (1991).
6. J. Penfold, E.M. Lee, R.K. Thomas, *Molecular Physics* **68**, 33-47 (1989).
7. M.N. Baibich, J.M. Broto, A. Fert, F. Nguyen Van Dau, F. Petroff, P. Etienne, G. Creuzet, A. Friedrich, J. Chazelas, *Phys. Rev. Lett.* **61**, 2472 (1988).
8. V.F. Sears, *Physics Report* **141**, 281 (1986).
9. X.L. Zhou, S.H. Chen, *Physics Reports* **257**, 223-348 (1995).
10. H. Glättli and M. Goldman, *Methods of experimental Physics, Vol 23C, Neutron Scattering* (Academic Press, Orlando, 1987).
11. S. Dietrich, A. Haase, *Physics Reports* **260**, 1-138 (1995).

12. V.F. Sears, *Methods of experimental Physics, Vol 23A, Neutron Scattering* (Academic Press, Orlando, 1987); V.F. Sears, *Neutron News* **3**, 26 (1992).
13. J. Lekner, *Theory of reflection of electromagnetic and particle waves* (Martinus Nijhoff, Dordrecht, 1987).
14. S.J. Blundell and J.A.C. Bland, *Phys. Rev. B* **46**, 3391 (1992).
15. C. Fermon, C. Miramond, F. Ott, G. Saux, *J. of Neutron Research* **4**, 251 (1996).
16. Pleshanov Z., *Physica B* **94**, 233-243 (1994).
17. L. Névot and P. Croce, *Revue de Physique Appliquée* **15**, 761 (1980).
18. C.F. Majkrzak, *Physica B* **221**, 342-356 (1996).
19. B. Farnoux, *Neutron Scattering in the 90', Conf. Proc. IAEA in Jülich*, 14-18 january 1985, 205-209, Vienna, 1985, X.D.
20. F. Mezei, *Z. Phys.* **255**, 146 (1972).
21. T.P. Russel, A. Menelle, W.A. Hamilton, G.S. Smith, S.K. Satija and C.F. Majkrzak, *Macromolecules* **24**, 5721-5726 (1991).
22. X. Zhao, W. Zhao, X. Zheng, M.H. Rafailovich, J. Sokolov, S.A. Schwarz, M.A.A. Pudensi, T.P. Russel, S.K.Kumar and L.J. Fetters, *Phys. Rev. Lett.* **69**, 776 (1992).
23. YY. Huang, G.P. Felcher and S.S.P. Parkin, *J. Magn. Magn. Mater.* **99**, 31-38 (1991).
24. A. Schreyer, J.F. Aukner, T. Zeidler, H. Zabel, C.F. Majkrzak, M. Schaefer and P. Gruenberg, *Euro. Phys. Lett.*, 595-600 (1995).
25. P. Böni, *Physica B* **234-236**, 1038-1043 (1997).
26. E. Train, C. Fermon, C. Chappert, A. Megy, P. Veillet and P. Beauvillain, *J. Magn. Magn. Mater.* **156**, 86 (1996).
27. O. Song, C.A. Ballentine, R.C. O'Handley, *Appl. Phys. Lett.* **64**, 2593 (1994).
28. F. Ott, C. Fermon, *Physica B* **234-236**, 522 (1997).
29. C Fermon, S. Gray, G. Legoff, V. Mathet, S. Mathieu, F. Ott, M. Viret and P. Warin, *Physica B* **241-243**, 1055 (1998).
30. G.P. Felcher, S. Adenwalla, V.O. De Haan and A.A. Van Well, *Nature* **377**, 409 (1995).
31. G.P. Felcher, S. Adenwalla, V.O. de Haan, A.A. van Well, *Physica B* **221**, 494-499 (1996).
32. C.F. Majkrzak and N.F. Berk, *Phys. Rev. B* **52**, 10827 (1995).
33. C.F. Majkrzak and N.F. Berk, *Phys. Rev. B* **58**, 15416 (1998).
34. J. Kasper , H. Leeb and R. Lipperheide, *Phys. Rev. Lett.* **80**, 2614-2617 (1998).

Part II

Applications

6 Statistical Physics at Crystal Surfaces

Alberto Pimpinelli

LASMEA, Université Blaise Pascal - Clermont 2, Les Cézeaux, 63177 Aubière Cedex, France
[1]

6.1 Surface Thermodynamics

6.1.1 Surface Free Energy

According to thermodynamics, physical properties can be deduced from the knowledge of the free energy. In this lecture, the surface free energy is introduced in a simplified way (disregarding, in particular, elasticity).

In order to create a surface, one has to break chemical bonds, and this costs energy. At finite temperature, the free energy has to be considered. It is easy to give a precise definition of the surface free energy: to break a crystal along a plane, it requires a work W. If L^2 is the area of the crystal section, the surface free energy for unit area or ***surface tension*** is $\sigma_0 = W/(2L^2)$, the factor of 2 coming from the two surfaces which are created this way. It is straightforward to see that the number of broken bonds per unit surface area varies with the surface orientation. In particular, in the broken-bond approximation compact surfaces are expected to have a larger surface tension than open ones: $\sigma_{(111)} > \sigma_{(001)} > \sigma_{(110)}$ (*Problem.* Prove that, for instance, $\sigma_{(111)} \approx 1.155\sigma_{(001)}$ for an fcc crystal, at low temperature.)

Given this dependence on orientation, if the surface is not a plane the total surface free energy may be expected to be the integral of the energies of all surface elements. This is only true for an incompressible solid of large enough size (the interested reader will find a discussion of this statement in [1]. If x, y and z are Cartesian coordinates and $z(x,y)$ is the height of the surface over the xy plane, the local surface tension is then a function $\sigma(p,q)$ of the partial derivatives

$$p = \partial z/\partial x,$$
$$q = \partial z/\partial y$$

representing the local slopes of the surface profile. The total free energy of a large, incompressible solid body containing N atoms is then:

$$F = F_0 + \int\int \sigma(p,q)\mathrm{d}S \qquad (6.1)$$

[1] Most of the material in this lecture is based on the book "Physics of crystal growth", A. Pimpinelli & J. Villain, Cambridge University Press (1998). The interested reader should refer directly to it for delving deeper into the different subjects quickly treated here.

where the first term on the right-hand side is the bulk free energy, and the second term is the surface free energy. The integration is made over the surface of the solid and dS is the surface element.

For an incompressible solid, one can equivalently use the (Helmholtz) free energy $F(T, V)$ or the Gibbs free energy (free enthalpy) $G(T, P) = F + PV$. The latter is known from thermodynamics to be equal to $\mu_0 N$ (in the thermodynamic limit $N \to \infty$ and in the absence of a surface). Processes at constant pressure are more adequately described if G is used. In the future, we shall mainly use the free energy F because, in the absence of external forces, it results directly from the interactions between the molecules of the solid. Sometimes, we will drop the word "free", especially at low temperature where energy and free energy are not very different.

In formula (6.1), the free energy is assumed to be independent of the local curvature. This is only true for a large sample. (*Problem.* Why? Is the assumption appropriate for membranes?) Another simplification comes from the fact that the surface itself may be difficult to define: its location may be rather imprecise. This point has been addressed by Gibbs [2]. Also, in formula (6.1) it is assumed that $z(x, y)$ is analytic, which implies that fluctuations on an atomic scale are ignored—or, better, averaged out (what is called ***coarse–graining*** in the jargon).

6.1.2 Step Free Energy

At zero temperature, a (001) or (111) surface is flat. At higher temperatures, thermal fluctuations will make atoms leave the surface plane and create holes and mounds. In other words, ***steps*** are present on the surface at any finite temperature. In the same way as the surface free energy, a step free energy may be defined. This is how to proceed: a bulk solid is cut into two pieces of cross section $L \times L$, but this time a step is also cut. The work needed to take apart the two pieces is proportional to the total surface exposed, which is now $L^2 + bL$, if b is the step heigth. The total free energy is thus

$$\sigma = \frac{W}{2L^2} = \sigma_0 + \gamma \frac{b}{L} \, . \tag{6.2}$$

The "excess" free energy due to the step, γ, is called the ***step free energy*** or ***step line tension.*** If more steps are present, (6.2) shows that the total free energy is proportional to the number of steps per unit length, or step density, as long as the steps are far enough from each other that their interactions can be neglected. Broken bond arguments show that the step free energy is a function of the step orientation.

6.1.3 Singularities of the Surface Tension

Let us suppose that the plane $z = 0$ is parallel to a high-symmetry orientation, for instance (001), and that the temperature is low. We will see the precise meaning of "low temperature" when addressing the roughening transition in Sect. 6.2. Of course, no experimental realisation of a flat high-symmetry surface is possible beyond a certain length. In other words, real surfaces are always slightly "misoriented" or "miscut" with respect to a given high-symmetry plane, so that they make a small angle θ with that orientation (a typical lower bound for θ is $\sim 0.1°$). As said above, if step-step interactions are neglected—which is licit, if steps are rare—the ***projected surface free energy*** $\phi = \sigma / \cos\theta$ is simply equal to the step free energy multiplied by the step density $|\tan\theta|$:

$$\phi(\theta) \simeq \phi(0) + \gamma|\tan\theta| = \phi(0) + \gamma\sqrt{p^2 + q^2}\,. \tag{6.3}$$

This shows that the surface free energy is non-analytic (as a function of the local slopes p and q). The line tension γ is a function of the step orientation $y' = \mathrm{d}y/\mathrm{d}x = -p/q$. As a matter of fact, it is an analytic function at all T. The singular behaviour of the surface free energy at low temperature is typical of the high-symmetry, low-index orientations, which are therefore called ***singular***. These appear as flat facets in the macroscopic equilibrium shape of the crystal. The linear size of a facet is proportional to γ. These statements are proved using the so-called ***Wulff construction***, a geometrical construction which relies the surface free energy of a crystal to its equilibrium shape [3]. Using the projected free energy ϕ is useful to derive the analytic equivalent of Wulff's contruction from a variational principle: the equilibrium shape of a crystal minimises the surface free energy for a given volume [4,3].

6.1.4 Surface Stiffness

It seems doubtful that equation (6.3) may hold at all temperatures. Steps can be straight as assumed in deriving (6.2), only at $T = 0$. At finite temperature, they must fluctuate—in what is called step meandering. It may well be that a step becomes so "delocalised" that the concept of step looses its meaning altogether, above a given temperature. Indeed, we will see in Sect. 6.2 that such a temperature exists, and the corresponding transition is the roughening transition; in the rough, high-temperature phase, the step line tension γ actually vanishes. What is the form of the surface free energy at high temperature? Our experience with critical phenomena suggests that it may be useful to assume for the free energy an analytic expansion *à la Landau* in the disordered, high temperature phase [4]. To do that, it is convenient to transform the surface integral in (6.1) into an integral over Cartesian coordinates x and y. For a closed surface, this requires that the surface be cut into a few pieces and different projection planes be used for the various pieces. Locally, i.e. for a given piece, it is convenient to introduce the projected surface free energy ϕ of equation (6.3),

$$\phi(p,q) = \sigma(p,q)\sqrt{1+p^2+q^2}$$

so that (6.1) can be rewritten as follows:

$$F = F_0 + \int\int \phi(p,q)\mathrm{d}x\mathrm{d}y \ . \tag{6.4}$$

Take now as reference the plane $z = 0$. Expanding ϕ to quadratic order in p and q, and noting that the linear terms in p and q can be made to vanish, as well as the mixed second derivatives $\sigma_{12} = \sigma_{21} = \partial^2\sigma/\partial p\partial q$, (6.1) and (6.3) become

$$F = F_0 + \frac{1}{2}\int\int \mathrm{d}x\mathrm{d}y \left[(\sigma+\sigma_{11})\,p^2 + (\sigma+\sigma_{22})\,q^2\right]$$

where $\sigma_1 = \partial\sigma/\partial p$, $\sigma_{11} = \partial^2\sigma/\partial p^2$, etc. Letting $\sigma_{11} = \sigma_{22} = \sigma''$, one obtains (within an immaterial additive constant):

$$\phi(p,q) \simeq \frac{1}{2}\tilde{\sigma}(p^2+q^2) \tag{6.5}$$

where we introduced the ***surface stiffness***

$$\tilde{\sigma} = \sigma + \sigma'' \ .$$

For obtaining (6.4) we started from an analytic surface free energy. In other words, we treated the solid as a continuum, completely disregarding the discrete structure of the crystalline lattice. Does it matter? The answer is no, above the roughening temperature, as we argue in Sect. 6.2.

6.1.5 Surface Chemical Potential

The chemical potential is the Gibbs free energy per particle (here, per atom), and it is defined from

$$\mu(T,P) = \left(\frac{\partial G}{\partial N}\right)_{T,P} , \tag{6.6a}$$

$$\mu(T,V) = \left(\frac{\partial F}{\partial N}\right)_{T,V} . \tag{6.6b}$$

The chemical potential determines, at equilibrium, the shape of a crystal. Indeed, it is also useful to describe the time evolution of a surface not too far from equilibrium. At equilibrium, the chemical potential of an infinite system must be constant everywhere (if more than one kind of particle are present, the chemical potential of each species must be constant):

$$\mu = \mu_0 \ .$$

How is this relation modified for a finite size system such as a crystal of fixed volume V? Let us locally modify the crystal surface so as to change isothermally and at constant volume the number of particles by a quantity δN. The most appropriate thermodynamic potential to use is the Helmoltz free energy. Then, μ is found from $(\delta F)_{T,V} = \mu \delta N$. Introducing the atomic volume $v = V/N$, let

$$N = \frac{1}{v} \int\int z(x,y) \mathrm{d}x \mathrm{d}y \, .$$

If $z(x,y)$ varies by δz, then

$$(\delta F)_{T,V} = \mu(T,V) \delta N = \frac{\mu}{v} \int\int \delta z(x,y) \mathrm{d}x \mathrm{d}y. \tag{6.7}$$

Varying (6.4) yields:

$$\delta F = \frac{\mu_0}{v} \int\int \delta z(x,y) \mathrm{dxdy} + \int\int \left(\frac{\partial \phi}{\partial \mathrm{p}} \delta \mathrm{p} + \frac{\partial \phi}{\partial \mathrm{q}} \delta \mathrm{q} \right) \mathrm{dxdy}. \tag{6.8}$$

The term within brackets in the second integral can be integrated by parts, provided ϕ is at least twice differentiable. Since $p = \partial z / \partial x$,

$$\int\int \left(\frac{\partial \phi}{\partial p} \delta p \right) \mathrm{d}x \mathrm{d}y = \int\int \left(\frac{\partial \phi}{\partial p} \frac{\partial}{\partial x} (\delta z) \right) \mathrm{d}x \mathrm{d}y$$

$$= \int\int \frac{\partial}{\partial x} \left(\frac{\partial \phi}{\partial p} \delta z \right) \mathrm{d}x \mathrm{d}y - \int\int \delta z \frac{\partial}{\partial x} \left(\frac{\partial \phi}{\partial p} \right) \mathrm{d}x \mathrm{d}y.$$

We only consider local variations δz, so that the first term vanishes. The second term in brackets in (6.8) can be handled in the same way, and we find:

$$\delta F = \int\int \delta z(x,y) \left[\frac{\mu_0}{v} - \frac{\partial}{\partial x} \left(\frac{\partial \phi}{\partial p} \right) - \frac{\partial}{\partial y} \left(\frac{\partial \phi}{\partial q} \right) \right] \mathrm{d}x \mathrm{d}y.$$

Finally, equating the preceding equation to (6.7) one gets (Herring 1953, Mullins 1963):

$$\mu = \mu_0 - v \left[\frac{\partial}{\partial x} \left(\frac{\partial \phi}{\partial p} \right) + \frac{\partial}{\partial y} \left(\frac{\partial \phi}{\partial q} \right) \right] . \tag{6.9}$$

In the following we will often choose our units in such a way that the molecular volume $v = V/N$ is 1. The reference chemical potential μ_0 is the chemical potential of the bulk solid, according to (6.1). At equilibrium, the chemical potential at the surface must also be equal to μ_0, and from (6.9) the surface must be flat.

If (6.4) holds, the chemical potential is $\mu + \delta\mu$, where $\delta\mu$ is, according to (6.9), given by

$$\delta \mu = -v \tilde{\sigma} \left(\frac{\partial^2 z}{\partial x^2} + \frac{\partial^2 z}{\partial y^2} \right) \tag{6.10}$$

where $\tilde{\sigma}$ is the surface stiffness. In the case of an isotropic surface tension, $\tilde{\sigma} = \sigma$. Locally, any convex surface $z(x,y)$ may be approximated as follows

$$z(x,y) \approx z_0 - \frac{1}{2}\left(\frac{x^2}{R_1} + \frac{y^2}{R_2}\right) , \tag{6.11}$$

where R_1 and R_2 are the two principal radii of curvature of the surface. Inserting (6.11) into (6.10) we find the Gibbs-Thomson relation

$$\delta\mu = v\tilde{\sigma}\left(\frac{1}{R_1} + \frac{1}{R_2}\right) . \tag{6.12}$$

In particular, for a sphere of radius R

$$\delta\mu = \frac{2v\sigma}{R}. \tag{6.13}$$

A similar expression yields the excess chemical potential in the vicinity of a curved step on the surface, as a function of the ***step stiffness*** $\tilde{\gamma} = \gamma(\theta) + \mathrm{d}^2\gamma/\mathrm{d}\theta^2$, where γ is the step line tension. Expliciting the atomic area a^2, and letting R be the local radius of curvature of the step, the relation reads

$$\delta\mu = a^2\frac{\tilde{\gamma}}{R} . \tag{6.14}$$

Note the factor 2 in (6.13)—a consequence of the 2 terms in (6.9).

6.2 Morphology of a Crystal Surface

6.2.1 Adatoms, Steps and Thermal Roughness of a Surface

At zero temperature, a high-symmetry surface—e.g. (100) or (111)—at equilibrium should be perfectly flat, i.e. it should contain no step. At low temperature, there are a few free atoms, or ***adatoms***, and vacancies. Their density depends on the energy one has to pay to create an adatom—and, as a consequence, a vacancy—from a terrace. In the approximation where bond energies are additive (***broken bond approximation***), and only nearest-neighbour interactions are considered, this energy is equal to $4E$ on an fcc–(001) surface and to 6ϕ on a (111) surface, if E is the bond energy.
At higher temperature, clusters of atoms start to appear. Such clusters are closed terraces bounded by steps, at equilibrium with a two-dimensional gas of adatoms. Indeed, adatoms are continuously absorbed at and released from steps. Atoms emission from steps requires an energy W_{ad}, much smaller than the energy cost for extracting an atom from a terrace—indeed, on an fcc surface the former is just half of the latter. Then, we expect the equilibrium adatom density n_{eq} to be given by a Gibbs formula

$$n_{\mathrm{eq}} = \exp\left(-\beta W_{\mathrm{ad}}\right) , \tag{6.15}$$

where $1/\beta = k_B T$ and k_B is the Boltzmann constant.
The step density (total step length per unit surface) increases with temperature. This increase is not easily seen directly by microscopy. Indeed, most of microscopy techniques work best at low temperature, where matter transport is hard and the surface does not easily attain thermal equilibrium. On the other hand, the atom scattering or x-ray diffraction signal does exhibit a dramatic change when temperature is increased and the surface roughens. Above some temperature, the lineshape, which is lorentzian at low temperature, undergoes a qualitative change. One can for instance measure the "specular" reflection, i.e. that whose reflection angle almost equals the incidence angle (***speculum*** is the Latin for mirror). The specular peak (as well as the Bragg peaks) is narrow for a smooth surface while a rough surface scatters radiation in all directions.

The interpretation of diffraction patterns from a hot surface is difficult [5–7] because the effect of atomic vibrations adds to roughness to broaden the reflected beam. It is however clear, from a quantitative analysis, that the total step length is greatly increased by heating.

The reason is essentially the following. At zero temperature, creating a step costs an energy (per unit length) W_1. Even if W_1 does not change much, the step entropy increases, so that the free energy decreases. A simple estimate is possible, if we consider on the (001) face of a cubic crystal a zig-zag step, whose average direction makes an angle of 45° with the bond directions. An approximate calculation is possible if we consider the configurations of a square-lattice random walk going from left to right, each step of the random walker being parallel to a lattice bond, and backward steps being forbidden. If the width of the system in the direction of the walk is uniformly equal to the bond length multiplied by $L\sqrt{2}$, all configurations have the same energy $2LW_1$. Since there are 2^{2L} configurations, the entropy is $2L \ln 2$ and the free energy per bond is

$$\gamma = W_1 - k_B T \ln 2 \,. \tag{6.16}$$

The free energy per unit length γ/a is called the ***line tension of steps***. Since a, the lattice parameter, is generally chosen as the length unit in this lecture notes, the term line tension is often employed for γ itself.

When γ is positive, one has to provide mechanical work to introduce a step into the surface. If the total step free energy $L\gamma$ vanishes, steps can appear spontaneously—they cost nothing. Thus, equation (6.16) tells us that the surface undergoes a transition at a temperature T_R approximately given by

$$T_R \approx \frac{W_1}{k_B \ln 2} \,. \tag{6.17}$$

This transition is called the ***roughening transition***.

6.2.2 The Roughening Transition

As seen above, the roughening transition temperature may be defined as the temperature at which the line tension of steps vanishes. According to the experiment [8] the line tension does vanish at some temperature, and its behaviour agrees with (6.16). Near T_R, the experimental curve bends away from the straight line predicted by formula (6.16): one can wonder whether it is an instrumental effect or a fundamental one. As will be seen in the next section, it is a fundamental effect. But before discussing that, we would like to make three remarks.

• 1. The roughening transition temperature depends on the surface orientation, i.e. it is different for a (111) and for a (1,1,19) surface. This point will be addressed in Sect. 6.2.6.

• 2. For a given surface orientation, the step line tension vanishes at the same temperature for all step orientations, so that T_R is independent of the step orientation. (*Problem*: prove this.)

• 3. Formula (6.16) suggests that the step line tension may become negative for $T > T_R$. Actually, it is not so, if a step is defined in a model-independent way. (*Problem*: could you think of the appropriate definition?) Indeed, the concept of a step looses its meaning above T_R.

6.2.3 Smooth and Rough Surfaces

The roughening transition is much more complicated than suggested by the discussion of Sect. 6.1.2. Indeed, thermally excited steps are not isolated objects, they are closed loops. In this Section, we shall try to give an idea of the what a rough surface really is. The reader will find more details about the roughening transition in the monographies by Balibar & Castaing (1985) [9], Van Beijeren & Nolden (1987) [10], Lapujoulade (1994) [11], Nozières (1991) [12] and Weeks (1980) [13].

As seen in the previous section, the concept of step is not useful above T_R. It is therefore appropriate to characterize roughness in an alternative way. Consider an infinite surface of average orientation perpendicular to the z axis. Let (x, y, z) be a point of the surface. The "height" z will be assumed to be a one-valued function of x and y, so that "overhangs" are excluded. Let $\mathbf{r}_\parallel = (x, y)$ be a point of the two-dimensional (x, y) space. We define the height-height correlation function

$$g(\mathbf{R}_\parallel) \equiv \left\langle \left[z(\mathbf{r}_\parallel) - z(\mathbf{r}_\parallel + \mathbf{R}_\parallel)\right]^2 \right\rangle . \qquad (6.18)$$

The interest of this function is that it has, if gravity is neglected, two qualitatively different behaviours above and below T_R:

$$\lim_{|\mathbf{R}_\parallel| \to \infty} g(|\mathbf{R}_\parallel|) \quad \text{finite for } T < T_R, \qquad (6.19a)$$

$$\lim_{|\mathbf{R}_{\|}|\to\infty} g(|\mathbf{R}_{\|}|) = \infty \quad \text{for } T > T_R \,. \tag{6.19b}$$

The finiteness of $g(|\mathbf{R}_{\|}|)$ at low temperature, in agreement with (6.19a), can be proved by a low-temperature expansion [3]. It is more difficult to prove (6.19b), and only a plausibility argument will be given. It is indeed reasonable to assume that, at sufficiently high temperature, say $k_BT \gg W_1$, the discreteness of the crystal lattice becomes negligible, so that the surface height $z(x,y)$ may be regarded as a differentiable function of x and y, as it is for a liquid. Thus, we just forget that we have a crystal, and we write the surface energy as if it were a continuous medium. The surface free energy F_{surf} is then simply proportional to the surface area, and the proportionality coefficient $\tilde{\sigma}$ is the surface stiffness (*cf.* 6.5). Introducing the local slopes p and q,

$$F_{\text{surf}} = \tilde{\sigma} \int\int \mathrm{d}x\mathrm{d}y\sqrt{1+p^2+q^2}$$

or, for small undulations (thermal fluctuations) of the surface,

$$F_{\text{surf}} = \text{Const} + \frac{\tilde{\sigma}}{2}\int\int \mathrm{d}x\mathrm{d}y\,(p^2+q^2) \,. \tag{6.20a}$$

However, in our world subject to gravity, thermal fluctuations of the surface acquire an additional energy from gravity. The effect of gravity is irrelevant below T_R (*Problem.* Why?). It is not so above the transition temperature. The energy of a column of matter of cross section $\mathrm{d}x\mathrm{d}y$, whose height is between z_1 and z, is

$$\rho g \mathrm{d}x\mathrm{d}y \int_{z_1}^{z} \zeta \mathrm{d}\zeta = \rho g \mathrm{d}x\mathrm{d}y(z^2 - z_1^2)/2$$

where ρ is the specific mass and g the gravity acceleration. The term containing z_1 is constant and will be omitted. The energy excess associated with surface shape fluctuations and resulting from both gravity and surface tension is

$$\delta F = -\frac{1}{2}\int\int \mathrm{d}x\mathrm{d}y\left[\rho g z^2 + \tilde{\sigma}\left(\frac{\partial z}{\partial x}^2 + \frac{\partial z}{\partial y}^2\right)\right] \,. \tag{6.20b}$$

The correlation function (6.18) is readily obtained by Fourier transforming (6.20b) and by using the equipartition theorem $\langle |\, z_q^2 \,|\rangle = k_BT/[\rho g + \tilde{\sigma}(q_x^2 + q_y^2)]$:

$$g(\mathbf{R}_{\|}) = 2k_BT\int_0^{1/a} \mathrm{d}q_x\mathrm{d}q_y \frac{1-\cos(\mathbf{q}\cdot\mathbf{R}_{\|})}{\rho g + \tilde{\sigma}(q_x^2+q_y^2)} \,. \tag{6.21}$$

An approximation of $g(\mathbf{R})$ may be obtained if the lower limit of integration is replaced by an appropriate cutoff, below which the numerator is almost zero. This allows us to replace the cosine by its average value 0, and one obtains

$$g(\mathbf{R}_{\|}) \approx 2\pi\frac{k_BT}{\tilde{\sigma}}\ln\left(\frac{\rho g + \tilde{\sigma}/a^2}{\rho g + \tilde{\sigma}/|\mathbf{R}_{\|}|^2}\right) \,. \tag{6.22}$$

If gravity can be neglected, which is possible for $|\mathbf{R}_{\|}| \ll \lambda = \sqrt{\sigma/\rho g}$, (6.22) goes as:

$$g(\mathbf{R}_{\|}) = \text{Const.} + (4\pi k_B T/\tilde{\sigma}) \ln |\mathbf{R}_{\|}| \quad . \tag{6.23}$$

This proves (6.19b). A surface is called ***rough*** if the height-height correlation function $g(\mathbf{R}_{\|})$ diverges as in (6.23), and ***smooth*** if there is no divergence.

Previously, we defined the roughening transition in terms of the vanishing of the step free energy γ. We should worry about the equivalence of the two definitions. Fortunately, they are equivalent. (*Problem.* Prove this equivalence. *Hint:* the proof of (6.23) relies on the use of (6.20b). If the step line tension is positive, equation (6.20b) cannot be true). Indeed, if $\gamma > 0$, the formation of a terrace of size L has a free energy cost which diverges with L. A diverging free energy means forbidden configurations—i.e., whose statistical weight vanishes—and the correlation function $g(\mathbf{R}_{\|})$ is finite for $|\mathbf{R}_{\|}| \to \infty$. Therefore, the roughening transition can be defined (at least on a high-symmetry surface) either by (6.19a,b) or by the vanishing of γ.

The role of gravity, which kills the roughening transition, becomes negligible at lengthscales shorter than the ***capillary length*** $\lambda = \sqrt{\tilde{\sigma}/\rho g}$. The order of magnitude of $\tilde{\sigma}$ is typically the energy of a chemical bond, i.e. $1eV$ per atom. The resulting value of λ is a few centimetres, i.e. much larger than the distance over which equilibrium can be reached at a crystal surface. For this reason, gravity is usually neglected in surface science.

6.2.4 Diffraction from a Rough Surface

Another important consequence of the peculiarity of the roughening transition is the form of diffraction peaks in a scattering experiment. The x-rays or atom scattering cross section has the form (see Chap. 2)

$$\frac{\mathrm{d}\sigma}{\mathrm{d}\Omega} = \frac{A}{q_z^2} \int\int \mathrm{d}X \mathrm{d}Y \left\langle \mathrm{e}^{-iq_z[h(X,Y)-h(0)]} \right\rangle \mathrm{e}^{-i(q_x X + q_y Y)} \quad , \tag{6.24}$$

where A depends on the electronic density of the surface atoms and $z = h(X,Y)$ describes the shape of the surface. The angular brackets denote thermal averaging over surface fluctuations. In order to compute this average, it is customary to make the assumption that fluctuations have a Gaussian probability distribution. With this assumption it is straightforward to show that (Chap. 2)

$$\Gamma(\mathbf{R}_{\|}) \equiv \left\langle \mathrm{e}^{-iq_z[h(R_{\|})-h(0)]} \right\rangle = \mathrm{e}^{-\frac{q_z^2}{2}\left\langle [h(R_{\|})-h(0)]^2 \right\rangle} \quad , \tag{6.25}$$

where we wrote $\mathbf{R}_{\|} = (X,Y)$. In the exponent at the right-hand side of equation (6.25) we recognize the height-height correlation function of the preceding section. The function $\Gamma(\mathbf{R})$ takes the name of ***pair correlation*** *function*. To appreciate the relation between the height-height and the pair correlation functions, the latter being the quantity directly measured in a scattering experiment, it is instructive to consider a few examples:

i) $g(\mathbf{R}_{||}) = C$. The pair correlation function is also a constant,

$$\Gamma(\mathbf{R}_{||}) = C' ,$$

and the scattered intensity obtained from (6.24) is a delta function.

ii) $g(\mathbf{R}_{||}) = C|\mathbf{R}_{||}|$. The pair correlation function is an exponential

$$\Gamma(\mathbf{R}_{||}) = \mathrm{e}^{-\frac{Cq_z^2}{2}|\mathbf{R}_{||}|} . \tag{6.26}$$

The scattering cross section allows us to find the scattered intensity I, which is a Lorentzian in this case:

$$I \sim \frac{2Cq_z^2}{C^2 q_z^4 + 4q_{||}^2} ,$$

where q_z and $q_{||}$ are the components of the momentum transfer wavevector respectively orthogonal and parallel to the average surface.

iii) $g(\mathbf{R}_{||}) = C \ln|\mathbf{R}_{||}|$. The pair correlation function has a power–law behaviour

$$\Gamma(\mathbf{R}_{||}) = \mathrm{e}^{-Cq_z^2 \ln|\mathbf{R}_{||}|} = |\mathbf{R}_{||}|^{-Cq_z^2} , \tag{6.27}$$

as well as the scattered intensity

$$I \sim q_{||}^{-2+Cq_z^2} .$$

The case of a vicinal (stepped) crystal surface will be treated here [14]. At low temperature $g(\mathbf{R}_{||})$ is a constant, and delta-function peaks are expected. Beyond T_R, the diffraction peaks are expected to acquire a power-law shape. However, a vicinal surface has two non equivalent directions, parallel and orthogonal to the steps. When computing the scattered intensity we must integrate only along a vector orthogonal to the steps. Then one finds

$$I \sim q_{||}^{-1+Cq_z^2} .$$

At the position of largest sensitivity, $q_z = \pi/a$, the intensity of the diffraction peaks behave as $I \sim q_{||}^{-1+C\pi^2/a^2}$, where C is actually a function of the temperature, $C = C(T)$. According to (6.23), at T_R the height-height correlation function has the form of case (iii) above, and $C(T_R)$ is equal to the universal value $2a^2/\pi^2$. Thus, the scattered intensity *at the transition temperature* T_R reads

$$I \sim q_{||}^{-1+2}$$

and the roughening temperature can be found as the temperature where

$$\tau = \frac{\mathrm{d}\,(\ln I)}{\mathrm{d}\,(\ln q_{||})} = 1 .$$

6.2.5 Capillary Waves

In the case of an isotropic surface, such as the surface of a liquid, (6.20b) describes thermal excitations of the surface, the so-called ***capillary waves***. As we have seen in Sect. 6.2.5, equation (6.22) implies that the ***reflected intensity*** of e.g. x-rays from the surface allows a direct measurement of the surface tension σ (which coincides with $\tilde{\sigma}$ for a liquid).

6.3 Surface Growth and Kinetic Roughening

6.3.1 Equilibrium with the Saturated Vapour

At any temperature, a crystal is in equilibrium with its saturated vapour at pressure P_{sat}. Assuming the vapour to be an ideal gas with density ρ_{sat}, kinetic theory dictates the impingement rate of atoms onto the surface [4,1]: $R_{\mathrm{imp}} = P_{\mathrm{sat}}/\sqrt{2\pi m k_B T}$.

Assuming that all the atoms are adsorbed—equivalently, that the sticking coefficient is unity, which is not necessarily true with molecular species—detailed balance gives the equilibrium adatom density n_{eq} once the saturated vapour pressure is known. If τ_{ev} is the average lifetime of an adatom before evaporation, then $R_{\mathrm{imp}} = R_{\mathrm{evap}} = n_{\mathrm{eq}}/\tau_{\mathrm{ev}}$, and

$$\frac{n_{\mathrm{eq}}}{\tau_{\mathrm{ev}}} = \frac{P_{\mathrm{sat}}}{\sqrt{2\pi m k_B T}} . \tag{6.28}$$

The right-hand side of this equation is called the ***evaporation rate*** of the crystal. The saturated vapour pressure for a given solid is obtained from the equation of state of an ideal gas; the detailed derivation is in [1]. The result is

$$P_{\mathrm{sat}} = (k_B T)^{\frac{5}{2}} \left(\frac{m}{2\pi\hbar^2}\right)^{\frac{3}{2}} \exp(-\beta W_{\mathrm{coh}}) . \tag{6.29}$$

Inserting (6.29) into (6.28) we obtain the evaporation rate in the form

$$\frac{n_{\mathrm{eq}}}{\tau_{\mathrm{ev}}} = m\frac{(k_B T)^2}{4\pi^2\hbar^3} \exp(-\beta W_{\mathrm{coh}}) . \tag{6.30}$$

The quantity in front of the exponential has dimension [length^{-2} time^{-1}]. Its temperature dependence is weak compared to the exponential, so that it is usually written as a T-independent coefficient τ_0^{-1}. Numerical values of $1/\tau_0$ are of order 10^{14} Å^{-2} s^{-1} for most elements, at room temperature.

6.3.2 Supersaturation and Vapour Deposition

In the equilibrium state, the chemical potential of the solid and of its vapour must be equal: $\mu_{\mathrm{solid}} = \mu_{\mathrm{vapour}} = \mu_{\mathrm{eq}}$. What happens if we increase the chemical potential in the vapour? The vapour atoms go where the chemical potential is lower, i.e. to the solid. The solid grows! The difference

$$\Delta\mu = \mu_{\text{vapour}} - \mu_{\text{eq}}$$

is called the ***supersaturation***. The chemical potential is not easily controlled by the experimentalist. It is easier to change the pressure. For a vapour treated as an ideal gas, the pressure can be written as

$$P = P_{\text{sat}} \exp(\beta\Delta\mu) .$$

Therefore, for $\beta\Delta\mu < 1$,

$$\beta\Delta\mu \approx \frac{P - P_{\text{sat}}}{P_{\text{sat}}} . \tag{6.31}$$

The growth rate $\dot{z}$ of a rough crystal surface is then simply determined by the balance between the atoms impinging from the vapour and those re-evaporating,

$$\dot{z} = R_{\text{imp}} - R_{\text{ev}} = (P - P_{\text{sat}})/\sqrt{2\pi m k_B T}, \tag{6.32}$$

where we used (6.28). From (6.29) we find

$$\dot{z} = \left(\frac{P}{P_{\text{sat}}} - 1\right) m \frac{(k_B T)^2}{4\pi^2\hbar^3} \exp(-\beta W_{\text{coh}}) . \tag{6.33}$$

From (6.33) we see that the growth rate is proportional to the super-saturation—when the latter is not too small, see Ref. [15] —and that it is thermally activated: a barrier must be overcome, which is given by the cohesion energy. In particular, one sees that the growth rate does not depend on the surface orientation. It is not always so, as it has been shown by [15]. However, (6.33) gives an upper limit to the growth rate of a crystal. Indeed, equation (6.33) is valid for a surface which is rough at equilibrium. Impinging atoms find plenty of favorable sites (kinks) for being incorporated. When the starting surface is parallel to a high-symmetry direction below the equilibrium roughening temperature, steps and kinks have to be created before the atoms can be incorporated. This is the phenomenon of nucleation.

6.3.3 Nucleation on a High Symmetry Substrate

On a close packed, step-free high-symmetry surface like (111) or (001), below its roughening temperature, atoms impinging from a vapour—as well as from an atomic or molecular beam—do not find energetically favourable sites to be incorporated into. Indeed, they first have to condense into aggregates, which will then grow by capturing other atoms. If the supersaturation $\Delta\mu$ is small, i.e. if the system is not far from equilibrium, atom condensation will be ruled by the free energy gain in forming a two-dimensional aggregate of size R,

$\Delta G_{\text{gain}} = -\pi \Delta\mu R^2$, which is opposed by the free energy cost of forming a step on the surface $\Delta G_{\text{cost}} = 2\pi\gamma R$. Adding the competing contributions one finds

$$\Delta G(R) = 2\pi\gamma R - \pi\Delta\mu R^2 \ . \tag{6.34}$$

This function of R has a maximum at $R = R_c = \gamma/\Delta\mu$. It is the phenomenon of ***nucleation***: aggregates of size smaller than the ***critical radius*** R_c can decrease their free energy by shrinking, while those whose radius is larger than R_c will grow further. The nucleation rate J_{nuc} is expected to be dominated by the probability of finding an aggregate at the ***nucleation barrier*** $\Delta G^* = \Delta G(R_c) = \pi\gamma^2/\Delta\mu$, i.e.

$$J_{\text{nuc}} \sim \exp\left(-\pi\frac{\beta\gamma^2}{\Delta\mu}\right) \ . \tag{6.35}$$

Small variations in the supersaturation are enormously amplified by the exponential dependence. The growth rate is now determined by the nucleation rate (6.35), and by the spreading velocity of the supercritical aggregates. Qualitatively, one expects to observe random nucleation of two-dimensional seeds, lateral spreading of the aggregates, and coalescence. The process will then start again on the freshly created layer. The growth rate will thus be an oscillating function of time, the period being equal to the formation time of a whole layer. The oscillation is a consequence of the discrete (layered) structure of a crystal, which is kept by its surface as long as it is below its roughening temperature.

However, if growth is continued, after a system-dependent time span the oscillations die out, and a quasi-stationary distribution of surface steps sets in. Indeed, the oscillations demand that each layer is started and completed in succession, with a high degree of temporal correlation. Randomness in deposition and nucleation destroy such correlations, and a disordered state, that we would be tempted to call rough, appears. In fact, it is found that after transients have died out, any growing surface is rough, with power-law correlations, as well in space as in time. This is ***kinetic roughening***.

6.3.4 Kink-Limited Growth Kinetics

Kinetic roughening is the outcome of the competition of two different mechanisms: randomness in deposition and nucleation, which, following the current jargon, we will call ***noise***, and matter transport processes. The effect of noise can be easily seen by picturing growth without matter transport at the growing interface. Atoms just stick where they hit, which happens at a rate F per unit surface and time. The number of deposited atoms in time t is thus $N \approx Ft$. The statistical fluctuation of this number is $\delta N \approx \sqrt{N} \approx \sqrt{Ft}$. One would thus expect the rms fluctuations of the surface height (or surface width),

$$w(t) = \sqrt{\langle [z(t) - z(0)]^2 \rangle} \tag{6.36}$$

to be proportional to δN, so that

$$w(t) \sim t^{1/2} . \tag{6.37}$$

This is an example of the power laws mentioned at the end of the previous section. Of course, the random deposition model without smoothening processes is an extreme example, though it may be appropriate to very low temperature, where surface mobilities are very slow. In real experimental situations, one seeks to obtain as flat a surface as possible. Indeed, the Gibbs-Thomson equation (6.12) states that an excess of chemical potential is stored where an excess of particle has accumulated: the surface will thus tend to relax by transferring these particles towards places with lower chemical potential. In a reference frame moving at the average growth velocity of the interface, F, the surface height $z(t)$ evolves according to the laws of linear thermodynamics:

$$\dot{z} = -\nu \delta \mu = \nu \tilde{\sigma} \left(\frac{\partial^2 z}{\partial x^2} + \frac{\partial^2 z}{\partial y^2} \right) \tag{6.38}$$

where the linearized expression (6.10) for the chemical potential has been used, and ν is a kinetic coefficient related to the rate of capture and emission of atoms by kinks, from and to the vapour.

The randomizing noise and the smoothing term (6.38) may be put together to obtain a Langevin equation for describing growth of a fluctuating interface in a vapour phase:

$$\frac{\partial z}{\partial t} = \nu \tilde{\sigma} \left(\frac{\partial^2 z}{\partial x^2} + \frac{\partial^2 z}{\partial y^2} \right) + \eta \tag{6.39}$$

where $\eta(\mathbf{x}, t)$ is a random function with the properties

$$\langle \eta \rangle = 0, \quad \langle \eta(\mathbf{x}, t) \eta(\mathbf{x}', t') \rangle = F \delta(\mathbf{x} - \mathbf{x}') \delta(t - t') . \tag{6.40}$$

6.3.5 Scaling

Since the equation (6.40) is linear, it is readily solved by Fourier tranformation [1]. Instead, we will perform a scaling analysis [16,1]. Rescale $\mathbf{x} = (x, y)$ by a factor λ, z by a factor λ^α, and the time by a factor λ^z. Equation (6.39) becomes

$$\lambda^{\alpha - z} \frac{\partial z}{\partial t} = \nu \tilde{\sigma} \lambda^{\alpha - 2} \left(\frac{\partial^2 z}{\partial x^2} + \frac{\partial^2 z}{\partial y^2} \right) + \lambda^{-1 - z/2} \eta \tag{6.41}$$

where the rescaling of the noise follows from the properties of δ-functions ($\delta(ax) = \delta(x)/a$).

Dividing both sides by $\lambda^{\alpha - z}$, we get

$$\frac{\partial z}{\partial t} = \nu \tilde{\sigma} \lambda^{z - 2} \left(\frac{\partial^2 z}{\partial x^2} + \frac{\partial^2 z}{\partial y^2} \right) + \lambda^{z/2 - 1 - \alpha} \eta \tag{6.42}$$

This equation coincides with (6.39) if $z = 2$ and $\alpha = 0$. What is the meaning of these exponents? To see this, we need playing a little with scaling relations. Consider the surface width (6.36). Under rescaling $z' = \lambda^\alpha z$, $t' = \lambda^z t$, w behaves as

$$w(t') = \sqrt{\langle [z'(t') - z'(0)]^2 \rangle} \approx \lambda^\alpha \sqrt{\langle [z(\lambda^z t) - z(0)]^2 \rangle} . \tag{6.43}$$

On the other hand, we expect that $w(t)$ be a power-law function of t as in (6.37). We introduce another exponent, β, such that

$$w(t) \sim t^\beta . \tag{6.44}$$

In critical phenomena, power-law behaviour at the critical point is a consequence of the absence of a characteristic lengthscale in the problem, the correlation length being infinite at T_c. We conclude that also in the case of growth of an interface no typical lengthscale exists, except for the lattice parameter a, which is immaterial at long wavelengths, and the size L of the system. Letting $\lambda = L$ in equation (6.43), we find

$$w(t) \approx L^\alpha \sqrt{\langle [z(L^z t) - z(0)]^2 \rangle} . \tag{6.45}$$

We see that for times t of the order of $t_{\text{crossover}} \sim L^z$, the surface width w reaches the time-independent (saturation) value

$$w_{\text{saturation}} \approx L^\alpha \sqrt{\langle [z(1) - z(0)]^2 \rangle} \sim L^\alpha . \tag{6.46}$$

To be coherent with (6.44), which implies $w_{\text{saturation}} \approx w(t_{\text{crossover}}) \sim t^\beta_{\text{crossover}} \sim L^{z\beta}$, equation (6.46) requires $L^\alpha = L^{z\beta}$, or

$$z = \alpha/\beta . \tag{6.47}$$

We see now how the exponents α, β, and z can be intrerpreted:

- α is the ***roughness exponent***, which characterizes the increase of the saturation value (the value at $t \gg t_{\text{crossover}}$) of the surface width with the system size L:

$$w_{\text{saturation}} \sim L^\alpha ;$$

- β is the ***growth exponent***, which characterizes the increase of the surface width with the time t (at $t \ll t_{\text{crossover}}$):

$$w(t) \sim t^\beta ;$$

- z is the ***dynamic exponent***, which characterizes the increase of the crossover time with the system size L:

$$t_{\text{crossover}} \sim L^z .$$

The equality (6.47) relates the three exponents. Random deposition, as seen at the beginning of Sect. 6.3.4, gives $\beta = 1/2$. In this extreme model, α and

ζ are not defined, due to the strict locality of the deposition process: the system's behaviour is completely insensitive to the system size. The resulting interface is completely—and maximally—uncorrelated. The growth model of equation (6.39), known as the Edwards-Wilkinson model[17], has $z = 2$ and $\alpha = \beta = 0$ in three dimensions—that is, for a two-dimensional surface. Indeed, the direct solution shows that $\alpha = \beta = 0$ means in reality logarithmically diverging correlations, like equation (6.23) of equilibrium roughening. Not all models behave like that, as seen in next section.

6.3.6 Surface-Diffusion-Limited Growth Kinetics

Another situation of interest is when the matter transport process charged to smoothen the growing surface is conservative: in this case, the evolution equation (6.38) is no longer valid, and it has to be replaced with

$$\dot{z} = -\mathrm{div}\mathbf{j}, \tag{6.48}$$

where $\mathbf{j}$ is the surface diffusion current. To find the latter, we invoke again the Gibbs-Thomson equation (6.10). However, we require now that the excess chemical potential due to a local atom excess is relaxed through diffusion along the interface. In other words, the current $\mathbf{j}$ is proportional to the gradient of the local chemical potential:

$$\mathbf{j} = -D\nabla\delta\mu \ ,$$

where D is a (collective) surface diffusion coefficient. From (6.10) we get

$$\mathbf{j} = D\tilde{\sigma}\nabla\left(\frac{\partial^2 z}{\partial x^2} + \frac{\partial^2 z}{\partial y^2}\right) ,$$

and inserting this into (6.48) yields, instead of (6.39):

$$\frac{\partial z}{\partial t} = -D\tilde{\sigma}\nabla^2(\nabla^2 z) + \eta \tag{6.49}$$

where we defined $\nabla^2 = (\partial^2/\partial x^2 + \partial^2/\partial y^2)$. The model of equation (6.49) is known as the Mullins model [18]. Performing the same rescaling as before, we find

$$\lambda^{\alpha-z}\frac{\partial z}{\partial t} = -D\tilde{\sigma}\lambda^{\alpha-4}\nabla^2(\nabla^2 z) + \lambda^{-1-z/2}\eta \ .$$

Dividing both sides by $\lambda^{\alpha-z}$, we get

$$\frac{\partial z}{\partial t} = -D\tilde{\sigma}\lambda^{z-4}\nabla^2(\nabla^2 z) + \lambda^{z/2-1-\alpha}\eta \ . \tag{6.50}$$

This equation coincides with (6.49) if $z = 4$ and $\alpha = 1$. The relation (6.47) yields now $\beta = 1/4$. Hence, the surface width w increases in this case faster ($w \sim t^{1/4}$), than at equilibrium or for kink-limited kinetics ($w \sim \ln t$).

Different physical situations will thus give different exponents. Again, this resembles very much the case of critical phenomena. Indeed, universality classes appear, where different models such as (6.39) or (6.49) find their place, characterised by different sets of values of α, β and z. Symmetry arguments, as well as physical considerations, dictate the form of the equation which rules the evolution of the interface, and thus the universality class where it belongs to. A lot more of details on this fascinating subject will be found in Refs. [16] and [1].

References

1. A. Pimpinelli, J. Villain *Physics of Cristal Growth* Aléa-Saclay series n° 4, Cambridge University Press, Cambridge (UK), 1998.
2. B. Caroli, C. Caroli, B. Roulet in *Solids far from equilibrium*, C. Godreche ed. (Cambridge University Press, 1991). *Instabilities of planar solidification fronts.*
3. J. Villain, A. Pimpinelli. *Physique de la croissance cristalline*, Coll. Aléa–Saclay, Eyrolles (Paris), 1995.
4. L. Landau, E. Lifshitz *Statistical Physics*, Pergamon Press, London, 1959.
5. G. Blatter *Surface Sci.* **145**, 419 (1984).
6. A.C. Levi *Surface Sci.* **137**, 385 (1984).
7. G. Armand, J.R. Manson *Phys. Rev. B* **37**, 4363 (1988).
8. F. Gallet, S. Balibar, E. Rolley *J. de Physique* **48**, 369 (1987).
9. S. Balibar, B. Castaing *Surface Sci. Reports* **5** (1985), 87.
10. H. van Beijeren, I. Nolden in *Structure and Dynamics of Surfaces* II, W. Schommers and P. von Blanckenhagen eds., Topics in Current Physics **43** (Springer, Berlin, 1987).
11. J. Lapujoulade *Surf. Sci. Rep.* **20**, 191 (1994).
12. P. Nozières in *Solids far from equilibrium*, C. Godréche ed. (Cambridge University Press, 1991). *Shape and growth of crystals.*
13. J.D. Weeks in *Ordering in Strongly Fluctuating Condensed Matter Systems* T. Riste ed. (Plenum, New York, 1980) p. 293. *The roughening transition.*
14. J. Villain, D. Grempel, J. Lapujolade *J. Phys. F* **15**, 809 (1985).
15. W.K. Burton, N. Cabrera, F. Frank *Phil. Trans. Roy. Soc.* **243**, 299 (1951).
16. A.-L. Barabási, H.E. Stanley *Fractal Concepts in Surface Growth* Cambridge University Press, Cambridge (UK), 1995.
17. Edwards, S.F., Wilkinson, D.R. *Proc. Roy. Soc. A 381*, 17 (1982).
18. Mullins, W.W. *J. Appl. Phys.* **30**, 77 (1959).

7 Experiments on Solid Surfaces

Jean-Marc Gay and Laurent Lapena

CRMC2 CNRS, Campus de Luminy, case 913, 13288 Marseille Cedex 9, France

The number of reflectivity studies has largely increased in the last years so that this technique is nowadays well developed using various x-ray sources ranging from sealed tubes and rotating anodes to last generation synchrotrons with setups adapted to liquid or solid surfaces [1,2]. This chapter is focussed on experimental solid surface studies. Let's just mention that solid surfaces allow more flexibility than liquid ones since they can be oriented in any direction without deviating the incident beam. In addition, some other questions about resolution with long range correlations on liquid surfaces are generally avoided with rough solid surfaces.

7.1 Experimental Techniques

7.1.1 Reflectivity Experiments

Measurement Setups and Procedures Setups for x-ray reflectivity experiments are now rather common. Figure 7.1 schematically shows a typical experimental system for standard reflectivity measurements which can use the more or less divergent beam emitted by a conventional or a synchrotron source. Slits and a monochromator produce a collimated monochromatic beam which impinges onto the sample surface under the incidence angle θ_{in}. Various monochromators are available depending on the desired resolution, and intensity. The sample is mounted on a goniometer with precise motors (the angular displacement accuracy is at least 1 mdeg) which control the sample surface (plane (x, y)) and the detector positions. Slits or an analyser crystal are set at the detector side to reduce the background and the divergence of the outgoing beam. The detector position is given by the θ_{sc} and ψ angles, polar angles in the plane of incidence (x, z) and out of this plane in the y-direction respectively (see Sect. 2.3.1). The components of the wave-vector transfer are then

$$\begin{cases} q_x = k_0(\cos\theta_{\text{sc}}\cos\psi - \cos\theta_{\text{in}}) \\ q_y = k_0 \cos\theta_{\text{in}}\sin\psi \\ q_z = k_0(\sin\theta_{\text{in}} + \sin\theta_{\text{sc}}) \end{cases} \quad (7.1)$$

Specular reflection is characterised by $\theta_{\text{sc}} = \theta_{\text{in}}$ and $\psi = 0$, that makes $\mathbf{q} = (0, 0, 2k\sin\theta_{\text{in}})$. The angle $\theta_{\text{in}} + \theta_{\text{sc}}$ is often also denoted 2θ, whereas the incident angle θ_{in} defined by the orientation of the surface is named ω.

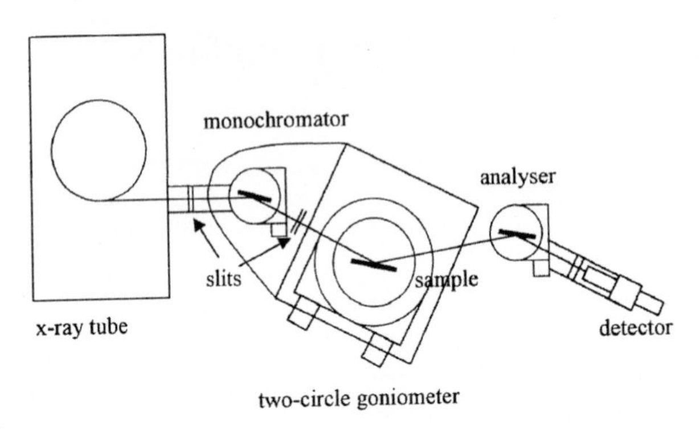

Fig. 7.1. Standard setup of a triple-axis diffractometer for specular reflectivity and off specular diffuse scattering. From [3]

The rotations ω and 2θ simply describe the position of the surface and the detector relatively to the incoming beam. For small scattering angles, the wave-vector transfer is

$$\mathbf{q} = (q_x \approx 2k_0\theta(\omega - \theta), q_y \approx k_o\psi, q_z \approx 2k_0\theta)$$

Different combinations of ω and 2θ are used for mapping the reciprocal space (q_x, q_z) as shown in Fig. 7.2:

- $\omega = 2\theta$ scan or specular scan, that keeps the condition $\theta_{\mathrm{sc}} = \theta_{\mathrm{in}}$. In this geometry, the detector measures the specularly reflected beam and maps the reciprocal space radially in the normal z-direction, ie $\mathbf{q} = (0, 0, q_z 2k_0\theta)$.
- $\omega = \theta + \Delta\theta_0 / 2\theta$ scan or longitudinal diffuse scan, with offset $\Delta\theta_0$. The reciprocal space is still radially mapped, but in a direction with an angle $\Delta\theta_0$ from the surface normal. This type of scan is useful for measuring the diffuse scattering contribution close to the specular peak. Subtracted from the measured specular reflectivity, it allows to get the true specular reflection.
- ω scan or rocking scan at fixed 2θ. The rotation is limited to the range $\omega = 0, \omega = 2\theta$. For small scattering angles, the reciprocal space is measured in the transverse q_x-direction of the accessible area at different q_z levels.
- 2θ - scan or detector scan at fixed ω. In this geometry, q_x and q_z are both varied with no limitation as soon as $2\theta > \omega$. In addition, a constant sample area is illuminated during the scan.
- ψ - scan at fixed $\omega = \theta$. In this geometry, the reciprocal space is measured in the q_y-direction normal to the incidence plane. This type of measurement is generally less used than the others because most reflectometers have no -motion out of the incidence plane even though it offers a full q_y range accessibility [4]. It is usually preferred to let slits widely opened in the y-direction leading to an effective integration over q_y.

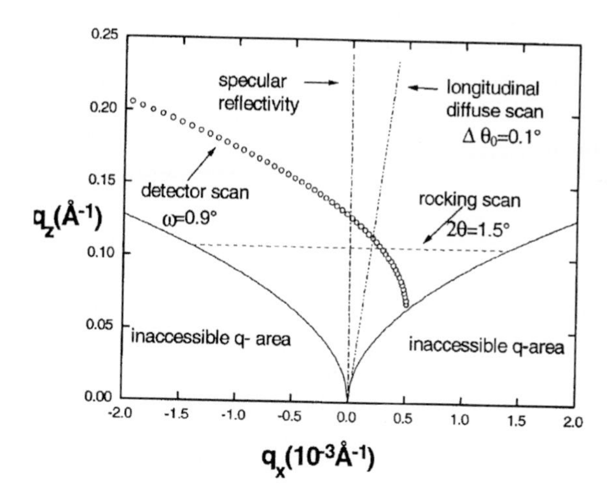

Fig. 7.2. Scans in reciprocal space with representation of the inaccessible areas. The line scans are shown for the $CuK\alpha_1$ radiation

Resolution The choice of slit widths or monochromator and analyser crystals is essential for setting the resolution of the measurements for given configurations [3,5–7]. The calculations of different resolution functions can be found in the literature. Considering the case of in-plane scattering ($\psi = 0$) with small scattering angles, one gets from Eqs. (7.1):

$$\begin{cases} \delta q_x = k_0(\theta_{\text{in}}^2 \Delta\theta_{\text{in}}^2 + \theta_{\text{sc}}^2 \Delta\theta_{\text{sc}}^2) \\ \delta q_z = k_0(\Delta\theta_{\text{in}}^2 + \Delta\theta_{\text{sc}}^2), \end{cases} \tag{7.2}$$

for a monochromatic radiation, and incoming and outgoing beams with angular divergence $\Delta\theta_{\text{in}}$ and acceptance $\Delta\theta_{\text{sc}}$ respectively. Assuming $\theta_{\text{in}} \approx \theta_{\text{sc}} \approx \theta$ and $\Delta\theta_{\text{in}} \approx \Delta\theta_{\text{sc}} \approx \Delta\theta$, one can estimate the resolutions as

$$\begin{cases} \delta q_x = q_z \Delta\theta \\ \delta q_z = 2k_0 \Delta\theta. \end{cases} \tag{7.3}$$

For the experimental results reported in the following section from measurements on a triple axis spectrometer using flat Ge(111) monochromator and analyser and the CuK_α radiation, the typical values for resolution are $\delta q_z \approx 2.10^{-3} \mathring{A}^{-1}$ and $\delta q_x \approx 2.10^{-4} q_z$.

The resolution functions determine the maximum length scales which can be coherently proben by the experimental measurements. For the above mentionned resolution, they are $z_{\max} \approx 10^3 \mathring{A}$ and $x_{\max} \approx 10^4/q_z \approx 10^7 \mathring{A}$.

Data Analysis Before any comparison with theoretical simulations of reflectivity, the data may have to be corrected for various geometrical factors depending on the measurement configuration [6,7]. Alternatively, the simulations may include the corrections and the data are considered as they are.

The specular reflectivity is defined (see Eqs. (3.5), (3.6)) as the ratio of the reflected intensity at the scattering angle 2θ to the intensity of the direct through beam. For very grazing incidence angles θ_{in}, the sample surface (length l along the x-direction) is almost parallel to the beam so that a fraction of the incoming beam (width b_{in}) does not illuminate the surface and cannot be consequently reflected. This geometrical effect is responsible for the bump experimentally observed at low angles. A plateau of total external reflection below the critical angle, as theoretically expected, is obtained upon renormalisation of the experimental data. Assuming a uniform rectangular flux distribution of the primary beam, the correction factor is simply expressed as a function of $\theta_{\text{in}}^0 = \arcsin(b_{\text{in}}/l)$ by:

$$\begin{cases} f_{\text{spec,in}} = \sin\theta_{\text{in}} / \sin\theta_{\text{in}}^0 & \text{for } \theta_{\text{in}} < \theta_{\text{in}}^0 \\ f_{\text{spec,in}} = 1 & \text{for } \theta_{\text{in}} \geq \theta_{\text{in}}^0 \end{cases} . \tag{7.4}$$

On the detector side, slits or the analyser crystal size may limit the measured beam width b_{sc}. A correction factor, independent of the incidence and exit angles θ_{in} and θ_{sc}, is given by:

$$\begin{cases} f_{\text{spec,sc}} = b_{\text{in}} / b_{\text{sc}}^0 & \text{if } b_{\text{sc}} < b_{\text{in}} \\ f_{\text{spec,sc}} = 1 & \text{if } b_{\text{sc}} \geq b_{\text{in}} \end{cases} . \tag{7.5}$$

The actual specular reflectivity is therefore deduced from the measured reflected intensity $I_{\text{spec}}(\theta_{\text{in}})$ at the incident angle $\theta_{\text{in}} = \theta$ using the formula:

$$R(\theta) = R(\theta_{\text{in}}) = f_{\text{spec,in}}(\theta_{\text{in}}) f_{\text{spec,sc}} \frac{I_{\text{spec}}(\theta_{\text{in}})}{I_0} ,$$

where I_0 is the intensity of the incident beam. On the other hand, the measured diffuse intensity is proportional to (i) the incident beam intensity, (ii) the illuminated area of the sample, and (iii) the resolution volume (see Sect. 4.7), whereas the calculation of the diffuse intensity is usually based on the expression of the scattering cross-section for a unit incident flux. The illuminated area which covers the full length l of the sample at low angles, decreases as soon as $\theta_{\text{in}} \geq \theta_{\text{in}}^0$. For the data normalisation, one has to consider a correction factor proportional to the actual illuminated area:

$$\begin{cases} f_{\text{diff,in}}(\theta_{\text{in}}) = 1 & \text{for } \theta_{\text{in}} < \theta_{\text{in}}^0 \\ f_{\text{diff,in}}(\theta_{\text{in}}) = \sin\theta_{\text{in}}^0 / \sin\theta_{\text{in}} & \text{for } \theta_{\text{in}} \geq \theta_{\text{in}}^0 \end{cases} . \tag{7.6}$$

This is the normalisation prefactor mentioned in formula (4.61).
Similar corrections have to be taken into account on the exit side depending on the slit widths or the size of the analyser crystal. A correction factor is

required when the width of the illuminated area (l for $\theta_{\text{in}} < \theta_{\text{in}}^0$, or b_l/θ_{in} for $\theta_{\text{in}} \geq \theta_{\text{in}}^0$) is larger than the area actually "seen" by the detector (width $b_{\text{sc}}/\sin\theta_{\text{sc}}$). The correction factor can be expressed by:

$$\begin{cases} f_{\text{diff,sc}}(\theta_{\text{in}}, \theta_{\text{sc}}) = 1 & \text{for } \theta_{\text{sc}} < \arcsin\left(\frac{b_{\text{sc}}}{l f_{\text{diff,in}}(\theta_{\text{in}})}\right) \\ f_{\text{diff,sc}}(\theta_{\text{in}}, \theta_{\text{sc}}) = \frac{b_{\text{sc}}}{l f_{\text{diff,in}}(\theta_{\text{in}})} & \text{for } \theta_{\text{sc}} < \arcsin\left(\frac{b_{\text{sc}}}{l f_{\text{diff,in}}(\theta_{\text{in}})}\right) \end{cases} . \tag{7.7}$$

The experimental diffuse intensity finally appears with the form:

$$I_{\text{diff}}(\theta_{\text{in}}, \theta_{\text{sc}}) = I_0 \; f_{\text{diff,in}}(\theta_{\text{in}}) \; f_{\text{diff,sc}}(\theta_{\text{in}}, \theta_{\text{sc}}) \int \frac{d\sigma}{d\Omega} d\Omega_{\text{sc}},$$

function of the scattering cross-section integrated over the angular resolution function of the detector.

Specular Reflectivity Once the true coherent specular reflectivity has been experimentally determined (see above the description and interest of the longitudinal diffuse scan), it is adjusted against a simulation with various parameters describing the investigated surface. Very often the surface of an homogeneous material is in fact made of a thin surface layer of different density resulting from various reasons: oxidation, mechanical treatment, inhomogeneous deposition, etc. As a consequence, the formalism for stratified media is the most commonly used even for single solid surfaces. Basically the thickness of the layer, its density and the surface and interfacial roughnesses are expected to come out from the specular reflectivity study. When diffuse scattering data are available, a simultaneous fit of all the data set is looked for, in order to get a full description of the sample configuration (see Sect. thereafter).

Exact theoretical descriptions of x-ray reflectivity from solid surfaces (of stratified media) are available for sharp surfaces and interfaces either from the matrix technique (see Chap. 3) or from the equivalent recursive approach initially developped by Parratt [8].

In reality, roughness of surfaces and interfaces can significantly alter the specularly reflected intensity. A rough interface can be seen as made of locally flat areas at different heights (see Chap. 2). A classical approximation considers a Gaussian height-distribution probability. The Névot-Croce factors [9] which depict the root-mean-square roughness of each interface can be easily introduced in the formalism derived for smooth interfaces (see Appendix 3.A). The analysis of the experimental data with this kind of model can be rather fast since the free parameters which describe each layer j of the considered system are restricted to the r.m.s roughnesses σ_j, the layer thicknesses d_j and densities δ_j. The final result of the analysis can then be shown as the density profile $\delta(z)$ of the investigated system. At the surface and at each interface, the profile looks like an error-function deduced from the gaussian probability of the height distribution. Each interface is treated independently from the others, that is expressed with the condition $\sigma_j \ll d_j$.

This approach does not hold when the r.m.s. roughness σ_j is on the same order of magnitude that the layer thickness d_j. For large roughnesses, the density profile does not show clearly identified plateaus corresponding to the different layers. The analysis is a little more complicated since one has first to guess the profile $\delta(z)$. With this initial guess, the investigated system is seen as a series of very thin layers p with sharp interfaces and density δ_p vaying from one to the other following the profile $\delta(z)$. This parametrisation of the system relies on a rather large number of density parameters which makes the calculation long, but still simply tractable with the exact Parratt formalism. A slightly different and faster procedure has been proposed by the group of Press [2]. This parametrisation is based on an " effective density model " which allows to consider the profile has made of individual layers even when the condition $\sigma_j \ll d_j$ is not fulfilled.

The above formalisms are based on an optical approach of light (not specially x-ray) scattering by stratified media. A different approach is also available for describing x-ray scattering. The so-called Born or kinematical approximation valid for the weak scattering regime does not hold when refraction effects are important and cannot be neglected, i.e. for angles close to the critical angle of total reflection. This drawback is usually overcome within the Distorded Wave Born Approximation (see Chap. 4). This theoretical framework is used for modelling both specular and off-specular diffuse scattering.

Off-Specular Diffuse Scattering Coherent specular and off-specular diffuse scattering are complementary for providing a complete set of parameters describing stratified media. In the simple case of a single layer on a semi-infinite substrate, besides the layer density, thickness and r.m.s. roughness parameters deduced from specular reflectivity, one can have access to a more detailed representation of the morphology of the sample through the lateral surface and interface height-height correlation functions, and the correlation function between the surface and the buried interface (see Sect. 2.2, and Sects. 8.2, 8.4, 8.5 for multilayers).

Many isotropic solid surfaces are self-affine so that the height-height correlation function C_{zz} can be simply expressed with three parameters (Eq. (2.26)): the r.m.s. roughness σ, the correlation length ξ which shows the scale on which the surface is rough and the Hurst parameter h related to the fractal dimension of the surface which describes how jagged or smooth it is [4,10].

7.1.2 Roughness Investigations with other Experimental Tools

Near Field Microscopy Near Field Microscopies have been now rather common tools for investigating the roughness of solid surfaces. Scanning Tunneling Microscopes (STM) and Atomic Force Microscopes (AFM) are particularly well suited for imaging surface morphology [11–13]. Combining a large number of line scans along the surface, they yield a detailed map of its roughness at various scales to some hundreds of μm^2. A statistical treatement of the images provides the power-spectral-density (PSD) of the surface from which can be determined the parameters σ, ξ, and h of the height-height correlation function mentionned in the previous section. Like x-ray reflectivity, near field microscopies are non-destructive techniques which can be used in rather fast measurements in laboratories. They are nevertheless local probes as compared to x-ray scattering. Studies with a satisfactory precision often require a very large amount of recorded data, that can be finally very time- and computer memory-consuming. Images of real space are always very appealing even if they only show the surface, ignoring the underlying interfaces.

Electron Microscopy Transmission Electron Microscopy (TEM) can be used for investigating the morphology of surface and interfaces of stratified solid materials [14]. This technique requires a delicate destructive sample preparation in order to get a thin slab cut normal to the surface. Different recipes are used depending on the investigated material which must be kept undammaged during the preparation process. Real space images of cross section TEM can clearly show the thickness of the layers and the morphology of the interfaces. R.m.s. roughness can be estimated from grey contrast profiles through the interfaces. More quantitative analysis of the TEM images would require calibrations against known standards. Information is always local.

7.2 Examples of Investigations of Solid Surfaces/Interfaces

7.2.1 Co/Glass - Self-Affine Gaussian Roughness

We report in this section the experimental study of a Co film, 150$\mathring{A}$ thick, deposited at room temperature on a glass substrate (see Sect. 6.2 for a description of the morphology of crystal surfaces).

X-Ray Reflectivity The x-ray scattering measurements were performed on a triple-axis diffractometer with flat Ge(111) monochromator and analyser crystals using the $CuK\alpha_1$ radiation emitted by a rotating anode operated at 10kW [15]. The detector was a standard NaI scintillator. Fig. 7.3a shows the specularly reflected intensity as a function of the scattering angle. The

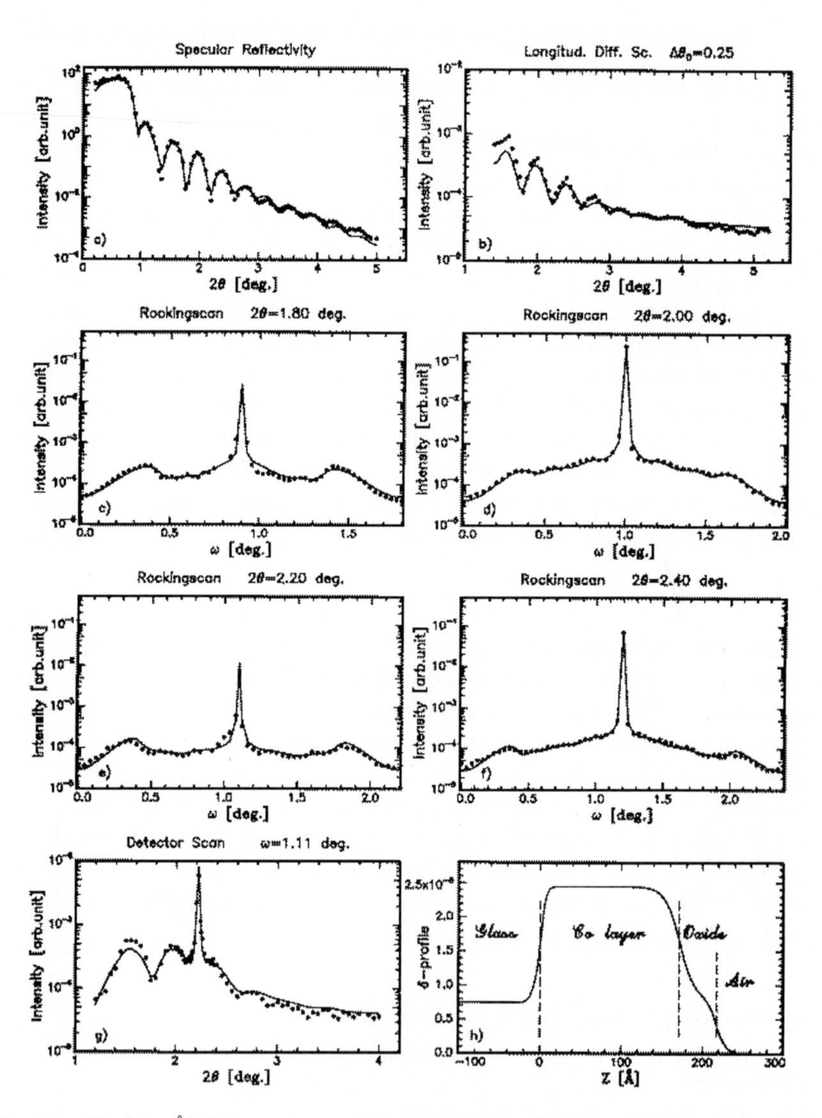

Fig. 7.3. Co (150 $\AA$) film on glass measured with the $CuK\alpha_1$ radiation (experimental data : symbols, and best fit : solid line). (a) Specular reflectivity. (b) Longitudinal diffuse scan with an offset $\Delta\theta_0 = 0.25°$. (c) to (f) Rocking scans at different scattering angles. (g) Detector scan with fixed incident angle ω. (h) δ-profile deduced from the adjustment of the experimental data

simulation has been calculated using the effective density model with two layers on top of a glass substrate (fixed density δ_0). The best fit parameters are given in Table 7.1. The corresponding δ-profile is represented in Fig. 7.3h. δ is proportional to the electron density of the medium (see Chap. 1). The surface layer is rather thin, that justifies the use of the effective density model. It is certainly made of cobalt oxide, since the sample was in air. The density of the cobalt layer is in quite good agreement with that of bulk cobalt. The r.m.s. roughness of the glass/Co interface is small as expected for the surface of a bare glass substrate.

Table 7.1. Co/glass. Summary of the different parameters deduced from the x-ray reflectivity, AFM and TEM studies. δ_j and d_j are the density and thickness of layer j respectively ($j = 0$ for the substrate). σ_j, ξ_j, h_j are the r.m.s., height-height correlation length and Hurst parameter describing the interfacial roughness between layer j and layer j+1. Vertical correlation between successive interfaces is considered with the α_{ij} coupling coefficient. In the adjustment of the x-ray reflectivity study, all the above parameters are free except δ_0. (*): fixed parameter

		X-Rays	AFM	TEM
Substrate (Glass)	$10^6\delta_0$	7.48(*)		
	σ_0 [Å]	5.1 ± 0.3	not accessible	6.8
	ξ_0 [Å]	1060 ± 500		
	h_0	0.5 ± 0.2		
	α_{01}	0.43 ± 0.05		
Co layer	$10^6\delta_1$	24.4 ± 0.5		
	d_1 [Å]	171.4 ± 0.5	not accessible	
	σ_1 [Å]	13.5 ± 0.3		
	ξ_1 [Å]	1200 ± 700		
	h_1	0.2 ± 0.1		
	α_{12}	1 ± 0.05		
Oxide surface layer	$10^6\delta_2$	8.9 ± 0.5		
	d_2 [Å]	45.8 ± 0.5		
	σ_2 [Å]	9.2 ± 0.6	9.1 ± 2.0	13.2
	ξ_2 [Å]	1200 ± 700	$1430 \pm 210/5000 \pm 60$	
	h_2	0.6 ± 0.3	0.8 ± 0.1	
	$d_1 + d_2$ [Å]	217.2 ± 1		215

Off-specular scattering has been measured in different modes shown in Fig. 7.3b-g. The longitudinal diffuse scan (Fig. 7.3b) has been used for extracting the true specular reflectivity reported above. Its oscillatory behaviour indicates correlated rough interfaces. A detector scan is also shown along with rocking curves at various scattering angles. All show the specular peak together with the diffuse scattering contribution. Yoneda wings (see Sect. 4.3.1) are observed on both sides of the rocking curves. They appear for incident and exit angles equal to the critical angle for total external reflection, for which refraction effect is clear. The off-specular diffuse scattering curves are fitted against a DWBA model with a self-affine roughness with gaussian probability. The best fit parameters can be found in Table 7.1. The adjustment performed with several diffuse scattering curves yields parameters for a complete description of the surface and interface morphology.

AFM Study of the Surface Roughness AFM images have been recorded on sample areas of $3.5 \times 3.5 \mu m^2$. Figure 7.4a shows a typical AFM image and Fig. 7.4b the height profile measured along an arbitrary line of the surface. A computational statistical analysis of the AFM images of the proben areas gives the r.m.s. roughness and the PSD. For isotropic self-affine rough surfaces described with the three parameters (σ, ξ and h), the PSD is expected to be almost constant for low q (ie large scale in direct space) and to decrease like $q^{-2(1+h)}$ for frequencies larger than a cutoff frequency associated to the correlation length. Figure 7.4c shows the experimental PSD with three regimes and two correlation lengths ξ_2 and ξ_2^*. The investigated sample surface is presumably more complicated than the simple self-affine model description. It is however worthwhile noting that the shortest correlation length and the Hurst parameter h agree pretty well with those deduced from the x-ray study (see Table 7.1).

TEM Study Cross section Transmission Electron Microscopy measurements have been performed on the same Co/glass sample. Figure 7.5a shows a TEM image on which can be seen the substate material, the intermediate layer (Co) and the surface transition layer. The image is quantitatively analysed by plotting the average grey level profile normal to the surface (Fig. 7.5b), which is representative of the density contrast through the sample. A smooth decrease of the profile at the surface prevents from distinguishing a surface oxide layer on top of the Co layer. After normalisation of the grey levels of the substrate and the Co layer, the profile can be compared to that deduced from the x-ray specular reflectivity work. The shape of the TEM and x-ray profiles are in rather good agreement indicating the same total thickness and similar r.m.s. roughnesses.

7.2.2 Si Homoepitaxy on Misoriented Si Substrate. Structured Roughness

Self affine surface models are very often well suited for statistical descriptions of isotropic growth-induced roughening of deposited films, as that presented in the above section. A quite different class of roughnesses is constituted with (quasi-)periodic undulations that make laterally structured rough surfaces/interfaces. For instance, such cases can result from growth on misori-

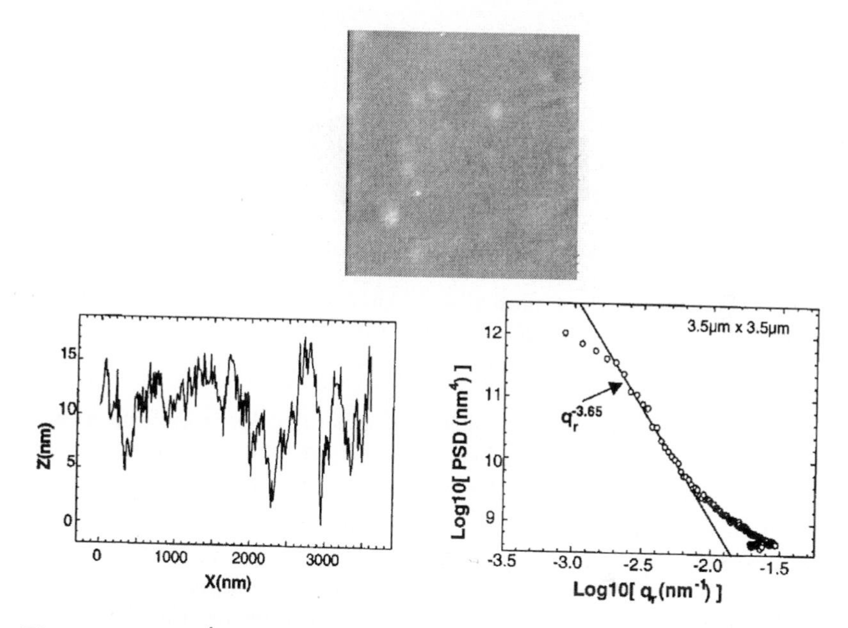

Fig. 7.4. Co (150Å) film on glass. (a) AFM image of the surface ($3.5 \times 3.5\mu$ m2), (b) height profile along an arbitrary line of the surface, (c) PSD from the AFM study

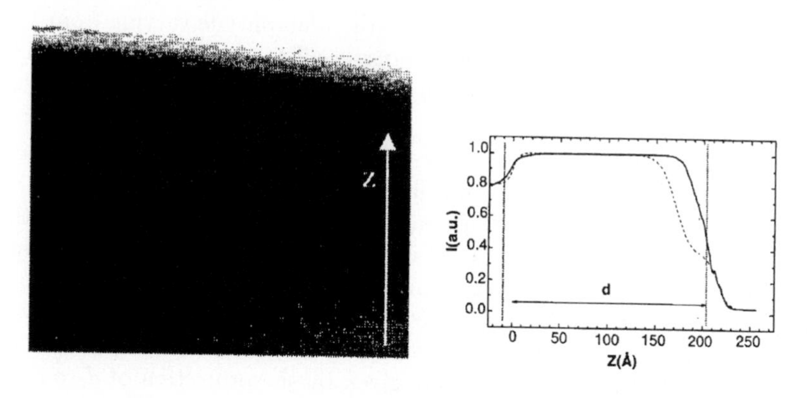

Fig. 7.5. Co (150Å) film on glass. (a) Cross section TEM image ($335 \times 335 Å^2$) and (b) normalised grey level profile normal to the surface (solid line) along with the *delta*-profile from the x-ray reflectivity study (dashed line)

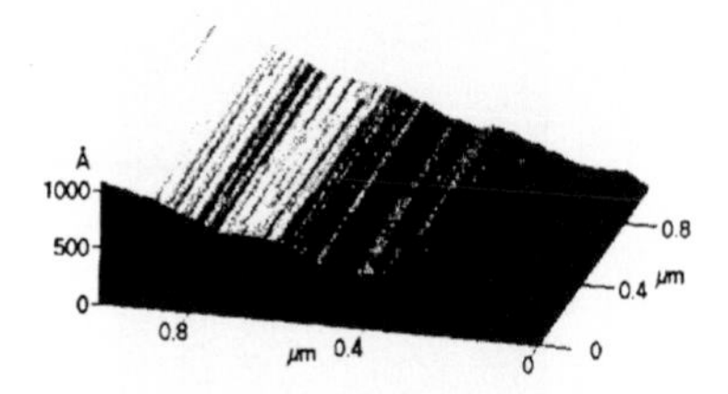

Fig. 7.6. AFM image of the surface of a Si film, 500 nm thick, deposited on misoriented Si(111) substrate. The image is $1.0 \times 1.0 \mu m^2$

ented surfaces with the influence of miscut-generated steps or from growth on specially designed surface gratings.

The work reported thereafter deals with a sample obtained by Si homoepitaxy on a Si Czochralski grown wafer. The substrate surface is misoriented by 10° around the[1-10] axis toward the [-1-12] axis with respect to the (111) plane. After cleaning, the initial substrate surface is composed of a regular array of triatomic steps with terraces about 5.5 nm wide and the 7×7 reconstruction. A total thickness of 500 nm of Si is deposited in a MBE chamber at 700° C with a rate of $0.15 nm.s^{-1}$ [16].

An AFM image of the Si(500nm)/Si sample surface is shown in Fig. 7.6. A long period (about 250 nm) undulation with an amplitude varying from 1 to 10 nm is clearly observed. X-ray specular and off-specular diffuse scattering investigations have been performed in order to provide a full description of the surface morphology.

The measured specular reflectivity can be easily modelled with a density profile made of a surface transition layer 6.6 nm thick on top of bulk Si. This thickness is in quite good agreement with the mean amplitude of the surface undulations seen in the AFM images. Off-specular diffuse scattering has been measured with rocking scans at different azimuthal orientations of the surface, ie at different angles α between the grazing incoming beam and the grooves of the surface undulations of period d_0 (see Fig. 7.7). The apparent period of the surface grating given by the projection of d_0 on the (x, z) scattering plane changes with α and gives rise to satellites in the rocking curves, the position of which allows the precise determination of d_0: 215 ± 5 nm. A careful examination of the satellite intensities shows that they are not symmetric. A simulation of the measured rocking curves is proposed (Fig. 7.8) in the simple frame of the kinematical approximation for the calculation of the structure factor of each satellite.

It is based upon an asymmetric surface profile shown in Fig. 7.8, the shape of which is adjusted to reproduce the experimental data. This profile of period 215 nm and amplitude 6.6 nm shows extended facets which make an angle of 2.6 ± 0.3 deg. with the average surface. This determination is

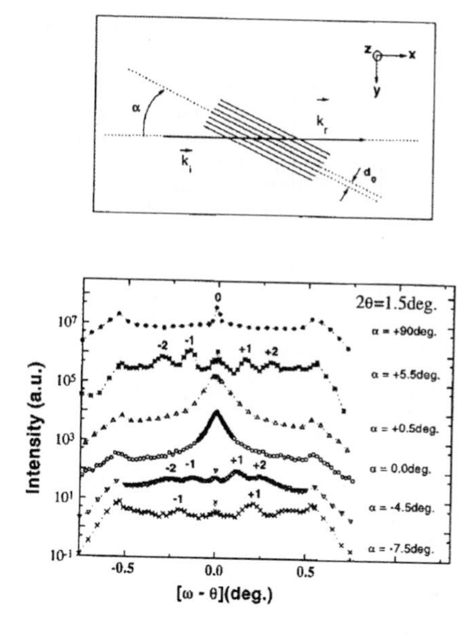

Fig. 7.7. Surface of a Si film, 500 nm thick, deposited on misoriented Si(111) substrate. (a) Azimuthal orientation of the surface grooves with respect to the incoming and reflected wave vectors. (b) X-ray off specular diffuse scattering measured with rocking scans recorded at $2\theta = 1.5deg.$ for various azimuthal angles α ($CuK\alpha_1$ wavelength)

quite consistent with cross section TEM images of this sample which reveal surface facets with an angle of 2.8 deg. with the average surface.

7.3 Conclusion

X-ray specular and off-specular diffuse scattering have now become rather common tools for investigating microscopic surface and interface morphology. Measurements can be performed on setups coupled with classical x-ray generators as well as synchroton light sources. The feasibility of the experiments and the associated information which can be deduced from the experimental data heavily depend on the quality of the investigated samples. Macroscopic faceting of a surface can completely obscure reflectivity. One has also to keep in mind that reflectivity decreases dramatically with r.m.s. roughness which consequently limits the measurable q-range. The extensive study of two types of roughness (self-affine fractal and structured grating surfaces) have been reported in this chapter as an illustration of experimental x-ray studies. When possible, a comparison is proposed with investigations with AFM or TEM

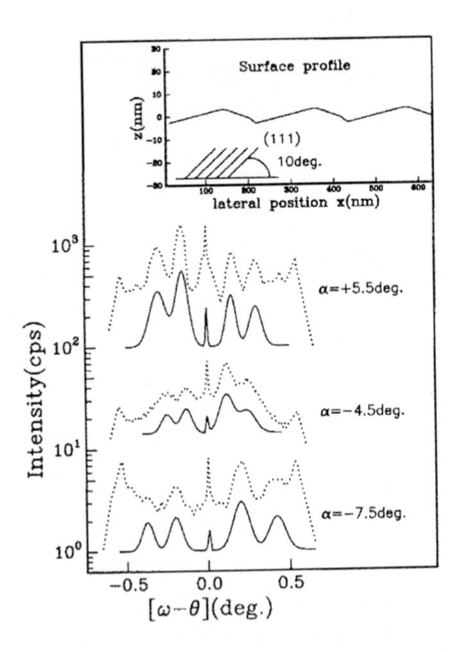

Fig. 7.8. Surface of a Si film, 500 nm thick, deposited on misoriented Si(111) substrate (see also Sects. 8.4 and 8.5). Simulation (solid line) of the experimental data (dashed line) with the surface profile shown in the inset

which give consistent parameters. X-ray reflectivity is nevertheless the only non-destructive tool for looking at buried interfaces.

References

1. Proceedings of the 4th Intern. Conf. on Surface X-ray and Neutron Scattering, G.P. Felcher and H. You Ed., North Holland, *Physica B* 221 (1996).
2. M. Tolan, in X-Ray Scattering from Soft-Matter Thin Films, Springer Verlag, 1998 in press.
3. L. Brügemann, R. Bloch, W. Press and M. Tolan, *Acta Cryst. A* **48**, 688 (1992).
4. T. Salditt, C. Brandt, T.H. Metzger, U. Klemradt and J. Peisl, *Phys. Rev. B* **51**, 5617 (1995).
5. W.H. de Jeu, J.D. Shindler and E.A.L. Mol, *J. Appl. Cryst.* **29**, 511 (1996).
6. M.F. Toney and D.G. Wiesler, *Acta Cryst. A* **49**, 624 (1993).
7. A. Gibaud, G. Vignaud and S.K. Sinha, *Acta Cryst. A* **49**, 642 (1993).
8. L.G. Parratt, *Phys. Rev. B* **95**, 359 (1954).
9. L. Névot et P. Croce, *Revue de Physique Appliquée* **15**, 761 (1980).
10. S.K. Sinha, E.B. Sirota, S. Garoff and H.B. Stanley, *Phys. Rev. B* **38**, 2297 (1988).
11. M.W. Mitchell and D.A. Bonnell, *J. Mater. Res.* **5**, 2244 (1990).
12. H. Rohrer, *Surface Science* **299-300**, 956 (1994).
13. C.F. Quate, *Surface Science* **299-300**, 980 (1994).
14. P. Schwander, C. Kisielowski, M. Seibt, F.H. Baumann, Y. Kim, and A. Ourmazd, *Phys. Rev. Lett.* **71**, 4150 (1993).
15. J.M. Gay, P. Stocker and F. Réthoré, *J. Appl. Phys.* **73**, 8169 (1993).
16. M. Ladevèze, I. Berbezier and F. Arnaud d'Avitaya, *Surface Science* **352-354**, 797 (1996).

8 X-ray Reflectivity by Rough Multilayers

Tilo Baumbach[1] and Petr Mikulik[2]

[1] Fraunhofer Institut Zerstörungsfreie Prüfverfahren, EADQ Dresden, Krügerstraße 22, D-01326 Dresden, Germany, *Present address:* European Synchrotron Radiation Facility BP 220, F-38043, Grenoble Cedex France,
[2] Laboratory of Thin Films and Nanostructures, Faculty of Science Masaryk University, Kotlářská 2, 611 37 Brno, Czech Republic

8.1 Introduction

One tendency in present material research is the increasing ability to structure solids in one, two and three dimensions at a sub-micrometer scale. Based on various material systems artificial *mesoscopic layered superstructures* such as multilayers, superlattices, layered gratings, quantum wires and dots have been fabricated successfully. This has opened new perspectives for manifold technological applications (e.g. for anticorrosion coating and hard coating, micro and optoelectronic devices, neutron and x-ray optical elements, magnetooptical recording).

The perfection of mesoscopic layered super-structures is characterised by

1. the perfection of the super structure (grating shape, periodicity, layer thickness ...),
2. the interface quality (roughness, graduated hetero-transition, interdiffusion ...),
3. crystalline properties (strain, defects, mosaicity ...).

Roughness is of crucial importance for the physical behaviour of interfaces. Roughness reduces the specular reflectivity of mirrors and wave guides for x-ray and neutron optics. Moreover it creates unintentional diffuse scattering. In magnetic layers it changes the interface magnetisation. Roughness promotes corrosion and influences the hardness of materials. It disturbs the electronic band structure in semiconductor devices. Interface roughness supports the generation of crystalline defects in layered structures. In multilayers already the roughness of the substrate or the buffer layer influences the quality of all subsequent layers. Depending on the growth process the roughness profile can be partially replicated from interface to interface.

Interface roughness is a random deviation of the layer shape from an ideally smooth plane. We consider here roughness with correlation properties of *mesoscopic* (sub-micrometer) scale. Irradiating a macroscopic area of the sample, surface sensitive x-ray scattering allows the investigation of the statistical behaviour of the roughness profile.

Interface roughness in multilayers can be studied by all surface sensitive x-ray scattering methods (x-ray reflection (XRR), grazing incidence diffraction

(GID), strongly asymmetric x-ray diffraction (SAXRD)) employing physical principles similar to the case of simple surfaces. They are based on

1. the reduction of the information depth at grazing angles of incidence and exit,
2. reflection of x-rays by the individual interfaces of a multilayer (ML) at small angles of incidence,
3. interference of the waves reflected by different interfaces,
4. diffuse scattering of x-rays by interface disturbances.

Specular x-ray reflection (SXR) as the most frequently used method studies the depth profile of the electron density. It detects the density gradient at the interface between two layers, where from we conclude on the r.m.s. roughness. Grazing incidence diffraction and strongly asymmetric x-ray diffraction detect interface roughness via the strain and the depth profile of the Fourier components of the electron density. The measurement of diffuse x-ray scattering (DXS) gives a clear evidence of interface roughness, distinguishing between roughness and graduated interfaces due to transition layers, interdiffusion or graduated hetero-transitions. Up to now DXS has frequently been observed in the XRR mode [1–10]. First measurements of DXS in the diffraction mode have been reported recently [11,?]. DXS by multilayers enables one to characterise the lateral correlation properties of interfaces similar to DXS by surfaces. Moreover it allows to detect vertical roughness replication from interface to interface. DXS at grazing incidence occurs under condition of simultaneous intense specular reflection. This gives rise to strong effects of multiple scattering [5,7,8,13,14,10,12]. That is why semi-dynamical methods such as the distorted wave Born approximation (DWBA) are more appropriate to explain the DXS features than kinematical treatments.

The paper intends to give an introduction into theoretical and experimental aspects of x-ray reflection by solid multilayers with rough interfaces, illustrated by various examples. We start in section 8.2 with a short presentation of rough multilayers and of the notations used in this chapter.

In section 8.3 we will introduce in the experimental set-up and usual experimental scans and in the following sections we apply the results of the chapters 3 and 4 on multilayered samples with different types of interface correlation properties. There we discuss typical features of the reflection curves and reciprocal space maps by various experimental examples. Afterwards, we mention the investigation of roughness by *surface sensitive diffraction methods* and at the end we study the reflectivity by intentionally laterally structured multilayers (*gratings*).

Throughout the chapter the *reciprocal space representation* of the optical potential and the scattering processes allows us to outline the scattering principles in a geometrical way. The basic principles of it are summarised in the appendix.

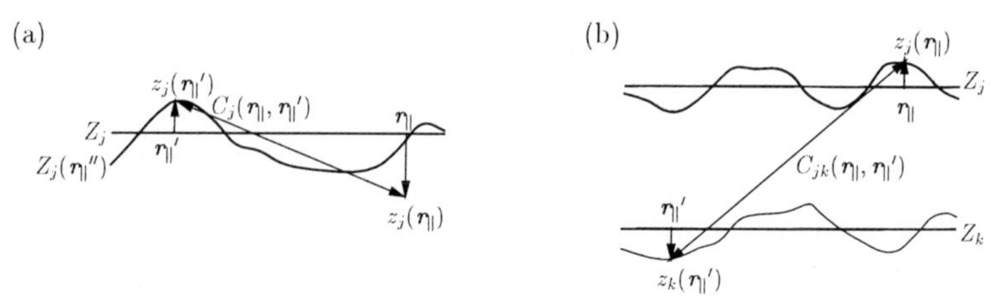

Fig. 8.2. Notation of the interface displacements and schematical representation of the correlation function of one (a) and of two interfaces (b)

one interface, i.e. for a substrate. In this section, we will treat the correlation properties between different interfaces of a multilayer. We introduce the two-dimensional probability density of *two* interfaces, Fig. 8.2(b),

$$p_2(z_j, z_k') = p\big(z_j(\boldsymbol{r}_\parallel), z_k(\boldsymbol{r}_\parallel{}')\big) , \tag{8.8}$$

and height-height correlation function

$$C_{jk}(\boldsymbol{r}_\parallel - \boldsymbol{r}_\parallel{}') = \big\langle z_j(\boldsymbol{r}_\parallel) z_k(\boldsymbol{r}_\parallel{}')\big\rangle . \tag{8.9}$$

Usually the perfection of interfaces in multilayers is essentially influenced by the quality of the substrate or buffer surface. The surface defects can be replicated in growth direction. Different replication behaviours have been observed, depending on the material system, layer setup and the growth conditions. The following replication model has been proposed in [15]: 1. during the growth of the jth layer, the roughness profile $z_{j+1}(\boldsymbol{r}_\parallel)$ of the lower interface is *partially replicated* and 2. other defects, an *intrinsic roughness* $\Delta_j(\boldsymbol{r}_\parallel)$, are induced by imperfections of the growth process

$$\begin{aligned} z_j(\boldsymbol{r}_\parallel) &= \Delta_j(\boldsymbol{r}_\parallel) + \int d\boldsymbol{r}_\parallel{}' \, z_{j+1}(\boldsymbol{r}_\parallel{}') \, a_j(\boldsymbol{r}_\parallel - \boldsymbol{r}_\parallel{}') \\ &= \Delta_j(\boldsymbol{r}_\parallel) + z_{j+1}(\boldsymbol{r}_\parallel) \otimes a_j(\boldsymbol{r}_\parallel) . \end{aligned} \tag{8.10}$$

where $\otimes$ denotes a convolution product. Here a non-random replication function $a_j(\boldsymbol{r}_\parallel)$ has been introduced, determining the "degree of memory" of the interface at the top for the roughnes profile at the bottom interface. If the replication function is zero, the upper interface of a layer "forgets" the interface profile at the layer bottom and its profile is entirely determined by the intrinsic roughness (*no replication*). *Identical profile replication* is achieved for zero intrinsic roughness and full replication ($a_j(\boldsymbol{r}_\parallel)$ equals the delta function). Other cases are discussed in detail in [15] and will win our interest within the discussion of the experimental results.

In later sections we will use the Fourier transformation of the interface correlation functions

$$\tilde{C}_{jk}(\boldsymbol{q}_\parallel) = \int d\boldsymbol{R}_\parallel \, C_{jk}(\boldsymbol{R}_\parallel) \, e^{i\boldsymbol{q}_\parallel \boldsymbol{R}_\parallel} = \big\langle \tilde{z}_j(\boldsymbol{q}_\parallel) \tilde{z}_k^*(\boldsymbol{q}_\parallel)\big\rangle \tag{8.11}$$

with

$$\tilde{z}_j(q_\parallel) = \tilde{\Delta}_j(q_\parallel) + \tilde{z}_{j+1}(q_\parallel)\,\tilde{a}_j(q_\parallel) \;. \tag{8.12}$$

In the following we neglect any statistical influence of the interface profile $z_{j+1}(r_\parallel)$ on the intrinsic roughness $\Delta_j(r_\parallel)$. Also the intrinsic roughness of different interfaces shall be statistically independent. Then we find the recursion formula for the Fourier transform of the correlation function

$$\tilde{C}_{jk}(q_\parallel) = \tilde{C}_{j+1,k+1}(q_\parallel)\,\tilde{a}_j(q_\parallel)\,\tilde{a}_k(q_\parallel) + \delta_{jk}\,\tilde{K}_j(q_\parallel) \;, \tag{8.13}$$

where $\tilde{K}_j(q_\parallel)$ is the Fourier transform of the correlation function of the intrinsic roughness

$$K_j(r_\parallel - r_\parallel{}') = \langle \Delta_j(r_\parallel)\,\Delta_j(r_\parallel{}')\rangle \;. \tag{8.14}$$

If we assume for all layers the same replication function $a(r_\parallel)$ and the same intrinsic roughness $\Delta(r_\parallel)$ (replicated substrate roughness $z_N(r_\parallel)$, for instance) we get the explicit expressions for the Fourier transforms of the *in-plane correlation function*

$$\tilde{C}_{jj}(q_\parallel) = \tilde{C}_{NN}(q_\parallel)\left[\tilde{a}(q_\parallel)\right]^{2(N-j)} + \tilde{K}(q_\parallel)\,\frac{\left[\tilde{a}(q_\parallel)\right]^{2(N-j-1)} - 1}{\left[\tilde{a}(q_\parallel)\right]^{2} - 1} \tag{8.15}$$

($\tilde{C}_{NN}(Q_\parallel)$ is the correlation function of the substrate) and of the *inter-plane correlation function*

$$\tilde{C}_{k\geq j}(q_\parallel) = \tilde{C}_{kk}(q_\parallel)\left[\tilde{a}(q_\parallel)\right]^{(k-j)} \;. \tag{8.16}$$

The physical meaning of the particular terms in (8.15) is obvious. The first term on the right hand side represents the influence of the substrate surface modified by the replication function, the second term is due to the intrinsic roughness of the layers beneath the layer j.

Knowing $\tilde{C}_{jj}(q_\parallel)$ we can calculate the mean square roughness σ_j^2 of the jth interface:

$$\sigma_j^2 = \langle z_j^2(r_\parallel)\rangle = \int dq_\parallel\,\tilde{C}_{jj}(q_\parallel) \;. \tag{8.17}$$

8.3 Setup of X-Ray Reflectivity Experiments

In this section we outline the experimental setup to investigate the fine structure of the reflected intensity pattern in vicinity of the origin of reciprocal space (000) under conditions of small angles of incidence and exit with respect to the sample surface.

8.3.1 Experimental Setup

A conventional x-ray reflectometer is drawn in Fig. 8.3. The x-ray source (a conventional x-ray tube or a synchrotron) emits a more or less divergent and polychromatic beam. The *monochromator* (a crystal or a multilayer mirror) and entrance slits produce a sufficiently monochromatic and parallel beam, hitting the sample surface under the incident angle θ_{in}. Its angular divergence is characterised by the spatial angle $\Delta\Omega_{\text{in}}$. The sample is mounted on a goniometer, which allows one to change the incident angle θ_{in} by the rotation ω. The x-rays are reflected (scattered) by the sample. The coherently reflected beam leaves the sample in specular direction (under the exit (final) angle $\theta_{\text{sc}} = \theta_{\text{sc}}$ in the plane of incidence). Due to roughness there occurs diffuse scattering into the upper half space of the sample. A detector rotates around the sample and measures the *flux* of photons (in units of counts per second) through the detector window, which defines the spatial angle interval $\Delta\Omega_{\text{det}}$ around a certain spatial angle Ω_{sc} (sufficiently defined by θ_{sc} in the coplanar case). If we suppose a perfectly monochromatic and parallel incident beam of intensity I_0 then the idealised flux through the detector window is related with the differential scattering cross section by

$$J = I_0 \int d\sigma = I_0 \int_{\Omega_{\text{sc}}-\Delta\Omega_{\text{det}}/2}^{\Omega_{\text{sc}}+\Delta\Omega_{\text{det}}/2} \left(\frac{d\sigma}{d\Omega}\right) d\Omega \,. \tag{8.18}$$

Taking the divergence and the intensity profile of the incident beam into account, we obtain

$$J = \int_{\Delta\Omega_{\text{in}}} d\Delta\Omega_{\text{in}}\, I_0(\Delta\Omega_{\text{in}}) \int_{\Omega_{\text{sc}}-\Delta\Omega_{\text{det}}/2}^{\Omega_{\text{sc}}+\Delta\Omega_{\text{det}}/2} \frac{d\sigma(\Omega_{\text{in}} + \Delta\Omega_{\text{in}}, \Omega)}{d\Omega}\, d\Omega \,. \tag{8.19}$$

Actually, in case of a large sample, the detector slits select another angular interval for each point on the illuminated sample area. That can be overcome replacing the detector slits by an analyser (also a perfect crystal or a multilayer mirror) in front of the detector similar a triple-crystal diffractometer (TCD). The monochromator is the "first crystal", the sample the "second

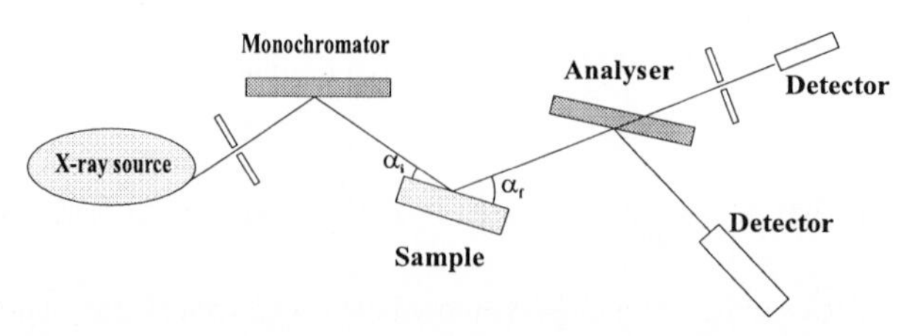

Fig. 8.3. Schematic setup of an x-ray reflectometer (source, monochromator, sample, slits and detector) and of a triple-crystal-like diffractometer (source, monochromator, sample, analyser and detector)

crystal" and the analyser the "third crystal". The flux measured by the TCD is

$$J = \int_{\Delta\Omega_{\rm in}} d\Delta\Omega_{\rm in}\, I_0(\Delta\Omega_{\rm in}) \int d\Omega\, \frac{d\sigma(\Omega_{\rm in} + \Delta\Omega_{\rm in}, \Omega)}{d\Omega}\, \mathcal{D}(\Omega - \Omega_{\rm sc}) \;, \quad (8.20)$$

where $\mathcal{D}(\Delta\Omega)$ is the reflectivity of the analyser.

8.3.2 Experimental Scans

Mapping the measured flux for different angles of incidence and exit we can plot the measured scattering pattern in angular space, $\mathcal{J}(\Omega_{\rm in}, \Omega_{\rm sc})$, or by three reciprocal space coordinates and one angular coordinate of the sample, e.g. $\mathcal{J}(\boldsymbol{k}_{\rm sc} - \boldsymbol{k}_{\rm in}, \theta_{\rm in})$. Restricting ourselves on coplanar reflection ($\boldsymbol{k}_{\rm sc}, \boldsymbol{k}_{\rm in}$ and the surface normal are in the same plane), the angular representation $\mathcal{J}(\theta_{\rm in}, \theta_{\rm sc})$ and the reciprocal space representation $\mathcal{J}(\boldsymbol{q})$ with the *scattering vector* $\boldsymbol{q} = \boldsymbol{k}_{\rm sc} - \boldsymbol{k}_{\rm in}$ are equivalent.

The principal rotations of a (coplanar) TCD are:

1. The rotation 2θ of the detector arrangement in the coplanar scattering plane around the sample: 2θ measures the *scattering angle* ($2\theta = \theta_{\rm in} + \theta_{\rm sc}$), the variation of 2θ changes $\theta_{\rm sc}$ ($\Delta 2\theta = \Delta\theta_{\rm sc}$).
2. The rotation ω of the sample around the same axis: $\theta_{\rm in} = \omega$, $\theta_{\rm sc} = 2\theta - \omega$, a variation of ω changes simultaneously $\theta_{\rm sc}$ and $\theta_{\rm in}$ ($\Delta\omega = \Delta\theta_{\rm in} = -\Delta\theta_{\rm sc}$).

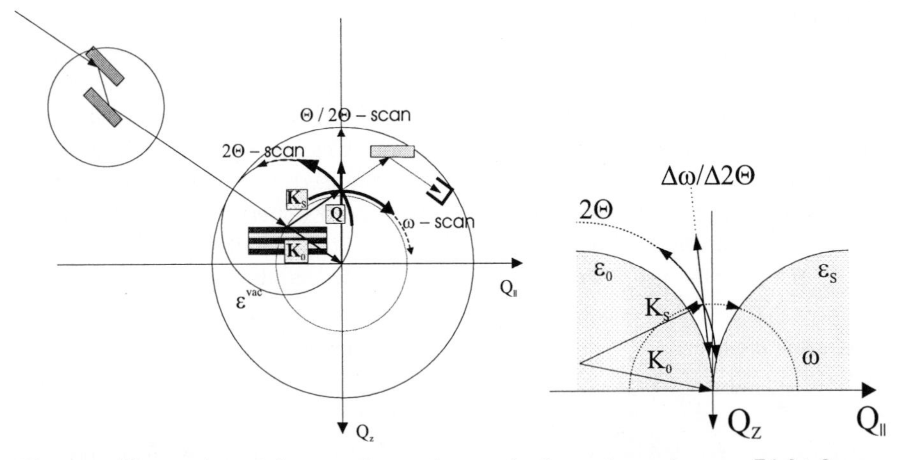

Fig. 8.4. Illustration of the experimental scans in the reciprocal space. Right figure shows the enlargement around its origin, where x-ray reflection takes place. The 2θ-scan (detector-scan) follows the Ewald circle of the incident wave. The ω-scan represents rocking scan, which is transversal for XRR. For $2\omega = 2\theta$ it is a q_z-scan with $q_{\parallel} = 0$ (specular scan)

Different experimental scans can be performed by coupling both rotations. In Fig. 8.4 the most usual scans are illustrated in real and reciprocal space. They are:

detector scan or 2θ-scan. The incident wave vector $\boldsymbol{k}_{\text{in}}$ opens out the Ewald sphere ϵ. If we keep the angle of incidence fixed (ω = const) and rotate the detector arrangement, we move in reciprocal space along the Ewald sphere ϵ.

ω-scan or constant q-scan. The ω-scan rotates the Ewald sphere around the origin of reciprocal space. Fixing the scattering angle 2θ, we fix the modulus of the scattering vector. Then the ω-scan represents a *constant q-scan* since we move in reciprocal space on a circle of radius $q = |\boldsymbol{q}|$ around the origin.

$\Delta\theta/\Delta 2\theta$-scan or radial scans. Rotating the sample and the detector arrangement in a ratio $\Delta\omega/\Delta 2\theta = 1/2$, we drive the TCD in reciprocal space in *radial direction* from the origin of reciprocal space.

$\theta/2\theta$-scan on the q_z axis or specular scan. This special radial scan with $\omega/2\theta = 1/2$ keeps the condition $\theta_{\text{in}} = \theta_{\text{sc}}$ and performs a q_z-scan at $q_x = 0$. This experimental mode is also called *specular scan*, since the detector selects always the specularly reflected beam.

q_x-scan and q_z-scan. These scans go parallel to the q_x and q_z axes at fixed q_z and q_x position, respectively.

Sometimes it is useful to measure a **reciprocal space map**, i.e., to measure the map of the scattered intensity by combining different scans, e.g. measuring a series of ω-scans (rocking-scans) in the interval from ω=0 to ω=2θ for varying 2θ. Using a position-sensitive detetor (PSD), one would detect PSD-spectra for different omega positions.

The angular region investigated by a reflection experiment is limited by the horizon of the sample. The limiting cases for grazing incidence ($\theta_{\text{in}} = 0$) and grazing exit ($\theta_{\text{sc}} = 0$) are illustrated in Fig. 8.5. The situation in reciprocal space is represented by the two limiting half spheres ϵ_0 and ϵ_{s}.

X-ray reflection experiments are usually realised at very small scattering angles. In Fig. 8.4(right) we show the introduced experimental scans in the x-ray reflection mode and their restrictions due to the sample horizon. Especially the ω-scans are narrowed down. In the accessible region of reflection,

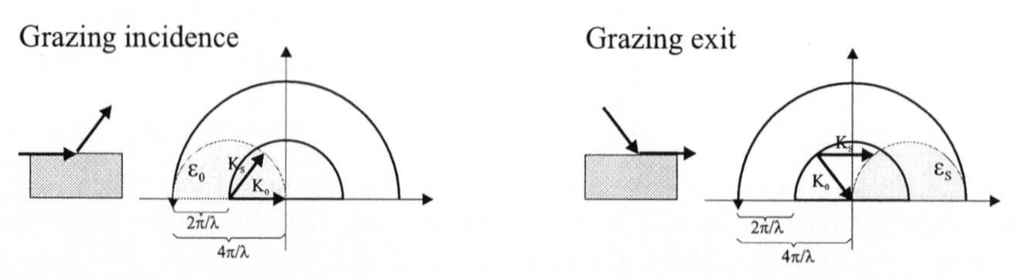

Fig. 8.5. Situation of grazing incidence (left) and grazing exit (right) in reciprocal and real space

i.e. near the origin of the reciprocal space, they perform approximately a transversal scan ($q_{\|}$-scan).

8.4 Specular X-Ray Reflection

In this section we discuss some theoretical and experimental examples of coherent specular x-ray reflection by layered structures with the aim to show *typical features* created by *different surface roughness point properties*. The coherent scattering intensity is concentrated along the specular rod. That means, the appropriate experimental scan is the specular or $\theta/2\theta$-scan.

8.4.1 Roughness with a Gaussian Interface Distribution Function

Single Surface The predominant number of samples have been successfully characterised assuming a Gaussian probability density of the interface roughness profile (see (2.19))

$$p_1(z) = \frac{1}{\sigma\sqrt{2\pi}}\, e^{-z^2/2\sigma^2} \,. \tag{8.21}$$

In this case, as shown in Chap. 3, Eq. (3.103), we obtain for a single surface (e.g. a substrate) the *amplitude ratio* of *dynamic reflection* [16,17]

$$r_{\text{dyn}}^{\text{coh}} = r_{\text{dyn}}^{\text{flat}}\, e^{-2k_{z,0}k_{z,1}\sigma^2} \tag{8.22}$$

with the amplitude ratio of the flat substrate being the dynamical Fresnel reflection coefficient of the substrate surface $r_{\text{dyn}}^{\text{flat}} = (k_{z,0}-k_{z,1})/(k_{z,0}+k_{z,1})$, see Eq. (3.68).

The ratio of *kinematical reflection* coefficients is (Eq. (3.104))

$$r_{\text{kin}}^{\text{coh}} = r_{\text{kin}}^{\text{flat}}\, e^{-2k_{z,0}^2\sigma^2} \tag{8.23}$$

with the kinematical Fresnel reflection coefficient of the surface $r_{\text{kin}}^{\text{flat}} = q_c^2/4q_z^2$, Eq. (3.91).

Both the kinematical and the dynamical Fresnel reflection coefficients are multiplied with a diminution factor containing the r.m.s. roughness σ in the exponent. The kinematical diminution factor decreases with the square of the scattering vector q_z, which is proportional to the angle of incidence. Its form resembles the static Debye-Waller factor. The dynamical diminution factor contains the product of the scattering vector in vacuum $q_{z,0}$ and that in the medium $q_{z,1}$. The angular dependence of the diminution factors in the dynamical and the kinematical theory differs substantially for small angles near the critical angle of total external reflection θ_c, see Fig. 8.6. Neglecting absorption, the scattering vector $q_{z,1}$ becomes purely imaginary below θ_c. Consequently there is no influence of roughness on the reflectivity in this

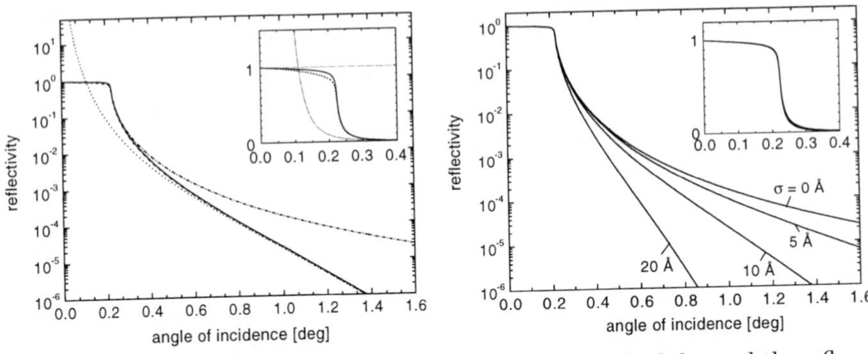

Fig. 8.6. The coherent reflectivity of a rough Si surface. In the left panel the reflectivity of a flat surface (dashed) is compared with that for the roughness $\sigma = 1\,\mathrm{nm}$, calculated by the "dynamical" theory (8.22) (full) and the kinematical theory (8.23) (dotted). The kinematical reflectivity diverges at grazing incidence. The "dynamical" curve coincides nearly with that of the flat surface below the critical angle θ_c. In the subfigure, the dashed line represents the coherent reflectivity of a rough surface calculated with *dynamical* Fresnel reflection coefficient and *kinematical* diminution factor. Thus the reflectivity decreases also below θ_c. In the right figure, influence of different roughness, calculated by dynamical formulae, is demonstrated. Close to θ_c (see subfigure), no essential change is observed

angular range within the dynamical description. At large incident angles both diminution factors coincide.

A more detailed discussion of both formulae (8.22) and (8.23) is given in [13]. There the contribution of the incoherent scattering to the specular direction has been studied by means of second order DWBA, showing its dependence on the lateral correlation length Λ. Concluding therefrom, the specularly reflected intensity can be described by the "dynamical" equation (8.22) for short Λ below $1\,\mu$m. For larger Λ the kinematical formula (8.23) with $r_{\mathrm{kin}}^{\mathrm{flat}}$ becomes more appropriate.

Surface roughness of numberless samples of amorphous, polycrystalline and mono-crystalline material systems has been studied by SXR. In Fig. 8.7 we plotted one experimental example, the reflectivity of a rough GaAs-substrate.

Multilayer Conventional SXR-simulation and fit programs are today based on a multilayer model with *independent r.m.s. roughness profiles* of each interface supposing a Gaussian probability density. This leads to effective Fresnel reflection and transmission coefficients (Eq. 3.103):

$$r_{j,j+1} = r_{j,j+1}^{\mathrm{flat}} e^{-2k_{z,j}k_{z,j+1}\sigma_{j+1}^2} \quad \text{and} \quad t_{j,j+1} = t_{j,j+1}^{\mathrm{flat}} e^{(k_{z,j}-k_{z,j+1})^2\sigma_{j+1}^2/2} \tag{8.24}$$

for each interface. The influence on the transmission function is rather small according to the small difference in the vertical scattering vector components

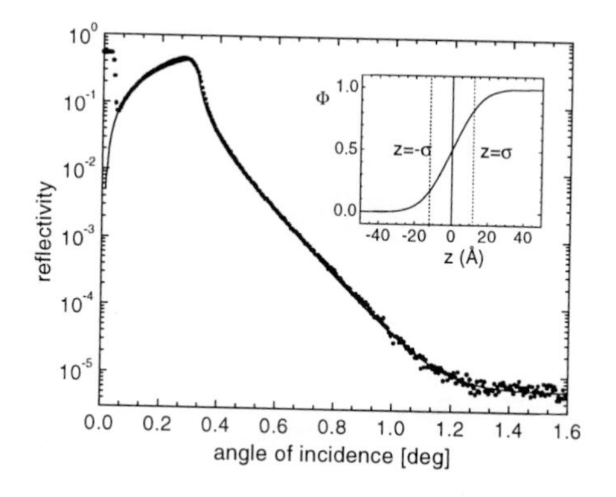

Fig. 8.7. Measured (points) and calculated (line) reflectivity curves of a GaAs substrate, $\sigma = 12$ Å [18]. In the inset the mean coverage of the surface is plotted

of the layers. However, the interface reflection is exponentially diminished by roughness, creating a strong change in the interference pattern. The effect of interface roughness versus surface roughness is shown in Fig. 8.8. The surface roughness mainly decreases the specular intensity of the whole curve progressively with q_z, where the interface roughness gives rise to a progressive dampening of the interference fringes (thickness oscillations). However, locally the variation in the Fresnel coefficients can cause more pronounced oscillations, too. In Fig. 8.9 we plotted the experimental and simulated curves of a magnetic rare earth/transition metal multilayer (Cr/$TbFe_2$/W on sapphire Al_2O_3), grown by laser ablation deposition. It shows a quite complicated non-regular interference pattern. A good agreement with the simulation was realised by considering a thin oxide film at the sample surface.

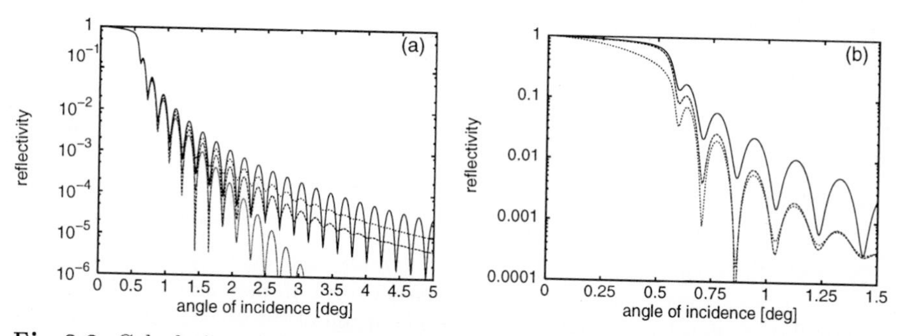

Fig. 8.8. Calculation of the specular reflectivity of a single layer (20 nm tungsten) on a substrate (sapphire) for different r.m.s. roughness and diminution factors. (a) Dynamical diminution factor. From the upper to the lower curve: without roughness, interface roughness 0.5 nm, surface roughness 0.5 nm, both surface and interface roughnesses 0.5 nm. Surface roughness yields a faster decay of the reflectivity, while interface roughness attenuates the peaks. (b) Different diminution factors. Surface roughness 1.2 nm and interface roughness 0.3 nm calculated for the kinematical "slow" roughness (lower curve), dynamical "rapid" roughness (middle curve), and without roughness (upper curve)

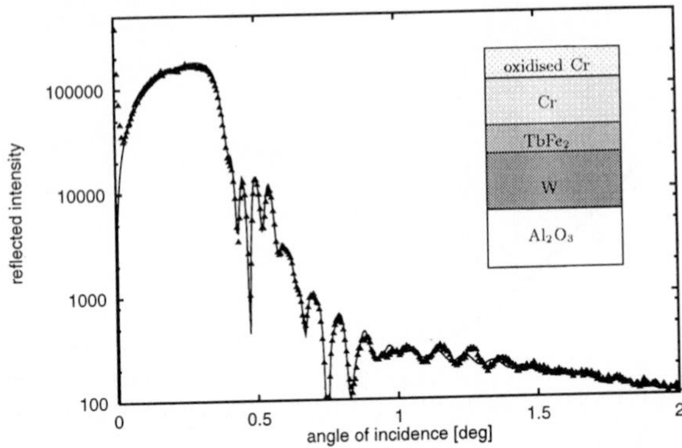

Fig. 8.9. Measurement (points) and the fit (full curve) of the specular reflectivity of a Cr/$TbFe_2$/W multilayer [19]. We determined the thicknesses (34.6 nm W, 4.8 nm $TbFe_2$, 50.5 nm Cr, 3 nm oxidised Cr) and the roughnesses (0.2 nm above sapphire, 2.0 nm W, 0.9 nm $TbFe_2$, 2.2 nm Cr)

Periodic Multilayer The main feature of the specular scans of a periodic multilayer are the multilayer Bragg peaks, giving evidence for the vertical periodicity, see Fig. 8.10 and Sect. 8.A.2.

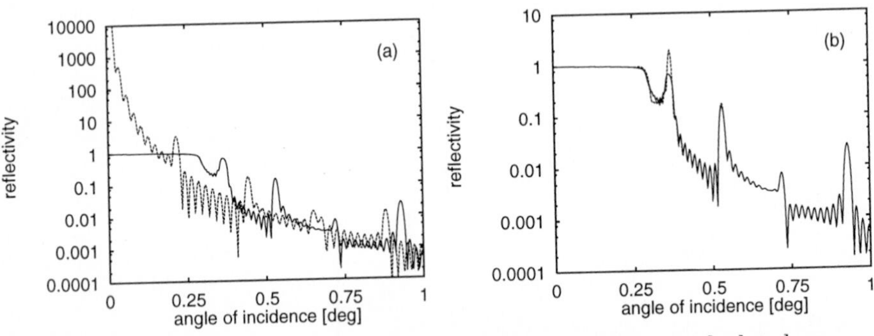

Fig. 8.10. Specular reflection by an "ideal" periodic multilayer—calculated curves for a [GaAs (13 nm) / AlAs (7 nm)] superlattice with 10 periods on a GaAs substrate, flat interfaces (no roughness). (a) Comparison of the dynamical theory (full curve) with the kinematical theory. The kinematical multilayer Bragg peaks correspond to the positions of the satellites of the (000) RLP. The curve diverges at low incident angles. The dynamical calculation shows the plateau of total external reflection below the critical angle. Due to refraction the multilayer Bragg peaks are shifted to larger angles. The first multilayer Bragg peak broadening is caused by multiple reflection (extinction effect). (b) Comparison of the dynamical theory with the semi-dynamical approximation (single-reflection approximation [18]). The satellite positions of all Bragg peaks coincide, also the shape and intensities except for the intense Bragg peaks

The intensity ratio of the Bragg peaks depends on the layer set-up within the multilayer period. The difference in the electron density determines the Fresnel coefficients, and the thickness ratio of the layers characterises the phase relations of the reflected waves of different interfaces. The laterally averaged gradual interface profile caused by interdiffusion or interface roughness leads to a damping mainly of the multilayer Bragg-peaks progressively with q_z, whereas the roughness of the sample surface reduces the intensity of the whole curve. This is demonstrated in Fig. 8.11.

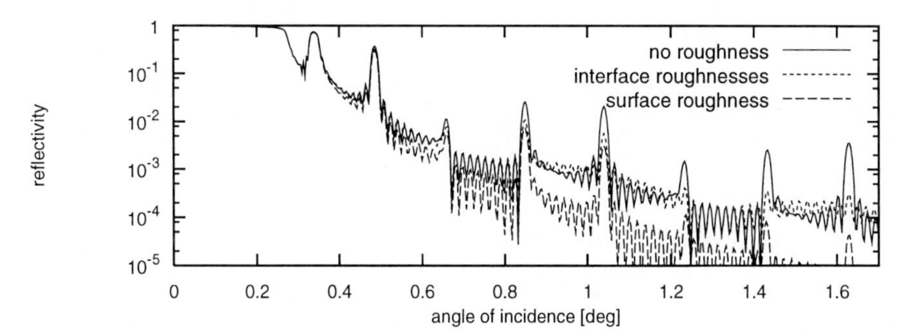

Fig. 8.11. Simulation of coherent reflectivity of a [GaAs (7 nm) / AlAs (15 nm)]10× periodic multilayer with no roughness (full curve) or 1 nm roughness of surface (dashed lower curve) or of all interfaces (dotted)

In Fig. 8.12 we plotted the measured SXR curves of an epitaxial CdTe / CdMnTe superlattice on a CdZnTe substrate. Due to the low contrast of the electron density of both layer materials the first order Bragg peak appears only as a very weak hump on the slope of the surface. The other Bragg peaks have a shape similar to a resonance line. From the best fit we obtain the mean compositional profile.

Increasing and decreasing roughness in multilayers The influence of roughness increasing or decreasing during the growth from the substrate towards the surface can be described by use of the roughness replication model introduced in Sect. 8.2.

We start the layer growth from a substrate with a Gaussian surface roughness profile,

$$C_{NN}(r_{\parallel} - r_{\parallel}') = \sigma_N^2 \, e^{-\left(\frac{|r_{\parallel} - r_{\parallel}'|}{\Lambda_N}\right)^2} . \tag{8.25}$$

For the non-random replication function in (8.10) we choose for all layers a Gaussian function

$$a(r_{\parallel} - r_{\parallel}') = \frac{1}{2\pi L^2} \, e^{-\frac{|r_{\parallel} - r_{\parallel}'|^2}{2L^2}} . \tag{8.26}$$

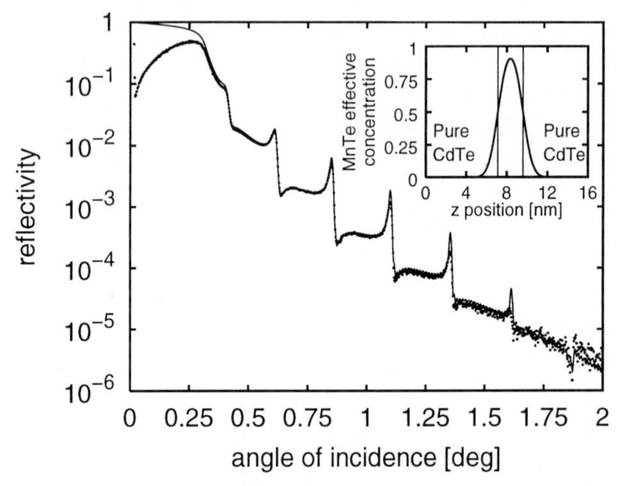

Fig. 8.12. Measured and calculated specular reflectivity of a [CdTe (14.2 nm) / CdMnTe (2.5 nm)]20× superlattice on CdZnTe [20]. In the subfigure, the roughness is represented by an effective MnTe concentration depth profile

The factor L determines the loss of memory from interface to interface. This choice arises from the aim to explain the different limiting cases of roughness replication models by one class of functions. It is not supported by any physical reason. However, the model allowed to describe measured curves of SXR and NSXR showing good agreement [15,8].

We assume the intrinsic correlation function (8.13) of all interfaces

$$K(\boldsymbol{r}_{\|} - \boldsymbol{r}_{\|}') = (\Delta\sigma)^2 \, e^{-\left(\frac{|\boldsymbol{r}_{\|} - \boldsymbol{r}_{\|}'|}{\Delta\Lambda}\right)^2} . \tag{8.27}$$

Now we continue like in Sect. 8.2. The Fourier transform of the *in-plane correlation function* is under these assumptions

$$\tilde{C}_{jj}(q) = \frac{1}{2}(\sigma_N \Lambda_N)^2 \, e^{-\frac{(q\Lambda_j')^2}{2}} + \frac{1}{2}(\Delta\sigma\Delta\Lambda)^2 \sum_{k=j}^{N-1} e^{-\frac{(q\Lambda_k')^2}{4}} , \tag{8.28}$$

where we have denoted

$$\Lambda_j' = \sqrt{\Lambda_N^2 + 4L^2(N-j)} . \tag{8.29}$$

The *inter-plane correlation* is then simply given by

$$\tilde{C}_{j \geq k}(q) = \tilde{C}_{jj}(q) \, e^{-\frac{(qL)^2}{2}(j-k)} . \tag{8.30}$$

We obtain the mean square roughness of the jth interface

$$\sigma_j^2 = \int dq \, \tilde{C}_{jj}(q) = \sigma_N^2 \frac{\Lambda_N^2}{\Lambda_j'^2} + (\Delta\sigma\Delta\Lambda)^2 \sum_{k=j}^{N-1} \frac{1}{\Lambda_k'^2} . \tag{8.31}$$

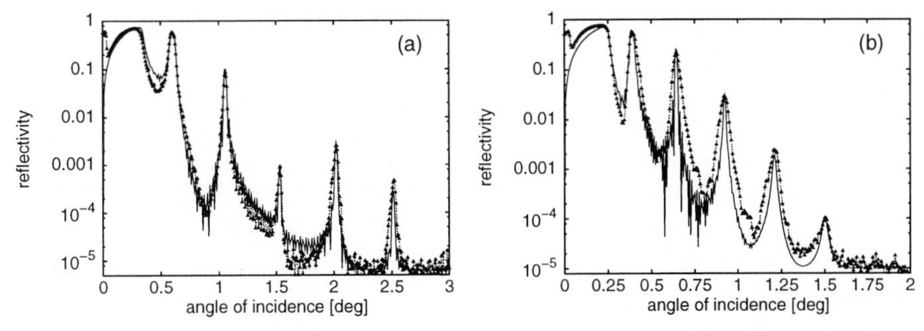

Fig. 8.13. Measurement (points) and simulation (full curve) of the specular reflectivity of a periodic Nb/Si multilayer of 10 periods [19]. (a) Sample A, fitted by the model of constant roughness, (b) sample B, fitted by the model of increasing roughness

Let us see what does it give for some limiting cases of the model:

1. *Identical interface roughness* is achieved with maximum replication and no intrinsic roughness: $L = 0$ and $\Delta\sigma = 0$. Consequently $\sigma_j = \sigma_N$, and all interfaces reproduce the profile of the substrate surface, $z_j(x) = z_N(x)$.
2. *Increasing roughness* towards the free surface is obtained by maximum replication and a non-zero intrinsic roughness ($L = 0$ and $\Delta\sigma > 0$). From (8.28) and (8.31) we find

$$\sigma_j^2 = \sigma_N^2 + (\Delta\sigma)^2 \, (N - j) \, , \tag{8.32}$$

describing the *roughening* during the growth.

3. *Partial replication and no intrinsic roughness* ($L > 0$ and $\Delta\sigma = 0$) leads to decreasing r.m.s. roughness towards the free surface (*smoothing* of the multilayer during growth), described by

$$\sigma_j^2 = \frac{\sigma_N^2}{1 + 4\left(\frac{L}{\Lambda_N}\right)^2 (N - j)} \, . \tag{8.33}$$

4. *No replication* occurs for diverging L, where $a(\boldsymbol{r}_\parallel - \boldsymbol{r}_\parallel{}')$ goes to zero. The roughness profile of each interface is independent.

We compare here the experimental example of two *periodic Si/Nb multilayers*, grown by magnetosputtering for superconductivity studies. The multilayer is deposited on a Si substrate with a thick SiO_2 layer and an Al buffer layer. The roughness of the buffer layer depends on its thickness and influences the quality of the interfaces. Two samples of different Al thickness have been investigated and the results are shown in Fig. 8.13. The multilayer periodicity generates the multilayer Bragg peaks or reflection satellites, which are dampened by interface roughness. The roughness of the substrate and the buffer layers has less influence on the reflection pattern. Sample A can be fitted by a roughness model of constant r.m.s. roughness for all interfaces. The peak widths of the first intense Bragg peak is broadened by extinction due to dynamical multiple scattering. For all higher order Bragg peaks we observe a narrower (kinematical) peak width. The satellite reflections of sample B are also rapidly damped, indicating a large interface roughness. Besides the widths of the peaks increases with q_z. That can not be explained by model 1. The satellite intensities and shape can be successfully reproduced by supposing increasing roughness according to (8.32). Due to their increased roughness, the upper layers near the surface contribute with decreasing effective Fresnel coefficients to the reflected wave. Within the Bragg position the contributions of all interfaces are still in phase, however, slightly away from the Bragg condition the contribution of interfaces near the substrate and those near the sample surface do not cancel completely, giving rise to the peak broadening.

8.4.2 Stepped Surfaces

The surface morphology of monocrystalline samples can also be described by a *discrete* surface probability distribution following the concept of terraces or small separated islands. In the simplest case, the *two-level surface* consists of randomly placed *islands* of uniform height d, so that the displacement $z(\boldsymbol{r}_{\|})$ has two possible values z_1 and $z_2 = d + z_1$ with the corresponding probabilities p_1 and $p_2 = 1 - p_1$, see Fig. 8.14 [21]. The surface probability distribution function $p(z)$ for this case writes

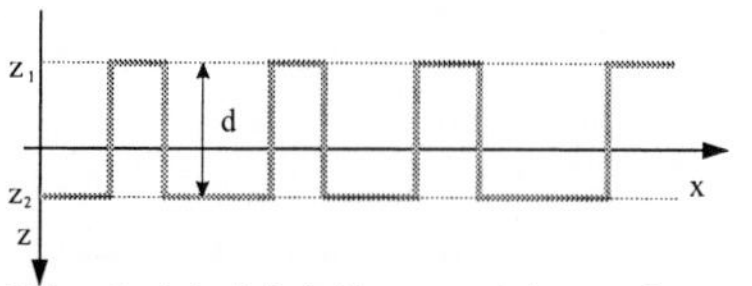

Fig. 8.14. Multilayer with random two-level islands

$$p(z) = p_1 \delta(z_1) + p_2 \delta(z_2) . \tag{8.34}$$

Since $\langle z(\boldsymbol{r}_{\|}) \rangle = 0$, then $Z_1 = -p_2 d$ and $Z_2 = p_1 d$. The mean square roughness is

$$\sigma^2 = p_1 Z_1^2 + p_2 Z_2^2 = p_1 p_2 d^2 \tag{8.35}$$

and the characteristic function (2.10) is

$$\chi(q_z) = e^{-iq_z d p_2} \left(p_1 + p_2 \, e^{iq_z d} \right) . \tag{8.36}$$

Putting this in the formulae for the reflected amplitude ratio of rough surfaces, we get the *amplitude ratio* of *kinematical specular reflection*

$$r_{\text{kin}}^{\text{coh}} = e^{-iq_z d p_2} \left(p_1 r_{0,1} + p_2 r_{1,2} e^{iq_z d} \right) . \tag{8.37}$$

A surface region perturbed in this way acts as a thin, homogeneous layer forming an upper and a lower interface with the Fresnel reflection coefficients

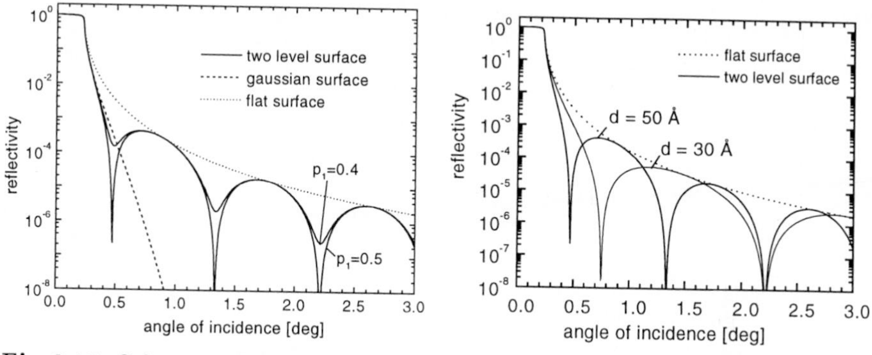

Fig. 8.15. Coherent reflectivity of a two level surface calculated within the kinematical theory for two values of the probability p_1 and the step height $d = 5\,\mathrm{nm}$ (left) and for two values of d and a symmetrical probability distribution $p_1 = p_2 = 0.5$ (right)

$p_1 r_{0,1}$ and $p_2 r_{1,2}$. They give rise to interference fringes which represent the height d (Fig. 8.15).

The example of a thin surface layer of porous silicon fits approximately this simple model, if its thickness is smaller than the vertical correlation lengths of the crystallites (Fig. 8.16(a)) [22]. Since the surface "layer" density is quite different from that of the substrate, we can observe two critical angles θ_1 and θ_2. The second one, θ_2, corresponds to silicon, the first one, θ_1, to the averaged surface region. Above θ_1 the wave can penetrate into the perturbed surface region, however total external reflection occurs at the "interface" with the non-perturbed region. That is why very intense fringes appear in this region between θ_1 and θ_2, which drop rapidly above θ_2. The whole curve is similar to that of a homogeneous layer of much less density or to that of a surface grating. In the fitted curves a small Gaussian deviation of the actual displacement around the z_1 and z_2 has been supposed, which leads to roughness diminution factors of the Fresnel reflection coefficients similar to (8.24).

8.4.3 Reflection by "Virtual Interfaces" Between Porous Layers

Porous silicon layers are fabricated by electrochemical etching in a monocrystalline silicon wafer. By a variation of the anode voltage, multilayers of modulated porosity can be produced. Following our division of the layer polarisability we can distinguish between the porous layer volume and the size of the layer of equal porosity. The interface between two layers of different porosities is not a microscopic laterally continuous and sharp interface between two media of different density, but an interface of two degrees of porosity. According to the coherent approach (used also in Sect. 3.4) we take for the coherent reflection an effective averaged refractive index into account.

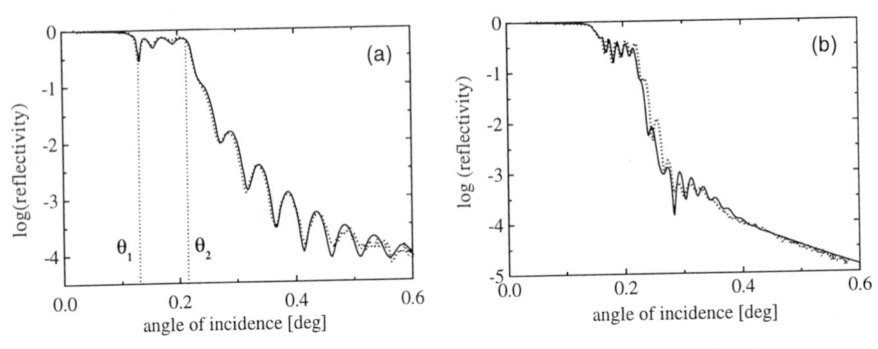

Fig. 8.16. Measured (full) and fitted (dashed) reflectivity curves of a thin porous silicon surface layer (a) and of a porous silicon double layer (b) on silicon substrate [22]. Positions θ_1, θ_2 are the critical angles of the porous layer and the substrate, respectively

Layers of statistically homogeneous porosity are assumed. We treat the slow "roughness" of the transition between two layers of different porosity by a Gaussian probability function. Same results are obtained by introducing a graduated transition of porosity from layer to layer. An experimental example is given in Fig. 8.16(b) for a double layer sample [22]. The thickness of the surface layer is much smaller than that of the buried layer. The fast oscillating fringes represent the total thickness. The fringe amplitude is modulated by a period, which corresponds approximately to the thickness of the surface layer. It has been found from the simulation that the interface between the two layers of different porosity is much sharper than the interface with the substrate (which is the end front of the etching process).

The occurrence of the modulation of thickness oscillations in Fig. 8.16(b) is a direct proof for the validity of the coherent scattering approach. Between the two porous layers there is nowhere a real roughly smooth lateral interface between two media. Nevertheless the x-rays are specularly reflected at this "microscopically non-existent interface" showing all features of the continuum theory of dynamical reflection by multilayers.

8.5 Non-Specular X-Ray Reflection

In this section we use the incoherent scattering approach (2) within the DWBA (Chap. 4) and derive some explicit expressions for the incoherent scattering cross section for x-ray reflection by rough multilayers. We discuss the main features of the scattering patterns illustrated by experimental examples. The representation of the scattering in reciprocal space allows a simple interpretation of the findings by the various scattering processes. We will treat samples with interfaces having a Gaussian roughness profile, diffuse scattering from terraced interfaces and finally non-coplanar diffuse scattering.

8.5.1 Interfaces with a Gaussian Roughness Profile

We will deal with interfaces having a Gaussian roughness profile. We start with the scattering from a single surface. Then we continue with a multilayer showing the effects of different roughness replication as well as dynamical scattering effects on reciprocal space maps.

Single Surface Firstly we will deal with *surfaces* of a gaussian probability distribution. The *pair probability distribution function* is in the stationary case (see [1,23–25], for instance)

$$p_2(z,z') = \frac{1}{2\pi\sqrt{\sigma^4 - C_{zz}^2(\boldsymbol{r}_\parallel - \boldsymbol{r}_\parallel')}} \exp\left\{ -\frac{z^2 + z'^2 - \frac{2zz'}{\sigma^2} C_{zz}(\boldsymbol{r}_\parallel - \boldsymbol{r}_\parallel')}{2\sigma^2[1 - \frac{1}{\sigma^4} C_{zz}^2(\boldsymbol{r}_\parallel - \boldsymbol{r}_\parallel')]} \right\} \tag{8.38}$$

with the two-dimensional characteristic function

$$\chi_{zz'}(q,q') = \langle e^{i(qz - q'z')} \rangle = e^{-\sigma^2(q^2 + q'^2)/2}\, e^{qq' C_{zz}(\boldsymbol{r}_\parallel - \boldsymbol{r}_\parallel')} \,. \tag{8.39}$$

One correlation function, which has been successfully applied to interpret the experimental findings, follows from similarities between the description of interfaces with fractal roughness properties and the Brownian motion, if we replace the lateral position by time. Supposing a behaviour like [1,24]

$$\left\langle \left[z(\boldsymbol{r}_\parallel) - z(\boldsymbol{r}_\parallel') \right]^2 \right\rangle = A \left| \boldsymbol{r}_\parallel - \boldsymbol{r}_\parallel' \right|^{2h} , \qquad 0 < h \leq 1 , \tag{8.40}$$

leads together with

$$\left\langle \left[z(\boldsymbol{r}_\parallel) - z(\boldsymbol{r}_\parallel') \right]^2 \right\rangle = 2\sigma^2 - 2C_{zz}(\boldsymbol{r}_\parallel - \boldsymbol{r}_\parallel') \tag{8.41}$$

to a correlation function, which only depends on the distance $|\boldsymbol{r}_\parallel - \boldsymbol{r}_\parallel'|$. The so-called *Hurst factor* h describes the jagged shape of the interface, determining the *fractal dimension* D of the interface, $D = 3 - h$. For $h = 1$ the fractal dimension is 2 and corresponds to the topological dimension of an interface (without a fractal structure). This function diverges for large distance $|\boldsymbol{r}_\parallel - \boldsymbol{r}_\parallel'|$. Thus it is suitable to introduce a *cut-off radius* ξ. Below ξ the correlation function shall approximately behave like (8.40), but above it should converge to zero. A function with such a behaviour is

$$C_{zz}(\boldsymbol{r}_\parallel, \boldsymbol{r}_\parallel') = \langle z(\boldsymbol{r}_\parallel) \cdot z(\boldsymbol{r}_\parallel') \rangle = \sigma^2\, e^{-\left(|\boldsymbol{r}_\parallel - \boldsymbol{r}_\parallel'|/\xi\right)^{2h}} \,. \tag{8.42}$$

The cut-off radius represents the *lateral correlation length of the interface.* Let us now determine the incoherent cross section for a surface with such properties. Using Eq. (4.41) we find for the incoherent scattering cross section of a single rough surface within the *full DWBA*

$$d\sigma_{\text{incoh}} = d\Omega\, \frac{k_0^4}{16\pi^2} \left| t_{0,1}^{\text{in}} \right|^2 \left| t_{0,1}^{\text{sc}} \right|^2 \tilde{Q}_1 \tag{8.43}$$

with the covariance function (4.D28)

$$\tilde{Q}_1 = A\,|n_1^2 - n_0^2|^2\,\frac{e^{-\frac{1}{2}\sigma^2(q_{z,1}^2+q_{z,1}^{*2})}}{|q_{z,1}|^2} \times \tag{8.44}$$
$$\times \int_A d(r_\| - r_\|{}')\, e^{i q_\| (r_\| - r_\|{}')} \left[e^{|q_{1z}|^2 C_{zz}(r_\| - r_\|{}')} - 1 \right] ,$$

where A is the area of integration, that means the illuminated surface of the sample. The result can be interpreted as follows: the incident wave transmits through the surface considered by the Fresnel transmission coefficients. This "distorted wave" is diffusely scattered by the surface disturbance. Thus the non-specularly reflected intensity depends on the r.m.s. roughness and is proportional to the Fourier transform of

$$\left[e^{|q_{1z}|^2 C_{zz}(r_\| - r_\|{}')} - 1 \right] .$$

Taking the correlation function (8.42), we have $C_{zz}(r_\| - r_\|{}') < \sigma^2$. For small roughness or small q_z fulfilling $(\sigma q_z)^2 \ll 1$, we can approximate (8.42) by the first two terms of its Taylor series and obtain finally

$$d\sigma_{\mathrm{incoh}} = d\Omega\,\frac{k_0^4 A|n_1^2 - n_0^2|^2}{16\pi^2}\,|t_{01}^{\mathrm{in}}|^2 |t_{01}^{\mathrm{sc}}|^2\, e^{-\frac{1}{2}\sigma^2\left(q_{z,1}^2+q_{z,1}^{*2}\right)}\,\tilde{C}(q_\|) , \tag{8.45}$$

i.e. an expression, which is proportional to the Fourier transform of the correlation function.

The according *kinematical* expressions are found by setting the transmission coefficients equal to 1 and substituting the scattering vectors in the medium $q_{z,1}$ by the scattering vectors in vacuum, q_z.

Multilayer with no vertical roughness replication In case of *independent roughness profiles* of all different interfaces we have the replication function $a_m(r_\|) = 0$ ($L \to \infty$ in (8.26)). There is no inter-plane correlation, that is why only the in-plane correlation functions have to be considered. We can proceed for each interface like in the case of a single surface described above. However, now we take four scattering processes (coresponding to downwards and upwards propagating incident and scattered waves), see (4.D27), into account instead of one in (8.43). Consequently, we consider 4×4 covariance functions for each interface. The incoherent scattering cross section adds up the contribution of all *single interfaces*

$$\left(\frac{d\sigma}{d\Omega}\right)_{\mathrm{incoh}} = \frac{k_0^4}{16\pi^2} \sum_{j=0}^{N} \sum_{\pm} \sum_{\pm} \sum_{\pm} \sum_{\pm} \tag{8.46}$$
$$U_j(\pm k_{\mathrm{in}\,z,j}, Z_j) U_j(\pm k_{\mathrm{sc}\,z,j}, Z_j) U_j(\pm k_{\mathrm{in}\,z,j}^*, Z_j) U_j(\pm k_{\mathrm{sc}\,z,j}^*, Z_j)$$
$$\tilde{Q}_{jj}(\pm k_{\mathrm{in}\,z,j} \pm k_{\mathrm{sc}\,z,j}, \pm k_{\mathrm{in}\,z,j} \pm k_{\mathrm{sc}\,z,j})$$

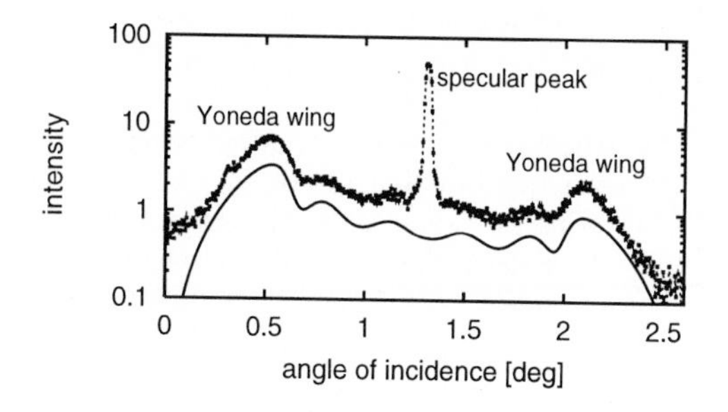

Fig. 8.17. Measurement (points) and fit (full line, shifted down 2×) of an ω-scan at $2\Theta = 2.63^{\circ}$ single layer W (11.1 nm) on Si substrate

with

$$\tilde{Q}_{jj}(q_z, q_z') = \frac{A\,|\chi_{0\,j+1} - \chi_{0\,j}|^2}{q_z\,(q_z')^*}\, e^{-\frac{1}{2}\sigma_j^2\left[q_z^2 + (q_z')^{*2}\right]} \times \tag{8.47}$$

$$\times \int_A d(r_\| - r_\|{}')\, e^{i q_\| (r_\| - r_\|{}')} \left(e^{q_z (q_z')^*\, C_{jj}(r_\| - r_\|{}')} - 1\right) ,$$

where we have used the polarisabilities $\chi_{0\,j+1} - \chi_{0,j} = n_{j+1}^2 - n_j^2$ instead of the optical indices. Assuming the same in-plane correlation functions for all interfaces the $\tilde{Q}_{jj}$ of different interfaces differ only by the scattering vectors and the differences of polarisability.

Figure 8.17 shows a measurement and fit of an ω-scan from a single layer sample.

Multilayer with partial vertical roughness replication In case of partial vertical roughness replication also the covariance functions of scattering at *different interfaces* have to be included. We get (cf. (4.45)–(4.46) and (4.D28)–(4.D29))

$$\left(\frac{d\sigma}{d\Omega}\right)_{\text{incoh}} = \frac{k_0^4}{16\pi^2} \sum_{j=0}^{N} \sum_{k=0}^{N} \sum_{\pm} \sum_{\pm} \sum_{\pm} \sum_{\pm} \tag{8.48}$$

$$U_j(\pm k_{\text{in}\,z,j}, Z_j) U_j(\pm k_{\text{sc}\,z,j}, Z_j) U_k(\pm k_{\text{in}\,z,k}^*, Z_k) U_k(\pm k_{\text{sc}\,z,k}^*, Z_k)$$

$$\tilde{Q}_{jk}(\pm k_{\text{in}\,z,j} \pm k_{\text{sc}\,z,j}, \pm k_{\text{in}\,z,k} \pm k_{\text{sc}\,z,k})$$

with the covariance function (see Fig. 8.18)

$$\tilde{Q}_{jk}(q_z, q_z') = \frac{A\,(\chi_{0\,j+1} - \chi_{0\,j})(\chi_{0\,k+1} - \chi_{0\,k})^*}{q_z\,(q_z')^*}\, e^{-\frac{1}{2}\left[\sigma_j^2 q_z^2 + \sigma_k^2 (q_z')^{*2}\right]} \times$$

$$\times \int_A d(r_\| - r_\|{}')\, e^{i q_\| (r_\| - r_\|{}')} \left(e^{q_z (q_z')^*\, C_{jk}(r_\| - r_\|{}')} - 1\right) . \tag{8.49}$$

Here σ_j and σ_k are the r.m.s. roughnesses of the corresponding interfaces determined by (8.31), C_{jk} are their inter-plane correlation functions. Restricting

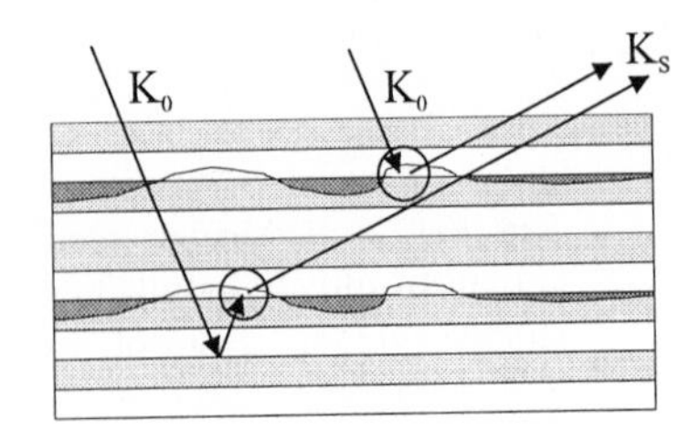

Fig. 8.18. Illustration considering the covariance function $\tilde{Q}_{jk}(q_z, q_z')$ of one scattering process q_z at the interface j and another scattering process q_z' at the interface k

ourselves on small roughness $(\sigma q_z)^2 \ll 1$, we can make approximations similar to (8.45) using the Fourier transform of the correlation functions $\tilde{C}_{jk}(\boldsymbol{Q}_\parallel)$ obtained in section 8.4, Eqs. (8.28) and (8.30).

The treatment of the corresponding expressions of the *simpler DWBA* for multilayers (p. 160), is straightforward. It neglects the influence of specular interface reflection on the diffuse scattering. Only the primary scattering processes are taken into account.

8.5.2 The Main Scattering Features of Non-Specular Reflection by Rough Multilayers

Let us give an overview of the main features in the non-specular reflected intensities and discuss their physical origin. The diffuse x-ray scattering (DXS) pattern is characterised by the *transmitted / reflected wave amplitudes* $U_j(\pm k_z)$ of the incident and final wave fields in the layers and by the 16 *covariances of the scattering processes*, $\tilde{Q}_{jk}(q_z, q_z')$ for each pair of interfaces j, k. We want to study the features of the DXS pattern under the aspect whether they are particularities of scattering by the *roughness profiles*, caused by the correlation properties, or of the *excited non-perturbed wave amplitudes*. With other words, we want to distinguish between effects of the random disturbance potential and of the non-perturbed potential. The latter effects do not depend on the statistical roughness properties, we call them *dynamical scattering effects*.

Resonant diffuse scattering First we investigate the influence of the interface roughness correlation. One essential characteristics caused by the interplane correlation is the so-called *resonant diffuse scattering* (RDS). We simplify the discussion of this phenomena by introducing a simpler model of vertical roughness correlation [26], where the inter-plane correlation function C_{jk} depends on the in-plane correlation function C_{ll}, $l = \max(j, k)$ of the lower interface, by

$$C_{jk}(r_\parallel - r_\parallel') = C_{ll}(r_\parallel - r_\parallel')\, e^{-|Z_j - Z_k|/\Lambda_\perp} \,. \tag{8.50}$$

In this phenomenological model the vertical correlation of the roughness profiles is limited by a *vertical correlation length* $\Lambda_\perp$. The model does not explain

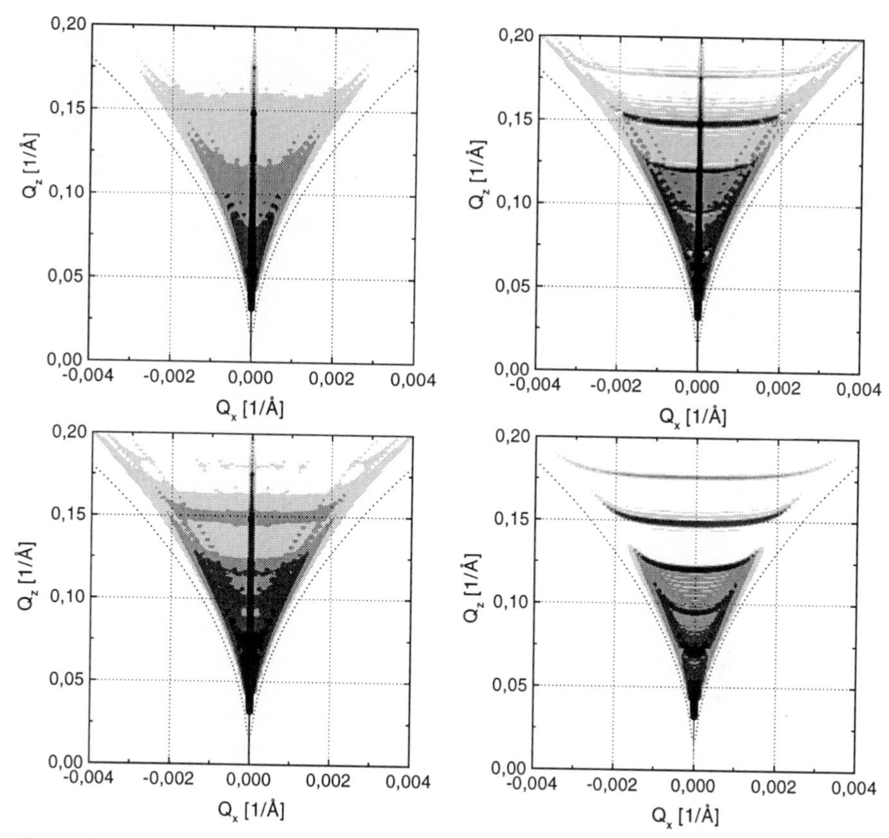

Fig. 8.19. Reciprocal space maps of the diffusely scattered intensity calculated for a [GaAs (7 nm) / AlAs (15 nm)]10× multilayer using the DWBA method and the simpler replication model (8.50) [18]. All the interfaces have the same r.m.s. roughness 1 nm, the correlation lengths 50 nm and different vertical correlation lengths $\Lambda_\perp$. Upper left panel: no replication, $\Lambda_\perp = 0$. Upper right panel: full replication, $\Lambda_\perp = \infty$. Bottom left panel: $\Lambda_\perp = 100$ nm. Bottom right panel: full replication, $\Lambda_\perp = \infty$, calculated by the simpler DWBA. The full lines represent the arcs of the Ewald spheres for the limiting cases of $\theta_{\rm in} = 0$ and $\theta_{\rm sc} = 0$. The RDS disappear, if the roughness profiles are not replicated (upper left panel). Bragg-like resonance lines are visible in all maps calculated by the full DWBA. They are not reproduced by the simpler DWBA (bottom right panel)

the effects of smoothening and roughening studied in section 8.4, since it neglects the interdependence of the r.m.s. roughness and the lateral correlation length (8.30). However, it makes the calculation and the discussion simpler. In Fig. 8.19 we see some calculated reciprocal space maps of the diffusely scattered intensity for a GaAs/AlAs superlattice assuming this vertical replication model. All the interfaces have the same r.m.s. roughness $\sigma = 1$ nm, and the *lateral* correlation length $\Lambda = 50$ nm. It shows the cases of *no repli-*

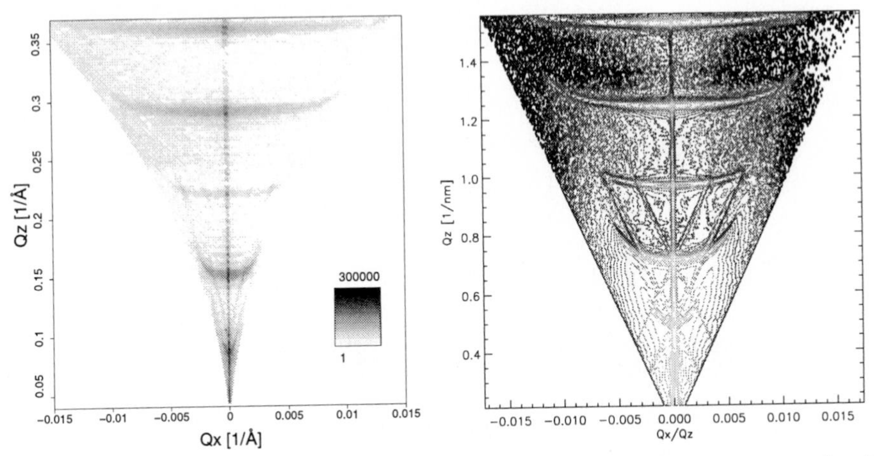

Fig. 8.20. Measured reciprocal space maps (top). Left map: periodic multilayer [Si (3.0 nm) / Nb (5.8 nm)] 10× starting from a rough Si substrate of $\sigma = 0.46$ nm and with interface roughness *decreasing* towards the free surface [19]. Right map: periodic multilayer with the setup corresponding to that of Fig. 8.19 with interface roughness *increasing* towards the free surface [8]. Left schema: the reciprocal space representation of diffuse scattering by a multilayer with interface roughness replication. The essential features are 1. the multilayer truncation rod through the RLP (000) with the multilayer satellite peaks and 2. horizontal sheets crossing the TR in the satellite positions

cation, partial replication and *full replication.* In the first case all interfaces scatter independently, the diffuse intensities of all individual interfaces superpose. The other two cases give rise to scattering with partial coherence, the resonant diffuse scattering. It occurs due to the vertical replication of the roughness profiles of different interfaces. The partial phase coherence of the waves diffusely scattered from different interfaces leads to a concentration of the scattered intensity in narrow sheets. These sheets of resonant diffuse scattering intersect the specular rod in the multilayer Bragg peaks. Neglecting refraction the sheets would be horizontally oriented with the centre fulfilling the one-dimensional Bragg conditions

$$q_z = k_0 \left(\sin\theta_{\text{in}} + \sin\theta_{\text{sc}}\right) = \frac{2\pi\, m}{D_{\text{ML}}} \,, \tag{8.51}$$

schematised in Fig. 8.20, which is the case in a kinematical treatment. Due to the angle dependent refraction of x-rays the sheets are curved forming

"RDS-bananas" following the modified Bragg law

$$\langle q_z \rangle_{\rm ML} = k_0 \left(\sqrt{\sin^2\theta_{\rm in} + \langle\chi_0\rangle_{\rm ML}} + \sqrt{\sin^2\theta_{\rm sc} + \langle\chi_0\rangle_{\rm ML}} \right) = \frac{2\pi m}{D_{\rm ML}} , \quad (8.52)$$

where $\langle\chi_0\rangle_{\rm ML} = \sum_{j=1}^{N} \chi_{0j}/D_{\rm ML}$ is the mean polarisability of the multilayer period and $\langle q_z\rangle_{\rm ML} = q_z(\theta_{\rm in}, \theta_{\rm sc}, \langle\chi_0\rangle_{\rm ML})$ the mean scattering vector in the medium. The length of the RDS-bananas in q_x direction is inversely proportional to some effective correlation length $\Lambda_{\rm eff}$ depending on the correlation length Λ_j of the interfaces. If all interfaces have the same correlation length, $\Lambda_{\rm eff}$ would equal Λ_j. The widths of the RDS-bananas in q_z direction represent the degree of replication. In the simple model it depends inversely on $\Lambda_\perp$ and for large $\Lambda_\perp$ on the total thickness of the multilayer. The sheets disappear if there is no vertical replication, $\Lambda_\perp = 0$, turning into a broad vertical maximum similar to that for a single surface. The RDS-bananas have no dynamical nature, their existence is not related with any kind of multiple scattering. They are also produced by the kinematical theory and by the simpler DWBA.

RDS has been experimentally observed at amorphous, polycrystalline as well as epitaxial multilayers as it is shown in Fig. 8.20. The RDS sheets are clearly visible, bent is due to the refraction. Their existence and narrow vertical width gives evidence for full roughness replication in both samples.

Dynamical scattering effects One typical dynamical feature is known from NSXR by rough surfaces. The so-called *Yoneda wings* arise if the incident or the exit angle equals the critical angle, $\theta_{\rm in/sc} = \theta_c$. The wings are generated by the enhancement of the transmitted wave amplitude at the inner sample surface, Figs. 4.4, 8.17. In the case of a single layer structure interference fringes can also be created due to the wave guide behaviour of the two interfaces in the layer structure. In general, this behaviour can produce *dynamical fringes* in ω-scans as well as in 2θ-scans.

In case of periodic multilayers we call them *Bragg-like resonance lines*, since the amplitudes of the reflected waves exhibit a maximum if the incident or exit wave fulfills the refraction-corrected Bragg-law

$$k_0 \sqrt{\sin^2\theta_{\rm in/sc} + \langle\chi_0\rangle} = \frac{\pi\, m_{\rm in/sc}}{D_{\rm ML}} , \qquad (8.53)$$

where $m_{\rm in}, m_{\rm sc}$ are integers. It is easy to proove that the zero order Bragg-like resonances are identical with the Yoneda wings. The resonance lines have a particular maximum, the so-called *Bragg-like peak (BL)*, where the incident and exit waves are simultaneously in Bragg condition and the Bragg-like

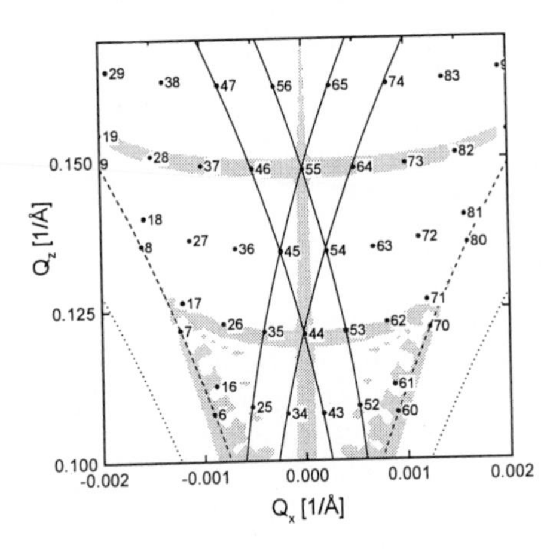

Fig. 8.21. The schema of the positions of the Bragg-like peaks (points) and the RDS-bananas (grey areas). The numbers denote the orders m_{in} and m_{sc} of the Bragg-like peaks according to (8.54). The dotted lines denote the positions of the Yoneda wings. The full lines are Bragg-like resonance lines, corresponding to $m_{\text{in}/sc} = 4$ and 5

resonances intersect. That is at the positions

$$\begin{aligned} Q_{z,m_{\text{in}}n_{\text{sc}}} &= \sqrt{(m_{\text{in}}2\pi/D)^2 - k_0^2\,\langle\chi_0\rangle} + \sqrt{(m_{\text{sc}}2\pi/D)^2 - k_0^2\,\langle\chi_0\rangle} \\ Q_{\|,m_{\text{in}}m_{\text{sc}}} &= \sqrt{k_0^2 - (m_{\text{in}}2\pi/D)^2 + k_0^2\,\langle\chi_0\rangle} + \sqrt{k_0^2 - (m_{\text{sc}}2\pi/D)^2 + k_0^2\,\langle\chi_0\rangle}. \end{aligned} \tag{8.54}$$

The existence of the Yoneda wings, dynamic fringes and Bragg-like peaks is of completely dynamical origin. They occur independent of the actual interface correlation function. However, their form and intensity is influenced by the interface correlation.

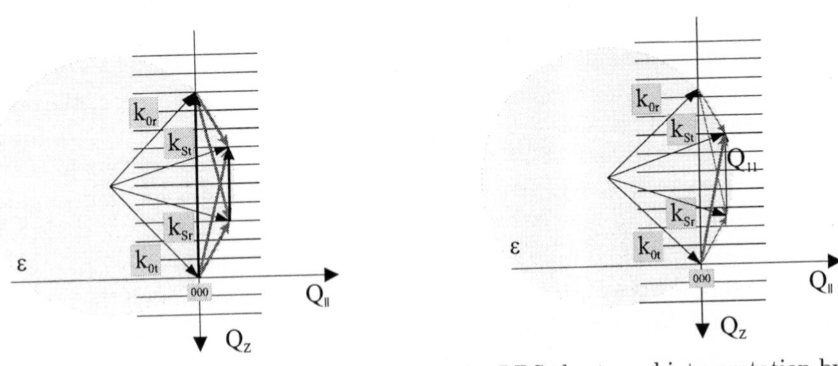

Fig. 8.22. Generation of Bragg-like peaks on the RDS-sheets and interpretation by the concept of Umweganregung. On the left side, both the incident and final non-perturbed state fulfill the Bragg condition (8.53). Simultaneously all four diffuse scattering processes are in the situation of resonant diffuse scattering (8.52). On the right side, the situation of RDS (8.52) is fulfilled for the primary scattering process. The incident wave is out of Bragg condition, consequently also the final state is out of Bragg condition. Additionally all three secondary diffuse scattering processes are out of resonance

In case of vertically replicated roughness we see with (8.52)–(8.54) that all Bragg-like peaks of an even number $m_{\rm in}+m_{\rm sc}$ are situated on RDS-sheets, Fig. 8.21. These Bragg-like peaks are very pronounced with respect to the others. That can be interpreted by the concept of *Umweganregung* (excitation of a reflection by another reflection), well known from x-ray diffraction and outlined in Fig. 8.22. In our experimental map of Fig. 8.20 the Yoneda wings and the Bragg-like resonances are well resolved. Along the RDS-sheets we observe intense Bragg-like peaks. All the features are reproduced by the calculation using the full DWBA treatment for multilayers.

Not always is it possible and necessary to measure a full well resolved map. In general ω-scans at different q_z and offset-scans or 2θ-scans are employed. Already one offset-scan or 2θ-scan is sufficient to give evidence for vertical replication.

8.5.3 Stepped Surfaces and Interfaces

The model of islands of nearly uniform height discussed in section 8.4.2 is the simplest case for a discrete stepped n-level surface. An infinite number of levels exist at a *terraced surface*, see Fig. 8.23, which is mostly the case of multilayers grown on slightly miscut substrates [27–29]. The miscut angle α equals the mean ratio of the step height $\langle h \rangle$ and the terrace widths $\langle L \rangle$: $\alpha = \langle h \rangle / \langle L \rangle$. The lateral correlation properties of such a stepped surface are determined by the conditional probability $p(\Delta x, z)$ giving the probability of displacement z for two surface points with the distance Δx. The two-dimensional characteristic function $\chi_{zz'}$ of such a stair-like surface can be described based on the approach of stationary random processes [21]. Using (8.47) one can calculate the covariance function $\tilde{Q}$ and with (8.46) the differential scattering cross section for the diffuse scattering by the stair-like surface. In Ref. [27] the gamma-distribution of order M has been supposed

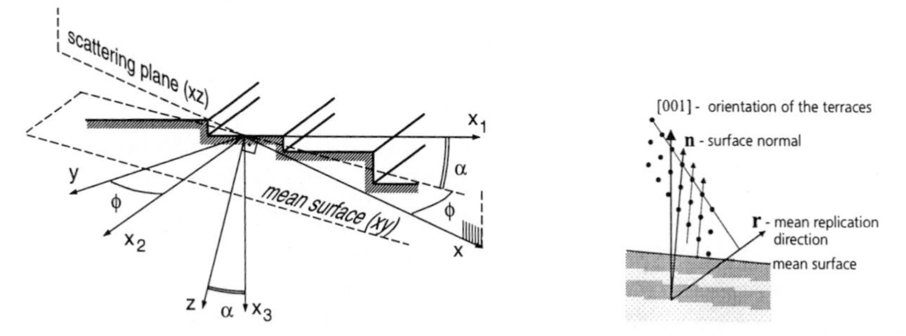

Fig. 8.23. (left) Model of a step-like surface. (right) Illustration of the stair-like interface pattern in the superlattice and of the corresponding fine structure in the reciprocal space

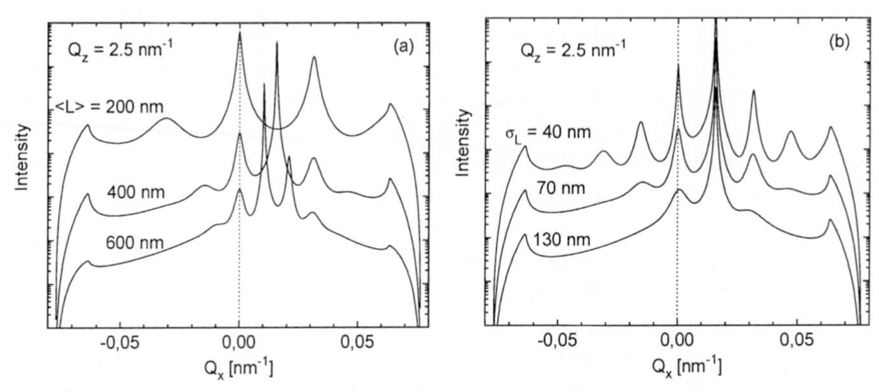

Fig. 8.24. ω-scans of a 3° miscut GaAs surface. (a) calculation for different terrace sizes, (b) for different dispersion of the terrace size [27]

for the distribution of the terrace widths L

$$p(L) = \frac{1}{\Gamma(M)} \left(\frac{M}{\langle L \rangle} \right)^M e^{-\frac{ML}{\langle L \rangle}} L^{M-1}, \tag{8.55}$$

with the dispersion of the distribution

$$\sigma_L^2 = \frac{\langle L \rangle^2}{M} . \tag{8.56}$$

The terrace length was described by a similar distribution. The step height between the terraces h was assumed to be normally distributed with the dispersion σ_h. For such a model the correlation function and the two-dimensional characteristic function have been calculated [27] and implemented in the expressions of the DWBA. The terrace size and its statistical distribution can be determined by transversal scans in reciprocal space or by ω-scans. In Fig. 8.24 the DXS intensity has been calculated for a terraced surface of GaAs with a slight miscut of 0.3°. Between the Yoneda wings there occur maxima, which are equidistant in reciprocal space and their distance is inversely proportional to the mean terrace size. The positions of these maxima correspond to the grating satellites of a mean surface grating with the lateral grating period D_G. The DXS peaks are broadened with increasing dispersion of the terrace lengths and of the step height.

Growing an epitaxial layer on a miscut substrate, the staircase profile can be replicated from the substrate/layer interface to the sample surface. In a superlattice on off-oriented substrates, the staircase profile can be replicated from interface to interface [28,29]. The direction of the replication may be inclined with respect to the growth direction (see Fig. 8.23). For simplicity we suppose first laterally uniform terrace lengths and perfect interface replication, giving the recursion formulae for the layer size functions

$$\Omega_j(\boldsymbol{r}) \approx \Omega_{j-2}(\boldsymbol{r} + D_{\mathrm{SL}}\hat{\boldsymbol{z}} + D_{\|}\hat{\boldsymbol{x}}) , \tag{8.57}$$

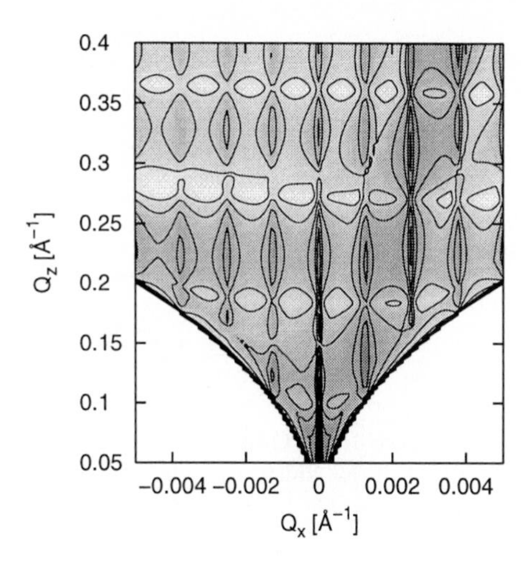

Fig. 8.25. Calculated map for a (7 nm GaAs / 15 nm AlAs)10× superlattice grown on a 0.5° miscut GaAs substrate. Averaged terrace distance is $\langle L \rangle$ =500 nm and interface steps are fully replicated at 40°.

where $D_{\|}$ is the lateral shift of the stair-like pattern during the growth of one superlattice period (here we assume a bilayer superlattice period). Such a two-dimensionally periodic morphological superstructure creates a two-dimensional fine structure, similar to later discussed multilayer surface gratings. In this case the whole reflected intensity would be concentrated along so-called grating truncation rods perpendicular to the sample surface, representing the lateral periodicity. Each truncation rod would contain the multilayer Bragg peaks due to the multilayer periodicity. An inclined replication direction of the interface profile creates inclined branches of multilayer Bragg peaks. All that is shown schematically in Fig. 8.23. In reality there will be a rather partial interface replication, characterised by an effective replication length $\Lambda_{\perp}$. In the Gaussian roughness model (discussed in Sect. 8.5.1) the vertical replication in the periodic multilayer caused horizontal bananas of resonant diffuse scattering, crossing the multilayer Bragg-peaks in the specular scan. In the present case of the lateral correlation of the interface steps similar horizontal sheets appear. However, they are, in addition, horizontally structured by lateral DXS maxima, which indicate the laterally and vertically correlated stair-like interfaces, see Fig. 8.25.

In result a two dimensionally structured pattern of resonant diffuse scattering is obtained with longitudinal DXS-satellites due to the superlattice periodicity and transversal DXS satellites

$$\delta q_x \approx \frac{2\pi}{\langle L \rangle_{\text{av}}} , \tag{8.58}$$

which represent the more or less periodic lateral morphological order of the interfaces. Both together form longitudinal stripes perpendicular to the mean sample surface, which remind to the grating truncation rods of multilayer surface gratings (see Sect. 8.7). Considering the q_z-dependence of the diffuse intensity one observes, that the envelope of the intensity follows with its

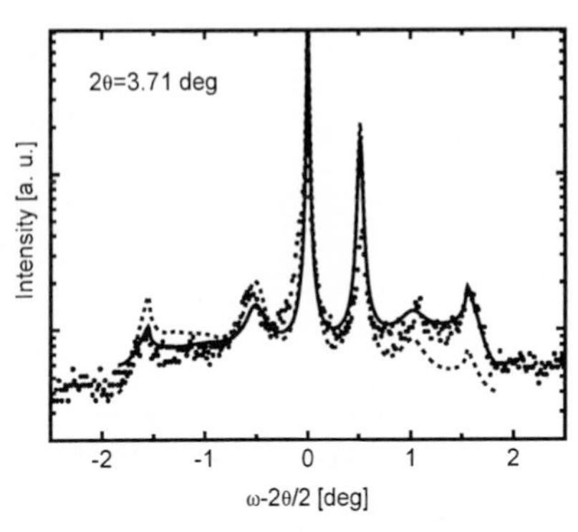

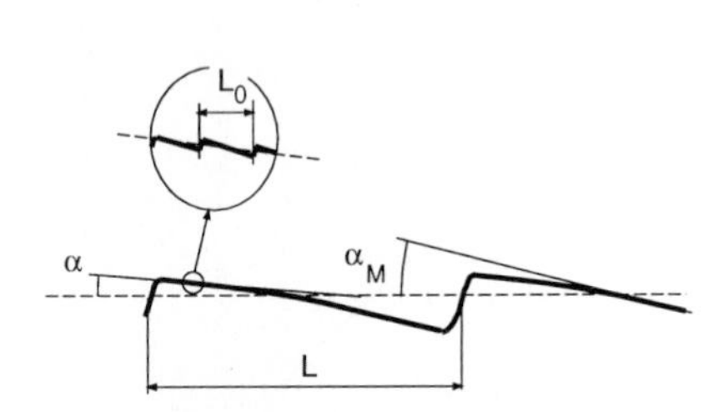

Fig. 8.26. Measured ω-scan of a GaInAs/GaAs/GaAsP/GaAs multilayer (dots) and its fit by the theory using a single type of steps (full), and two sets of the steps (dashed) [27]. The left-hand figure shows a possible microscopic structure of terraces

maximum the direction of the terrace orientation. However, the simultaneous existence of large terraces formed by *step bunching* and atomic scale micro-terraces can modify the DXS pattern (see Fig. 8.26) [27–29].

The investigation of step-like interface morphology by interface sensitive *diffraction* methods is briefly discussed in Sect. 8.6.

8.5.4 Non-Coplanar NSXR

XRR in *coplanar* geometry is most common and simple to realise with conventional diffractometers and reflectometers. The intensity distribution is resolved in the q_x/q_z-plane which contains the surface normal. The region in the q_x/q_z-plane accessible by coplanar reflection geometry is restricted by the Ewald spheres for the limiting cases of grazing incidence and grazing exit, which represent the horizon of the sample surface. Especially for small values of q_z the measurable lateral momentum transfer decreases and consequently the information is cut about roughness with small lateral dimensions of nanoscopic scale.

By use of a *non-coplanar* scattering geometry this limitation has been overcome [30,31,10]. The equipment requires monochromatic beam collimated an in two directions, which can be provided by synchrotron radiation sources. First experiments used the setup of a small angle scattering instrument with a well collimated beam and a two dimensional position sensitive detector. Other setups are based on surface diffraction instruments, working usually in a strongly non-coplanar (grazing incidence diffraction) geometry, see Fig. 8.27. The detection of the diffusely scattered intensity up to a parallel momentum transfer of $1\,\text{Å}^{-1}$ enables to study the correlation properties up to a few Å. The diffusely scattered intensity is usually drawn in a double logarithmic scale. Fitting the asymptotic intensity decay with increasing $Q_{||}$ by a power law, the Hurst factor introduced in section 8.5.1 can be determined with good

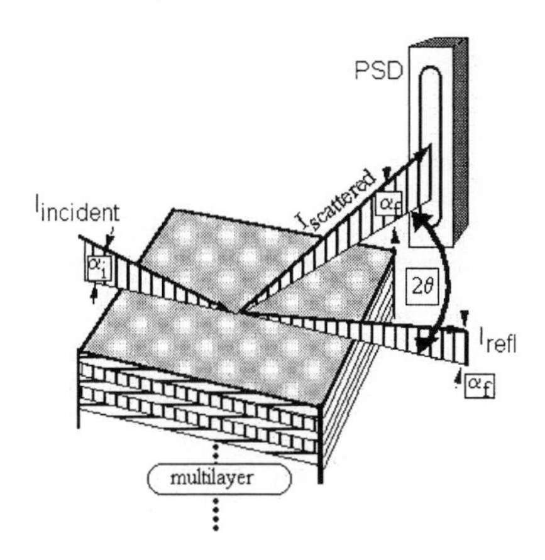

Fig. 8.27. Schema of non-coplanar x-ray reflectivity setup [30]

precision, wherefrom one can conclude on the validity of different growth models.

Fig. 8.28(a) shows measured θ_{sc}-scans of an amorphous W/Si superlattice for different $q_{||}$. They cross the RDS-sheets indicated by roman numbers. For increasing $q_{||}$ the width of the RDS-sheets increases and finally the resonant diffuse scattering disappears, indicating a reduction of the vertical replication length L for the higher frequencies of the roughness profile. In Fig. 8.28(b), the decrease of the intensity of the first RDS-sheet is plotted. The measurements prove the validity of a logarithmic scaling behavior as predicted by the Edward-Wilkinson equation [32].

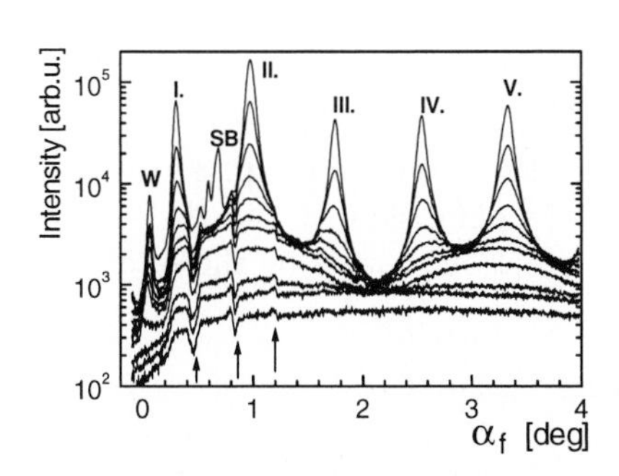

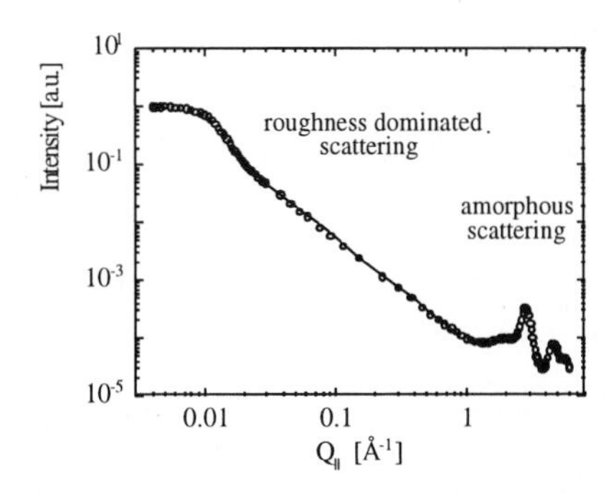

Fig. 8.28. Measurement of non-specular x-ray reflectivity of an amorphous W/Si superlattice. (a) θ_{sc}-scans for different $q_{||}$ [10]. Intersections with the RDS-sheets are indicated by roman numbers. (b) Intensity profile of the first RDS-sheet [30]

8.6 Interface Roughness in Surface Sensitive Diffraction Methods

In case of epitaxial multilayers surface and interface roughness can also be studied by surface sensitive x-ray *diffraction* methods such as *grazing incidence diffraction* (GID) and *strongly asymmetric x-ray diffraction* (SAXRD). Beside reflection at the interfaces there occurs diffraction by the layer lattices. The principles of diffraction by rough multilayers are similar to those described in more detail for x-ray reflection. All used theoretical treatments can be extended.

The polarisability of each layer can be developed in a Fourier series after its reciprocal lattice vectors

$$\chi_j^{\text{layer}}(\boldsymbol{r}) = \sum_{g} \tilde{\chi}_{g,j}(\boldsymbol{r})\, e^{-i\boldsymbol{g}\boldsymbol{r}} \,. \tag{8.59}$$

Measuring the intensity pattern of a Bragg-reflection with the reciprocal lattice vector $\boldsymbol{h}$ in a conventional diffraction geometry (so-called *two-beam case*) only the Fourier components with the indices $\boldsymbol{h}$, $-\boldsymbol{h}$ and 0 are of importance.

Crystal truncation rods through each reciprocal lattice point characterise the structure amplitude of a crystalline layer. All truncation rods of a periodic multilayer contain the fine structure of equidistant superlattice satellites similar to the schema in Fig. 8.20.

The non-perturbed wave field of diffraction by a planar epitaxial multilayer under conditions of grazing incidence consists in each plane layer of 8 plane waves for each polarisation

$$E_j^{\text{pl}}(\boldsymbol{r}) = \sum_{n=1}^{4} \left[T_j^n\, e^{-i\boldsymbol{k}_{0\|}\boldsymbol{r}_\|} e^{-ik^n_{0z,j}(z-Z_j)} + R_j^n\, e^{-i\boldsymbol{k}_{h\|}\boldsymbol{r}_\|} e^{-ik^n_{hz,j}(z-Z_j)} \right] \Omega_j^{\text{pl}}(z) \tag{8.60}$$

with the rough interface shape function $\Omega_j^{\text{pl}}(z) = H(z-[Z_j+z_j]) - H(z-Z_j)$. For superlattices with rough interfaces the layer disturbance includes the variation of the Fourier components of the polarisability and the lattice displacement $\Delta\boldsymbol{u}(\boldsymbol{r})$ due to the lattice deformation created by the interface roughness profile. In layer j,

$$\Delta\chi_j^{B_{\text{layer}}}(\boldsymbol{r}) = \sum_{g=0,-\boldsymbol{h},\boldsymbol{h}} \Delta\chi_{g,j}(\boldsymbol{r})\, e^{-i\boldsymbol{g}\boldsymbol{r}} \qquad \text{with} \tag{8.61}$$

$$\Delta\chi_{g,j}(\boldsymbol{r}) = \left[\chi_{g,j}\left(e^{i\boldsymbol{g}\Delta\boldsymbol{u}(\boldsymbol{r})} - 1\right) + \Delta\chi_{g,j}\, e^{i\boldsymbol{g}\Delta\boldsymbol{u}(\boldsymbol{r})}\right] e^{i\boldsymbol{g}\boldsymbol{u}_0(z)}\, \Omega_j^{\text{pl}}(z) \,.$$

Similar to x-ray reflection by the rough interfaces the disturbances give rise to diffuse scattering. The number of possible diffuse scattering processes between two non-perturbed states at one interface, see (4.D23), increases up to 64.

Fortunately a certain number of them is almost negligible. If the roughness profile is replicated, the diffusely scattered intensity is concentrated in horizontal sheets of resonant diffuse scattering crossing the crystal truncation rods in the position of the diffraction satellites. Their origin arises now from partially coherent *diffraction and reflection* by the interface disturbances. For weak strain the covariance functions are formally quite similar to (8.49) found for x-ray reflection

$$\tilde{Q}_{jk}^{mnop} = \frac{A\,\Delta\chi_{g,j}(\Delta\chi_{g',k})^*}{\delta q_{z,j}^{mn}(\delta q_{z,k}^{op})^*} \int_s d(r_\| - r_\|')\, e^{iq_\|(r_\| - r_\|')} \times \tag{8.62}$$
$$\times \left[\chi_{z_j,z_k}(\delta q_{z,j}^{mn}, (\delta q_{z,k}^{op})^*) - \chi_{z_j}(\delta q_{z,j}^{mn})\chi_{z_k}((\delta q_{z,k}^{op})^*)\right] ,$$

however now with the *reduced* scattering vectors of the corresponding scattering process in the layers, which depend on the *local* reciprocal lattice vectors in the layers by $\delta q_{z,j} = q_{z,j} - g_{z,j}$.

In Fig. 8.29 the scattering geometry in reciprocal space and the corresponding experimental results of strongly asymmetric diffraction by a GaAs/AlAs superlattice are shown. The measured sheets of resonant diffuse scattering (RDS) of the diffraction mode are clearly visible. It is an advantage of the AXRD measurements, that the RDS sheets are not limited by the sam-

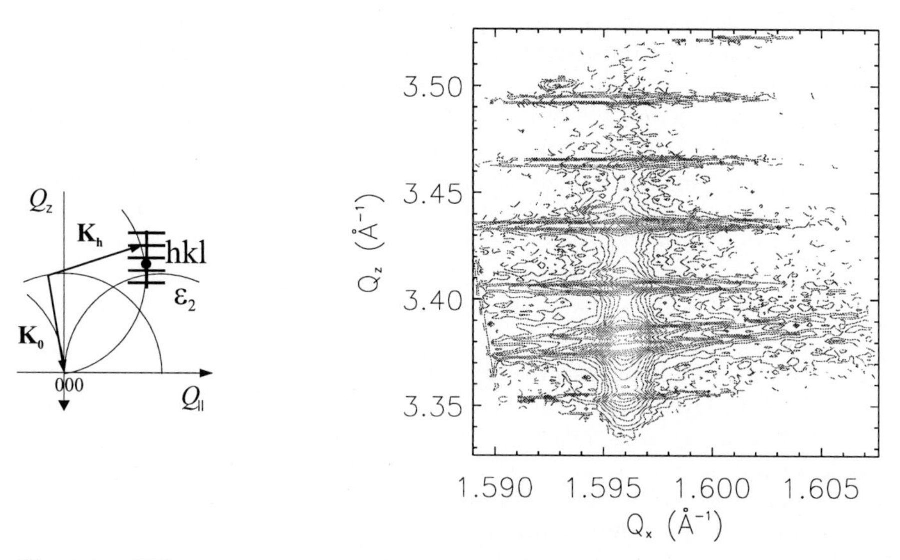

Fig. 8.29. Diffuse x-ray scattering by rough interfaces in the strongly asymmetric diffraction mode. Left: schematic situation in reciprocal space. Right: reciprocal space map of (113) diffraction of a GaAs/AlAs superlattice for $\lambda = 1.47$ Å. The coherent crystal truncation rod (CTR) is crossed by horizontal RDS sheets, indicating correlated roughness. The sheets are laterally not limited by the experimental geometry

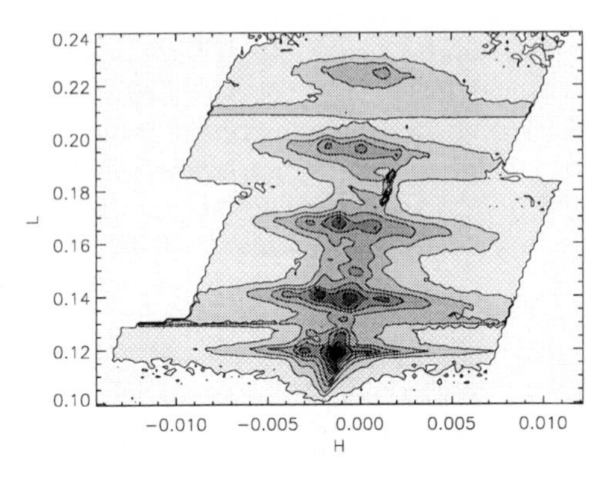

Fig. 8.30. Diffuse x-ray scattering by stepped interfaces in the grazing incidence diffraction mode. Reciprocal space map of $(0\bar{2}0)$ diffraction of a GaInAs/GaAs/GaAsP/GaAs multilayer on a 2° off-oriented [001]-substrate. The axes are normalised in crystallographic units (HKL)

ple horizon, in contrast to coplanar XRR. So the full range of momentum transfer can be detected in a coplanar scattering geometry.

The application of x-ray diffraction methods is limited on epitaxial structures. On the other hand, x-ray reflection experiments are less successful for many semiconductor systems due to the missing contrast in the electron density modulation. Thus the choice of suitable Bragg-reflections allows increasing the contrast between the layers in the diffraction mode.

GID, a non-coplanar surface sensitive diffraction method, was successfully applied for the measurement of RDS by rough multilayers in Ref. [12].

Beside Gaussian roughness correlation behaviour, the *step-like interface morphology* was also investigated by various diffraction methods. In Fig. 8.30 we show the measured 200-reciprocal space map of a GaInAs/GaAs/GaAsP/GaAs-superlattice on a 2° off-oriented GaAs substrate, measured by grazing incidence diffraction. This reflection is highly sensitive for the morphological ordering, since the scattering contrast of the corresponding Fourier components of the susceptibility is much larger than that in the above discussed reflection mode. Similarly to Figs. 8.23 and 8.25, the diffuse scattering is concentrated in stripes, resonant diffuse scattering along so-called grating truncation rods, which are *perpendicular to the averaged surface.* The grating rods are therefore inclined with respect to the *crystallographic* orientation, which is simultaneously the orientation of the terraces. Each grating rod contains multilayer Bragg-peaks. The Bragg-peaks of the same vertical order but of different grating rods form branches which are inclined with respect to the sample surface according to the inclination of the morphological interface replication via the surface normal. The envelope maximum of the diffuse scattering follows the 001-direction, which is the orientation of the terraces.

8.7 X-Ray Reflection from Multilayer Gratings

In this section we discuss the calculation of the x-ray reflection from *multilayer gratings* (MLGs), Fig. 8.31. Gratings are etched into planar multilayers

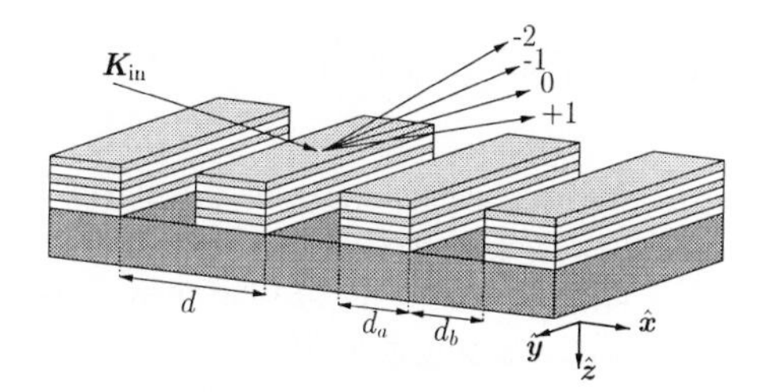

Fig. 8.31. A sketch of a multilayer grating with a fan consisting of four diffracted-reflected waves

so that their lateral structure is formed by wires distributed equidistantly with period d along the surface. We focus the present study mainly on the short-period gratings with the periodicity at about micrometers, which are of most interest in semiconductor physics.

The part etched out (dips between wires) can be several hundreds nanometers deep. Thus these structures can be considered as a special case of huge deterministic roughness or as an artificial lateral one-dimensional crystals contrary to the crystals periodic in all three directions. Thus the reflectivity from gratings can be treated by *approximate as well as rigorous methods* [33–35,?,37,38,?,39], thus making possible to treat and compare the adequateness of various approximations. In this section, we formulate the approximate perturbative treatment by the kinematical theory and by DWBA and compare them to the exact dynamical calculation. We determine region of validity of DWBA and we show that the correct choice of the eigenstates can lead to good results even when the perturbed potential is present in the most volume of the sample, contrary to the small roughness of interfaces.

8.7.1 Theoretical Treatments

MLG possesses the translation symmetry so that it is fully sufficient to determine its susceptibility $\chi(\boldsymbol{r})$ in one period $(-\frac{d}{2} \leq x \leq \frac{d}{2})$ only. Therefore we first describe it for any of the layer j. The period consists of two parts (wires) named $\underline{a}_j$ and $\underline{b}_j$ (for the case of an etched grating, one of the parts is the air). We denote their susceptibilities χ_j^a, χ_j^b and their widths $d_j^a = \Gamma_j\, d$, $d_j^b = (1 - \Gamma_j)\, d$ with $0 \leq \Gamma_j \leq 1$.

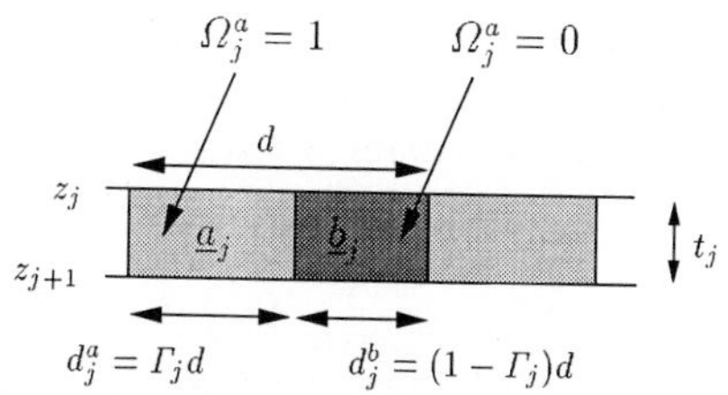

Fig. 8.32. Notation of the variables describing a laterally structured layer

We introduce the *shape function* $\Omega_j^{a1}(\boldsymbol{r})$ of the material $\underline{a}_j$ in the period. It equals unity inside the volume occupied by the material $\underline{a}_j$ and it is zero elsewhere, see Fig. 8.32. Then the susceptibility of one period is

$$\chi_j^1(\boldsymbol{r}) = \chi_j^a \Omega_j^{a1}(x,z) + \chi_j^b\left(1 - \Omega_j^{a1}(x,z)\right) . \tag{8.63}$$

By $\tilde{\Omega}_j^{a1,h}(q_z)$ we denote the two-dimensional Fourier transform of the shape function of one period.

Because of the presence of two types of interfaces, horizontal and vertical ones, different theories treat the respective reflectivities using different approximations. Further, we mean by the single-scattering approaches those related to the lateral diffraction case. We treat separately the perturbative (single-scattering) and rigorous dynamical (multiple-scattering) theories.

Perturbative treatments MLG is periodic along the axis $\hat{\boldsymbol{x}}$ with the lateral periodicity d, Fig. 8.31. Then the scattering potential $V(\boldsymbol{r})$ of the sample, as defined in Sect. 8.2, can be given as a convolution of the scattering potential of one period $V^1(\boldsymbol{r})$ (defined on the interval $-\frac{d}{2} \leq x \leq \frac{d}{2}$) with a periodic arrangement of δ-functions

$$V(\boldsymbol{r}) = V^1(\boldsymbol{r}) \otimes \sum_n \delta(x - nd) \ . \tag{8.64}$$

Its Fourier transform is product of two terms (A denotes the sample area)

$$\tilde{V}(\boldsymbol{q}) = \int d\boldsymbol{r}\, V(\boldsymbol{r})\, e^{i\boldsymbol{q}\boldsymbol{r}} = \frac{A}{d}\, \tilde{V}^1(q_x, q_z) \cdot \sum_{h=\frac{2\pi}{d}m} \delta_{q_x,h}\, \delta_{q_y,0} \ . \tag{8.65}$$

The second (summation) term expresses the reciprocal lattice of the grating, which are *grating truncation rods (GTRs)* in Q_z direction positioned equidistantly along the axis q_x at points $q_x = h_m = \frac{2\pi}{d}m$, where m is an integer (see Fig. 8.33). The first term $\tilde{V}^1$ is the Fourier transform of the potential in one period, behaving like an envelop function for the wave fields associated to the GTRs.

In a multilayer grating, the potential of one period $V^1(\boldsymbol{r})$ is the sum of the potentials of individual layers $V_j^1(\boldsymbol{r})$, and similarly for their two-dimensional Fourier transforms $\tilde{V}_j^1(h, q_z)$. The latter separates into the zeroth component proportional to the laterally averaged susceptibility and to the discrete Fourier components proportional to the susceptibility contrast

$$\tilde{V}_j^1(h, q_z) = \begin{cases} -K^2\, d \int dz\, \chi_{0j}(z)\, e^{iq_z z} & \text{for } h = 0 \ , \\ -K^2\, (\chi_j^a - \chi_j^b)\, \tilde{\Omega}_j^{a1,h}(q_z) & \text{for } h \neq 0 \ . \end{cases} \tag{8.66}$$

Now we will consider the scattering from the sample we characterised generally above. We first determine the directions $\boldsymbol{K}_h$ of scattered waves. We use the principles for the Ewald construction, discussed in Appendix 8.A, which state that the wave vector end-points lay at the intersection of the sample reciprocal lattice and the Ewald sphere of the incident wave. Thus the incident wave is scattered into the fan of reflected and transmitted waves associated to each GTR, see Figs. 8.31 and 8.35.

Further, we calculate the reflection amplitudes. Scattering potential of MLG is *deterministic* and thus the reflection amplitude of all GTRs comes from the *coherent scattering* only (even though into non-specular directions). It is expressed similarly to the coherent specular reflection amplitude of rough MLs calculated by DWBA. Using the formalism from Appendix 4.D, the amplitude at the sample surface is $R^h(\boldsymbol{K}_h) = T_{0h}/2iK_{hz}A$. The scattering matrix element $T_{0h} = \langle E_h|V(\boldsymbol{r})|E_0\rangle$ can be decomposed into the sum over the individual layer contributions τ_j^h. The sample reflectivity along GTR h is finally $|R^h|^2\, K_{hz}/K_z$.

The reflection amplitude then depends on the approximation used in the evaluation of the scattering matrix element. We discuss briefly the calculation by the kinematical theory and by the first-order DWBA applying the approach of Sects. 4.D.2 and 4.D.3, respectively.

Kinematical calculation Kinematical theory is equivalent to the first Born approximation [36,37] thus calculating the scattering process as the single-scattering transition of the incident vacuum plane wave $|E_0\rangle = e^{-i\boldsymbol{K}_0\boldsymbol{r}}$ into the diffracted vacuum plane wave $|E_h\rangle = e^{-i\boldsymbol{K}_h\boldsymbol{r}}$, see Fig. 8.35(a). The scattering matrix element for one period and one layer is proportional to the Fourier transform of the layer potential in one period (with the scattering vector $\boldsymbol{Q}_h = \boldsymbol{K}_h - \boldsymbol{K}_0$)

$$\tau_j^h = \langle e^{-i\boldsymbol{K}_h\boldsymbol{r}}|V_j^1(\boldsymbol{r})|e^{-i\boldsymbol{K}_0\boldsymbol{r}}\rangle = \tilde{V}_j^1(h, Q_{hz})\ . \tag{8.67}$$

According to (8.66), we can see that the Fourier transform for h=0 is determined by the profile of laterally averaged susceptibility. Thus the *specular reflectivity* profile coincides with a kinematical reflection from laterally averaged planar multilayer and the specular reflectivity curve exhibits the same features as those calculated in the framework of the kinematical theory and the stationary phase method (SPM) [19,39]. (SPM helps to avoid

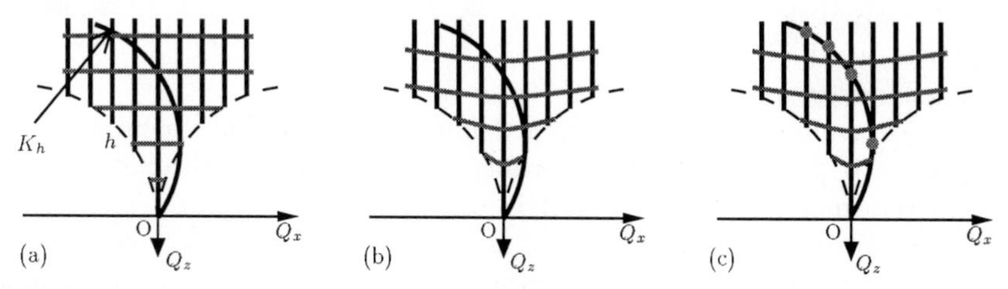

Fig. 8.33. Schematical drawing of the reciprocal space maxima of a laterally periodic grating etched into a periodic multilayer. The "Bragg" sheets are parallel to the q_x axis in the kinematical treatment (a), whereas they are curved and shifted upwards in the DWBA (b) and dynamical (c) calculations due to refraction. In addition, the subfigure (c) illustrates the multiple scattering interaction among wave fields of the simultaneously excited GTRs which is taken into account within the dynamical theory

the Fraunhofer approximation which is not suitable for laterally extended samples.)

Considering the intensity of the *non-specular truncation rods* ($h\neq 0$), the scattering matrix contribution is

$$\tau_j^h = -k_0^2(\chi_j^a - \chi_j^b) \cdot \tilde{\Omega}_j^{a1,h}(Q_{hz}) \ . \tag{8.68}$$

By calculating the kinematical scattering integral by the stationary phase method we generalise the *kinematical Fresnel reflection coefficient for lateral diffraction case*

$$r_{j,j+1}^{h,\mathrm{kin}} = \frac{k_0^2\left(\tilde{\chi}_j^h - \tilde{\chi}_{j+1}^h\right)}{2K_{hz}Q_{hz}} \ . \tag{8.69}$$

For specular reflection it perfectly coincides with the kinematical Fresnel reflection coefficient for planar multilayers $r_{j,j+1}^{0,\mathrm{kin}} = k_0^2\left(\chi_{0,j} - \chi_{0,j+1}\right)/Q_z^2$, cf. (3.91), as we said above.

As all the kinematical theories, also in the present case the effects of absorption and refraction are not comprised. Thus the kinematical intensity is much larger than unity below the critical angle and it diverges for the specular scan at the origin of the reciprocal space. Further, the kinematical period of oscillations of a MLG converges slowly to that calculated by a theory including the refraction.

Let us figure out the positions of maxima of a periodic multilayer grating using a reciprocal space schema, Fig. 8.33. They lay on the intersections of the grating truncation rods (reciprocal lattice of the grating represents the lateral periodicity) and the sheets passing through the ML maxima on the specular truncation rod (which represents the vertical periodicity).

Calculation by DWBA

We follow the basis of the DWBA as treated for the roughness and we split the MLG potential $V(\boldsymbol{r})$ into two parts, see Fig. 8.34. We choose the *ideal (unperturbed) potential* $V^A(\boldsymbol{r})$ as that of a planar laterally averaged multilayer and thus calculating the eigenstates $|E_{\boldsymbol{K}}^A\rangle$, see (4.D19), according to (3.47). For the simplicity of the further treatment we restrict ourselves to the rectangular gratings only [40]. From (8.65) and (8.66) it follows that the ideal potential V_j^A is constant in each etched layer, $V_j^A(\boldsymbol{r}) = \tilde{V}_j^1(0,0)/dt_j$, whilst the *perturbed potential* $V^B(\boldsymbol{r}) = V(\boldsymbol{r}) - V^A(\boldsymbol{r})$ is the sum of non-zero Fourier components, $V_j^B(\boldsymbol{r}) = \sum_{h\neq 0}\tilde{V}_j^1(h,0)e^{ihx}/dt_j$.

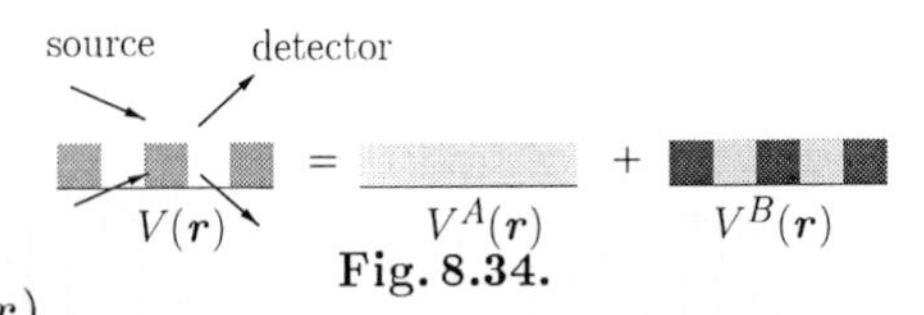

Fig. 8.34.

Consequently the scattering element of the perturbed potential does not intervene into the *specular* term

$$\tau_j^{h=0} = \langle E_0^A|V_j^A|E_0\rangle + \langle E_0^A|V_j^B|E_0^A\rangle \ . \tag{8.70}$$

The specular reflection amplitude from the whole MLG then equals the (dynamically) calculated reflection from the laterally averaged multilayer. From this it clearly follows that this DWBA considers multiple scattering between the horizontal interfaces of averaged layers by using the dynamical Fresnel reflection coefficients, but neglects the influence of multiple scattering by the vertical side walls.

The amplitude of the wave scattered into a *non-specular* GTR $h \neq 0$ is

$$\tau_j^h = \langle E_h^A | V_j^B | E_0^A \rangle \ . \tag{8.71}$$

The contribution of each laterally structured layer consists of four terms

$$\tau_j^h = -K^2 \Big(T_{k_{h,j}} S_j^{11} T_{k_{0,j}} + T_{k_{h,j}} S_j^{12} R_{k_{0,j}} + R_{k_{h,j}} S_j^{21} T_{k_{0,j}} + R_{k_{h,j}} S_j^{22} R_{k_{0,j}} \Big) \tag{8.72}$$

where the amplitudes T_{k_j}, R_{k_j} are equal to $U_j(\pm k_{z,j})$ in (3.48) and the layer structure factor (4.D13) is $S_j^{mn} = S_j(\boldsymbol{q}_j^{mn}) = \left(\chi_j^a - \chi_j^b\right) \tilde{\Omega}^{a1}_{q^{mn}_{x,j}}(-q^{mn}_{z,j})$. The four scattering wave vectors $\boldsymbol{q}_j^{11}, \ldots, \boldsymbol{q}_j^{22}$ are defined as in the case of diffuse scattering, se (4.D23) and Fig. 8.40. We draw them in the reciprocal space schema in Fig. 8.35(b) while demonstrating there the single-scattering character of the diffraction from the incident to the diffracted wave fields.

Because the eigenstates of the ideal potential are calculated using the usual dynamical matrix formalism for specular reflectivity from a planar multilayer, thus the effects of absorption and refraction are taken into account. Then the maxima of a periodic multilayer grating, Fig. 8.33(b), lay on the intersection of the truncation rods and the refraction-curved sheets passing through the maxima on the specular truncation rod.

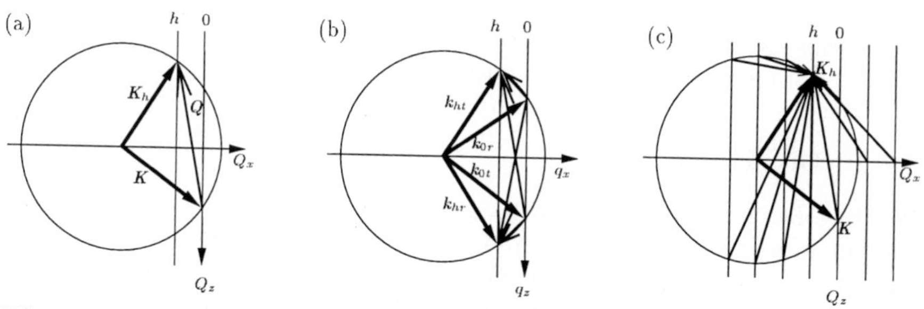

Fig. 8.35. Single-scattering approaches, i.e. kinematical (a) and DWBA (b), calculate the diffracted field as a single-scattering process from GTR 0 to a GTR h, while the multiple-scattering approaches (c) take the contributions from all the GTRs into account

Multiple scattering treatment by the dynamical theory The dynamical theory treats the reflection from a multilayer grating by rigorously solving the wave equation under the condition of a one-dimensional periodicity

$$\chi(\boldsymbol{r}) = \chi_0(z) + \sum_h \tilde{\chi}_h(z)\, e^{-ihx} \,. \tag{8.73}$$

There are miscellaneous approaches found in the literature, reviewed e.g. in [19,39]. Their formulation comes from the optics of visible light [33], while they have been applied in XRR only for surface gratings [38] using integral Rayleigh-Maystre formulae. XRR from multilayered gratings is studied deeply in [19,39] using matrix modal method. Dynamical theory takes into account the multiple scattering among the wave fields (each consisting of pair of a transmitted and reflected wave), which are associated to all truncation rods, including the real as well as evanescent GTRs as shown in Fig. 8.35(c).

Using a convenient matrix formalism similar to that for planar multilayers, the *generalisation of the Fresnel coefficients for lateral diffraction case*, compare (3.68) and (3.70), has been found [19,39]

$$r^{hg}_{j,j+1} = \frac{k^h_{z,j} - k^g_{z,j+1}}{k^h_{z,j} + k^g_{z,j+1}} \quad \text{and} \quad t^{hg}_{j,j+1} = \frac{2k^h_{z,j}}{k^h_{z,j+1} + k^g_{z,j+1}} \,. \tag{8.74}$$

Here, the indices h and g relate the transmission and reflection processes to simultaneous diffraction between wave fields of two GTRs h and g. Wave vectors $\boldsymbol{k}_h$ of scattered waves do not point to a *spherical* Ewald sphere, but to a so-called *dispersion surface* like in dynamical theory of x-ray diffraction.

In the dynamical theory the energy is conserved. Therefore a strong wave field corresponding to a certain GTR can influence significantly the intensity profile of another GTR. This may be the case, for instance, in the angular region where the wave field of the first GTR changes from evanescent to real (near the intersection of the Ewald sphere with the GTR +1, see Fig. 8.33(c)). There the specular intensity can be enhanced with respect to the specular intensity of an averaged planar multilayer.

8.7.2 Discussion

For the following discussion we will consider short period rectangular gratings (period around one micrometer) and the wavelength at about one Ångstrom. Since we already mentioned that the kinematical theory does not involve the refraction, which is of crucial importance in XRR, we will further devote our discussion to the comparison of DWBA to the dynamical theory. We choose the ratio Γ of the wire width with respect to the period one half. Then we can find truncation rods of three types:

Specular truncation rod ($h = 0$). Here, the DWBA and dynamical theory give the same profiles, except for the known angular region of the enhanced interaction with GTR +1 as discussed earlier.

Weak, kinematically forbidden truncation rods (h is even). The associated Fourier coefficients are zero, and therefore single-scattering theories, including

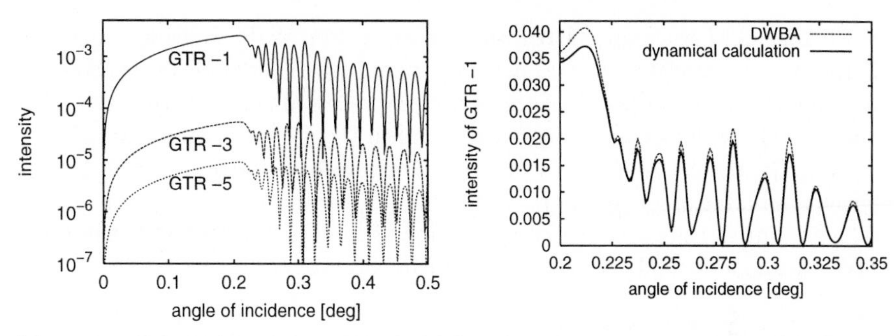

Fig. 8.36. Calculation of the odd-order GTRs for a GaAs surface grating (thickness 300 nm) for period of (a) 0.8 μm and (b) 5 μm for wavelength 1.54 Å. In the former case, DWBA gives the same results as the dynamical theory. In the latter case the multiple scattering starts to be important and DWBA of the first order gives only approximative result

the kinematical one and DWBA, predict zero intensity for them. Thus these GTRs are excited by multiple scattering in the etched layers and consequently their profile can be calculated by the dynamical theory or by higher-order DWBA.

Strong truncation rods (h is odd). Here, both DWBA and dynamical theory coincide, see Fig. 8.36(a). The good coincidence depends on the force of the dynamical interaction between diffracted wave fields. There are more GTRs excited in the Ewald sphere of the incident wave for large periods or small wavelengths, thus the dynamical effects will be enhanced and DWBA starts to be only approximative, see Fig. 8.36(b). We found possible to formulate a condition separating the two cases using a two-beam approximation of the dynamical theory [19].

8.7.3 Reflectivity from Rough Multilayer Gratings

The influence of interface roughness on grating reflectivity can be studied within all three theoretical treatments discussed earlier. Within the matrix approach of the dynamical theory [19,39], the generalised Fresnel coefficients (8.74) corrected for roughness were found formally similar to those for rough planar multilayers (8.24)

$$r^{hg}_{j,j+1} = r^{hg,\mathrm{flat}}_{j,j+1}\, e^{-2k^h_{z,j}k^g_{z,j+1}\sigma^2_{j+1}} \text{ and } t^{hg}_{j,j+1} = t^{hg,\mathrm{flat}}_{j,j+1}\, e^{(k^h_{z,j}-k^g_{z,j+1})^2\sigma^2_{j+1}/2} . \tag{8.75}$$

Roughness in gratings decreases the scattered intensity for the incidence angles even *below* the critical angle. Furthermore, there is different sensitivity to the surface and interface roughnesses for *weak and strong GTRs*, respectively. Finally we can notice that the kinematical reflection coefficients (8.69) are attenuated by $e^{-Q^2_{hz}\sigma^2_{j+1}/2}$ similarly to (3.104).

In Fig. 8.37 we show XRR results of a periodic W/Si multilayer grating. Structural parameters (lateral periodicity and wires width, layer thicknesses and interface roughnesses) of the sample were obtained by fitting the measured GTR profiles employing the dynamical theory for rough gratings.

Finally, the calculation (by DWBA) of the diffuse scattering from MLGs, such as simulation of the map in Fig. 8.37(c), is even more tricky procedure which requires the preliminary calculation of the eigenstates either using the DWBA for perfect MLG or the dynamical theory.

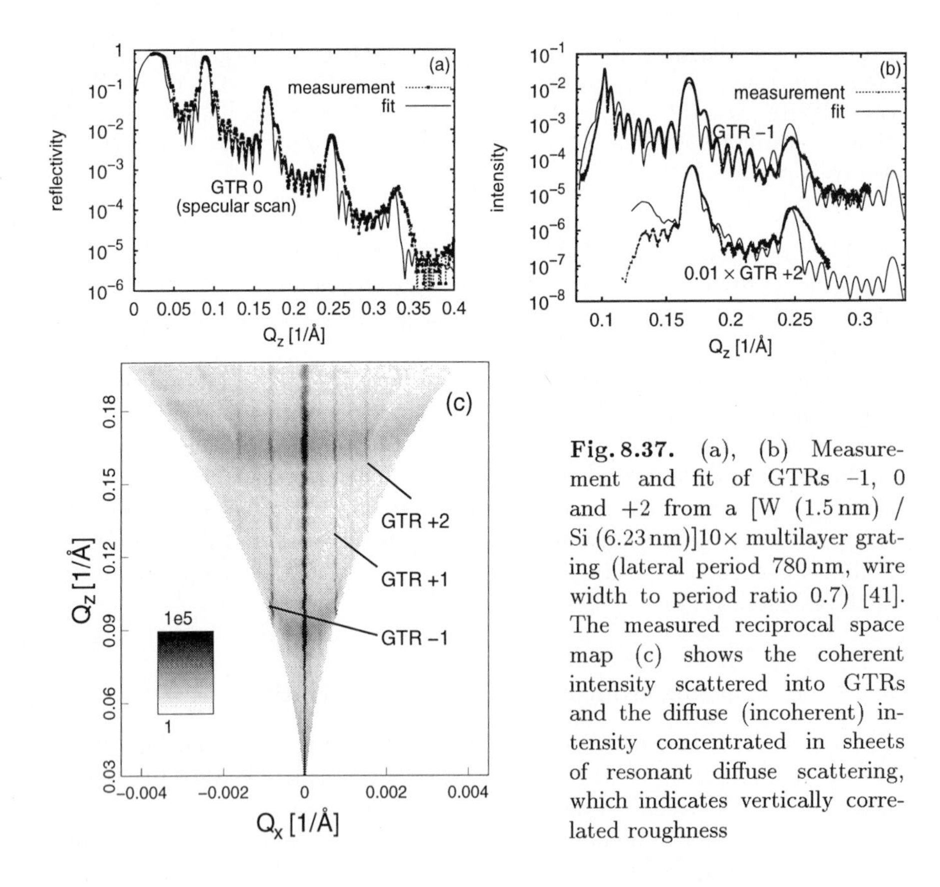

Fig. 8.37. (a), (b) Measurement and fit of GTRs −1, 0 and +2 from a [W (1.5 nm) / Si (6.23 nm)]10× multilayer grating (lateral period 780 nm, wire width to period ratio 0.7) [41]. The measured reciprocal space map (c) shows the coherent intensity scattered into GTRs and the diffuse (incoherent) intensity concentrated in sheets of resonant diffuse scattering, which indicates vertically correlated roughness

Acknowledgments

The work was supported by the Deutsche Forschungsgemeinschaft (grant BA1642/1-1), by the Lise Meitner Fellowship of FWF, Austria (project M428-PHY) and by the grant VS 96102 of the Ministry of Education of the Czech Republic.

8.A Appendix: Reciprocal Space Constructions for Reflectivity

In some of the previous chapters in the book, the *reciprocal space* representation was used for drawing the experimental scattering geometry: experimental scans and inaccessible regions for coplanar reflectivity (Figs. 7.2, 8.4 and 8.5). In addition, throughout this chapter we use the reciprocal space to describe graphically the scattering events of x-ray reflection. Since this approach may not be common to the reader who is not accustomed to that representation, we give here some schematic interpretations of the reflection by multilayers in reciprocal space, which help in finding the intuition for an easy understanding of the scattering features in a simple geometrical way. We start by the interpretation of fundamental laws of reflection and refraction at interfaces. We relate the reflection curves of thin films and periodic multilayers to their particular reciprocal space features and discuss multiple scattering as it is considered within the treatment by a DWBA.

The idea to represent x-ray scattering by reciprocal space constructions has been introduced by P.P. Ewald in the early stage of the dynamical theory of x-ray diffraction. The goal is to relate the directions of the scattered waves and the symmetry of the sample represented by the Fourier transform of the crystal lattice and/or the shape function of the scatterers. *Ewald (reciprocal space) construction* visualizes two basic physical principles:

1. Energy conservation. x-ray reflection is an elastic scattering process, conserving the wave vector length. Then the end-points of all scattered waves can lay only on the *Ewald sphere* of the radius of the wave vector length, Fig. 8.38(a).
2. Momentum conservation except of a reciprocal lattice vector if the diffraction condition is fulfilled. This reflects the symmetry properties of the sample.

In this book we use Ewald construction for the illustration of the reflection by layers and multilayers, including the wave vectors in the vacuum and in the medium.

8.A.1 Reflection from Planar Surfaces and Interfaces

Let us discuss the reflection and refraction laws in reciprocal space, Fig. 8.38, by use of Ewald construction. The wave propagation in the vacuum and in the media is determined by the different length of the wave vectors. In case of a homogeneous half space of a slightly absorbing medium with a flat surface or of planar layers with smooth interfaces their reciprocal space structure is defined by a so-called *truncation rod* passing through the origin and normal to the surface (i.e., it usually coincides with the axis $\boldsymbol{q}_z$). We call the truncation rod through the origin of the reciprocal space here *specular rod*, since it defines

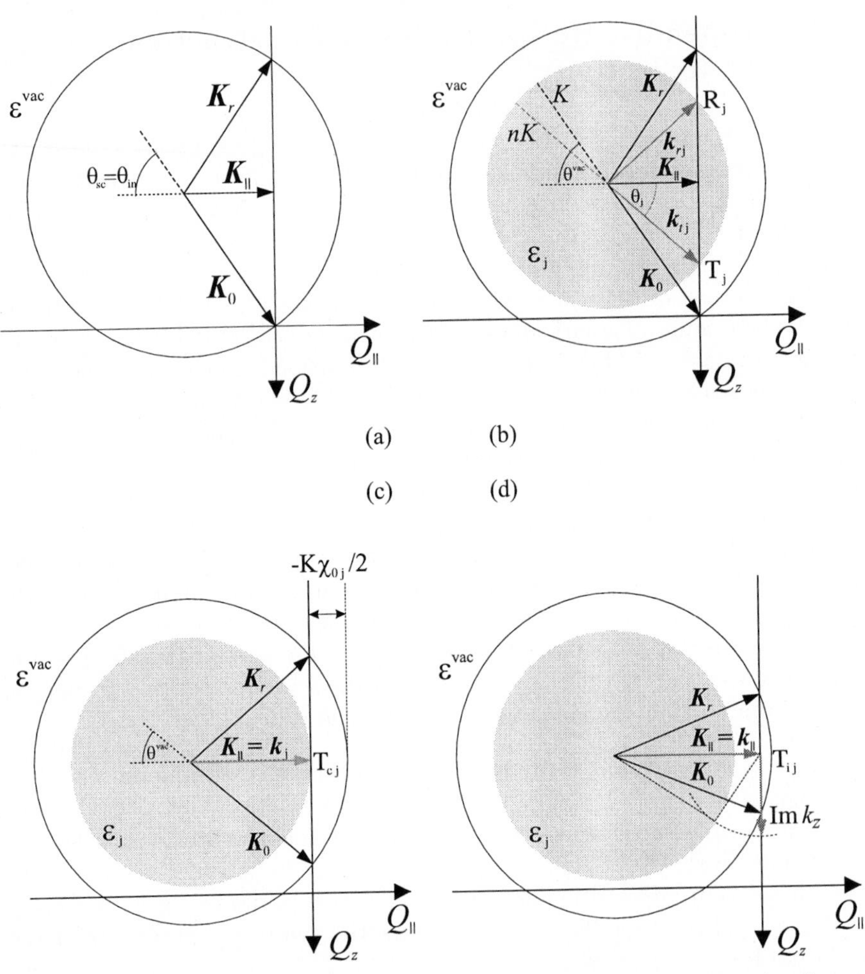

Fig. 8.38. Graphical representation of the laws of reflection and refraction by an interface by means of the Ewald construction. (a) The law of reflection, (b)–(d) Snell's law: (b) above, (c) at and (d) below the critical angle. Below the critical angle the lateral component of $\boldsymbol{k}$ is larger than the radius of the Ewald sphere of the medium j thus it has purely imaginary k_z component (neglecting absorption) and the wave is called evanescent

the conditions for specular reflection. It intersects the vacuum Ewald sphere $\epsilon^{\rm vac}$ of the incident wave $\boldsymbol{k}_0$ in two points, which pin down the wave vectors of the reflected wave $\boldsymbol{k}_r$ and of the transmitted wave in the vacuum, see subfigure (a). Therefrom we obtain the *the law of reflection*—the reflected wave makes the same angle with the surface as the incident one. The Ewald construction with the specular rod represents the symmetry of the sample and of the scattering process, which permits a momentum transfer only along the $\boldsymbol{q}_z$ direction (along the surface normal).

Inside a layer j of a multilayer (or in a substrate) the wave vectors are determined

1. by the *dispersion relation* $k_j = n_j k_0$ giving the radius of the Ewald sphere within the medium ϵ_j,
2. by the continuity of the lateral wave vector components at the interface.

These two conditions lead to the *Snell's law* (also refraction law) for the transmitted wave as outlined in subfigures (b)–(d). The tie points T_j and R_j of the transmitted and reflected wave in the layer j, respectively, are located at the intersections of the specular rod and the "inner" Ewald sphere ϵ_j. For x-rays is $n < 1$ ($\chi < 0$), thus three distinct cases may happen in each layer. Case (b) marks the refraction law above the *critical angle*: two waves, reflected $\boldsymbol{k}_{rj}$ and transmitted $\boldsymbol{k}_{tj}$, propage in the layer. The case (c) visualises the situation at the critical angle for total external reflection in the layer. There is one tie point T_{cj} only and the wave in the layer propagates parallel to the interface, $\boldsymbol{k}_j = \boldsymbol{k}_{\|}$. Case (d) interprets the generation of the *evanescent wave* in the layer, propagating parallel to the interface and exponentially damped perpendicularly to it.

According to the Fresnel formulae, see (3.68) and (3.70), the reflected and transmitted wave amplitudes depend exclusively on the complex wave vectors of the media bordering the interface.

8.A.2 Periodic Multilayer

Reciprocal lattice of a periodic multilayer, Fig. 8.39(a), is a set of points positioned equidistantly along the q_z axis, subfigure (c). Thus the "super-periodicity" in real space causes a periodic fine structure along the specular rod, and we find so-called *multilayer Bragg peaks* on the specular reflectivity curve, see Fig. 8.11 for instance.

Following from Fig. 8.38 the refraction in the layers causes a shift of the actual multilayer multilayer Bragg peaks with respect to the position of the reciprocal lattice points. This is shown by the comparison between the kinematical the dynamical reflection curve of a smooth multilayer in Fig. 8.10. The position of the kinematical Bragg-peaks coincide exactly with the reciprocal lattice points.

The finite total multilayer thickness gives rise to additional side maxima, so-called *Kiessig fringes* between the multilayer Bragg peaks (not shown in the figure). There are $p-2$ maxima in between two Bragg peaks for a flat multilayer with p periods.

Reciprocal lattice of a laterally periodic multilayer grating etched into a planar periodic multilayer is shown in Fig. 8.33. The lateral periodicity gives rise to a grating rod pattern. The grating rods are equidistantly positioned along the direction of patterning with the specular rod in the center.

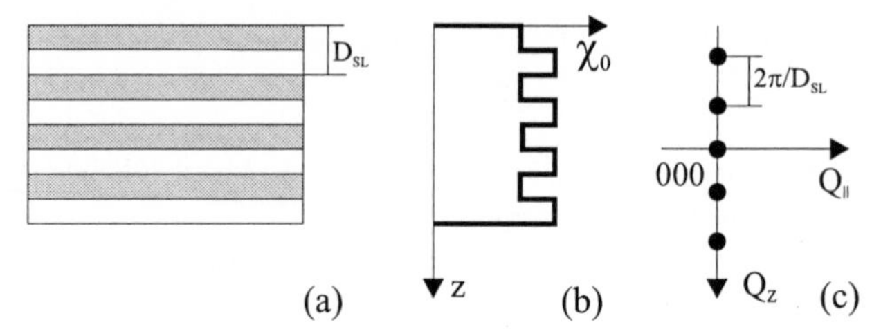

Fig. 8.39. Schematic set-up of a periodic multilayer: (a) in real space, (b) the polarizability profile, (c) in reciprocal space

8.A.3 Reciprocal Space Representation of DWBA

The formulae for the calculation of the first order DWBA have been derived in Chap. 4. Here, we show the graphical representation of the corresponding scattering events. Each of the two eigenstates of the unperturbed potential V^A consists of a transmitted and reflected wave $T = U(+k_z)$, $R = U(-k_z)$. The four wave vector transfers $\boldsymbol{q}^{11}, .., \boldsymbol{q}^{22}$ defined by (4.D23) and corresponding to $(\boldsymbol{k}_{\mathrm{sc}\|} - \boldsymbol{k}_{\mathrm{in}\|}, \pm k_{\mathrm{sc},z} \pm k_{\mathrm{in},z})$ in (4.41),(4.46) or (8.48), are represented in the reciprocal space by the four intervening scattering processes. They are schematically drawn in Fig. 8.40. We call the first (transmission-transmission) term the *primary* scattering process $\boldsymbol{q}_j^{11}$, since it is directly excited by the incident wave and it corresponds to the measured scattering vector in vacuum $\boldsymbol{q} = \boldsymbol{k}_{\mathrm{sc}} - \boldsymbol{k}_{\mathrm{in}}$. The other three terms are *secondary* scattering processes. They are of purely dynamical nature, called *Umweganregung* (detour or non-direct excitation), which occurs exclusively due to multiple scattering (direct or non-direct excitations).

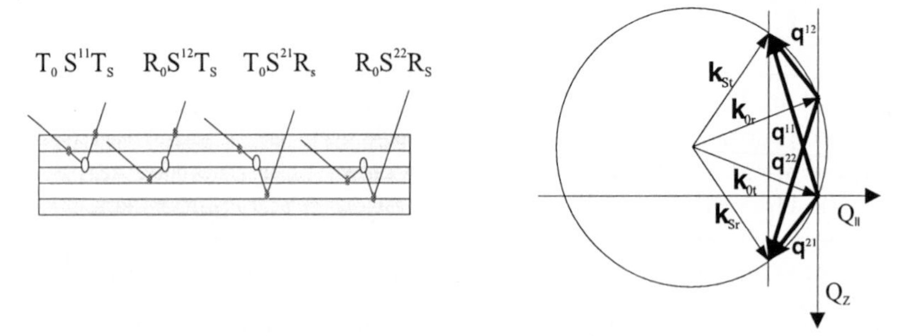

Fig. 8.40. Schematic representation of the four x-ray reflection processes in real space (left) and in the reciprocal space (right) of the first order DWBA. The full circles denote the dynamical reflection and transmission in the ideal multilayer, open circles indicate the diffuse scattering due to the interface roughness. The process with the indices 11 is the primary scattering process, described also by the kinematical approximation. The other three are processes of Umweganregung

The division of the perturbed potential V^B into the layer disturbances V_j^B allowed to represent the scattering in terms of structure factors S_j, Eq. (4.D13), an advantage usually reserved for the kinematical theory. The contribution of one scattering process in a single layer to the amplitude reflected by the whole sample depends on the structure factor of the layer disturbance and on the amplitudes of the participating waves.

Reciprocal space representation of the scattering processes in the Born approximation, DWBA and dynamical theory for reflection by gratings is shown in Fig. 8.35.

References

1. S.K. Sinha, E.B. Sirota, S. Garoff, and H.B. Stanley, , *Phys. Rev. B* **38**, 2297 (1988).
2. D.G. Stearns, , *J. Appl. Phys.* **65**, 491 (1989).
3. J.B. Kortright, , *J. Appl. Phys.* **70**, 3620 (1991).
4. D.E. Savage, J. Kleiner, H. Schimke, Y.H. Phang, T. Jankowski, J. Jacobs, R. Kariotis, and M.G. Lagally, , *J. Appl. Phys.* **69**, 1411 (1991).
5. D.K.G. de Boer, , *Phys. Rev. B* **44**, 498 (1991).
6. J. Daillant and O. Bélorgey, , *J. Chem. Phys.* **97**, 5824 (1992).
7. V. Holý, J. Kuběna, I. Ohlídal, K. Lischka, and W. Plotz, , *Phys. Rev. B* **47**, 15896 (1993).
8. V. Holý and T. Baumbach, , *Phys. Rev. B* **49**, 10668 (1994).
9. J.-P. Schlomka, M. Tolan, L. Schwalowsky, O.H. Seeck, J. Stettner, and W. Press, , *Phys. Rev. B* **51**, 2311 (1995).
10. T. Salditt, D. Lott, T.H. Metzger, J. Peisl, G. Vignaud, P. Høghøj, J.O. Schärpf, P. Hinze, and R. Lauer, , *Phys. Rev. B* **54**, 5860 (1996).
11. V. Holý, G.T. Baumbach, and M. Bessière, , *J. Phys. D* **28**, A220 (1995).
12. S.A. Stepanov, E.A. Kondrashkina, M. Schmidbauer, R. Köhler, and J.-U. Pfeiffer, , *Phys. Rev. B* **54**, 8150 (1996).
13. D.K.G. de Boer, , *Phys. Rev. B* **49**, 5817 (1994).
14. G.T. Baumbach, V. Holý, U. Pietsch, and M. Gailhanou, , *Physica B* **198**, 249 (1994).
15. E. Spiller, D. Stearns, and M. Krumrey, , *J. Appl. Phys.* **74**, 107 (1993).
16. P. Croce and L. Névot, , *Revue Phys. Appl.* **11**, 113 (1976).
17. L. Névot and P. Croce, , *Revue Phys. Appl.* **15**, 761 (1980).
18. V. Holý, U. Pietsch, and G.T. Baumbach. *High-Resolution X-Ray Scattering from Thin Films and Multilayers.* Springer-Verlag, Berlin, 1999.
19. P. Mikulík. PhD thesis, Université Joseph Fourier (Grenoble) and Masaryk University (Brno), 1997.
20. J. Eymery, J.M. Hartmann, and G.T. Baumbach, , *J. Cryst. Growth* **184**, 109 (1998).
21. P.R. Pukite, C.S. Lent, and P.I. Cohen, , *Surf. Sci.* **161**, 39 (1985).
22. D. Buttard, G. Dolino, D. Bellet, T. Baumbach, and F. Rieutord, , *submitted to Appl. Phys. Lett.*, (1998).
23. J. Krim and G. Palasantzas, , *Int. J. Mod. Phys.* **9**, 599 (1995).
24. G. Palasantzas and J. Krim, , *Phys. Rev. B* **48**, 2873 (1993).

25. G. Palasantzas, , *Phys. Rev. B* **48**, 14472 (1993).
26. Z.H. Ming, A. Krol, Y.L. Soo, Y.H. Kao, J.S. Park, and K.L. Wang, , *Phys. Rev. B* **47**, 16373 (1993).
27. V. Holý, C. Giannini, L. Tapfer, T. Marschner, and W. Stolz, , *Phys. Rev. B* **55**, 55 (1997).
28. V. Holý, A.A. Darhuber, J. Stangl, G. Bauer, J.F. Nützel, and G. Abstreiter, , *Il Nuovo Cimento* **19D**, 419 (1997).
29. V. Holý, A.A. Darhuber, J. Stangl, G. Bauer, J.F. Nützel, and G. Abstreiter, , *Semicond. Sci. & Technol.* **13**, 590 (1998).
30. T. Salditt, T.H. Metzger, and J. Peisl, , *Phys. Rev. Lett.* **73**, 2228 (1994).
31. T. Salditt, T.H. Metzger, Ch. Brandt, U. Klemradt, and J. Peisl, , *Phys. Rev. B* **51**, 5617 (1995).
32. S.F. Edwards and D.R. Wilkinson, , *Proc. R. Soc. London, Ser. A* **381**, 17 (1982).
33. D. Maystre. Rigorous vector theories of diffraction gratings. In E. Wolf, editor, *Progress in Optics XXI*. North-Holland, Amsterdam, 1984.
34. M. Nevière, , *J. Opt. Soc. Am. A* **11**, 1835 (1994).
35. A. Sammar, J.-M.André, and B. Pardo, , *Optics Communications* **86**, 245 (1991).
36. A. Sammar and J.-M. André, , *J. Opt. Soc. Am. A* **10**, 2324 (1993).
37. S. Bac, P. Troussel, A. Sammar, P. Guerin, F.-R. Ladan, J.-M.André, D. Schirmann, and R. Barchewitz, , *X-ray Sci. Technol.* **5**, 161 (1995).
38. M. Tolan, W. Press, F. Brinkop, and J.P. Kotthaus, , *Phys. Rev. B* **51**, 2239 (1995).
39. P. Mikulík and T. Baumbach, , *Phys. Rev. B* **59**, (1999).
40. P. Mikulík and T. Baumbach, , *Physica B* **248**, 381 (1998).
41. M. Jergel, P. Mikulík, E. Majková, Š. Luby, R. Senderák, and E. Pinčík, , *submitted to J. Appl. Phys.*, (1998).

9 Reflectivity of Liquid Surfaces and Interfaces

Jean Daillant

Service de Physique de l'Etat Condensé, Orme des Merisiers, CEA Saclay, 91191 Gif sur Yvette Cedex, France

The aim of this chapter is to make a presentation of reflectivity experiments on liquid surfaces and interfaces. This is a field where reflectivity techniques are widely used, in particular because the range of available techniques is relatively less important than for solid surfaces (no high-vacuum techniques, no scanning tunneling microscopy, atomic force microscopy is difficult). Reflectivity experiments on liquid surfaces present specific features both experimentally and conceptually. Experimentally, the liquid surface is always horizontal, and therefore requires adapted experimental setups. Morevover, the subtraction of the high background scattering in the bulk liquid phases imposes severe constraints on the experiments.

Conceptually, the distinctive property of liquid surfaces is the low q divergence of the height-height correlation functions which makes the separation of specular reflectivity and diffuse scattering impossible. On the other hand, analytical expressions for the height-height correlation functions are available, at least in the capillary regime when the physics is governed by surface tension (see chapter 6 for the surface energy of solid surfaces). This allows a thorough analysis of x-ray surface scattering methods through exact model calculations which can be interesting even for readers having no particular interest in liquid surfaces.

After a short introduction to liquid interfaces, we shall comment on the specific features of liquid surface reflectivity studies and then give some examples.

9.1 Statistical Description of Liquid Surfaces

In this section, we consider the microscopic structure of the interfacial region as it can be determined by a reflectivity or surface scattering experiment.

A first approach to the structure of liquid surfaces was initiated by van der Waals and consists in describing the liquid-vapour interfacial region as a region from smooth transition from the liquid density to that of the gas [1]. A complete description of van der Waals and related theories is given in Ref. [2]. The principle of such density functional theories is to minimise a free energy (or grand potential) functional taking into account both the local free energy of the fluid at a given density and temperature, and the effect

of density gradients (as a square gradient term in the most simple version of the theory). The minimisation of this functional yields the liquid-gas interface density profile and the surface tension which is the surface excess of the grand potential (or of the free energy if the Gibbs dividing surface is chosen [3]), i.e. the total free energy minus that of bulk liquid and gaseous phases extended up to an arbitrary dividing surface [2].
The alternative capillary wave model of Buff, Lovett, and Stillinger [4] assumes a step-like profile for the liquid-vapour interface. Then all the structural information about the interface is contained in its profile $z(\mathbf{r}_{\|})$, or, since only a statistical description is meaningful (see chapter 2), in the height correlations which are assumed to result from the propagation of capillary waves (i.e. surface deformation modes). There is now a good experimental evidence using different techniques that this model gives an accurate description of the liquid surface structure for in-plane lengthscales larger than one micron, and also describes the mean surface roughness better than the van der Waals theory. We will therefore limit the discussion below to a description of this model.

9.1.1 Capillary Waves

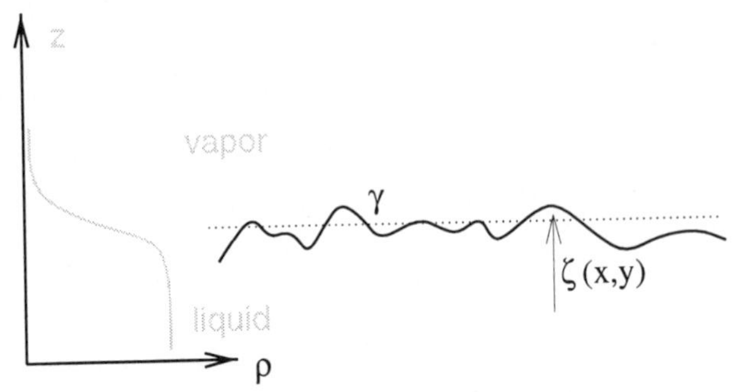

Fig. 9.1. Capillary waves. γ is the surface tension and $z(x,y)$ the interface height. The corresponding average density profile is given on the left

Let $z(\mathbf{r}_{\|})$ be the interface height in $\mathbf{r}_{\|}$ (Fig. 9.1). The interface fluctuations are characterised by a spectrum $\langle z(\mathbf{q}_{\|})z^*(\mathbf{q}_{\|})\rangle$ for the wavevector $\mathbf{q}_{\|}$.

$$z(\mathbf{r}_{\|}) = \sum_q z(\mathbf{q}_{\|})e^{i\mathbf{q}_{\|}\cdot\mathbf{r}_{\|}} . \tag{9.1}$$

The work necessary to deform the interface $z(\mathbf{r}_{\|})$ is:

$$W = \int\int_A dxdy \left[\int_0^{z(x,y)} \Delta\rho\, gzdz + \gamma\left(\sqrt{1+(\partial z/\partial x)^2+(\partial z/\partial y)^2} - 1\right)\right] \tag{9.2}$$

ρ is density and γ the surface tension. Developping the square root and Fourier transforming, one obtains:

$$W = \gamma A + \frac{1}{2} \sum_{\mathbf{q}_{\|}} \sum_{\mathbf{q'}_{\|}} z(\mathbf{q}_{\|}) z(\mathbf{q'}_{\|}) \int_A d^2\mathbf{r} e^{i((\mathbf{q}_{\|}+\mathbf{q'}_{\|}).\mathbf{r}_{\|})} \left[\Delta\rho\, g - \gamma \mathbf{q}_{\|} . \mathbf{q'}_{\|} \right], \tag{9.3}$$

where the modes $\mathbf{q}_{\|}$ are multiples of $2\pi/L$, with L the surface dimension. The terms $\mathbf{q'}_{\|} \neq -\mathbf{q}_{\|}$ vanish and

$$W = \gamma A \left\{ \frac{1}{2} \sum_{\mathbf{q}_{\|}} z(\mathbf{q}_{\|}) z(-\mathbf{q}_{\|}) \left[\frac{\Delta\rho\, g}{\gamma} + \mathbf{q}_{\|}^2 \right] \right\}. \tag{9.4}$$

The length $\sqrt{\gamma/\Delta\rho g}$, on the order of $1mm$ is the so-called capillary length. The equipartition of energy (Gaussian Hamiltonian) among the degrees of freedom of the system in thermal equilibrium gives:[1]

$$\langle z(\mathbf{q}_{\|}) z(-\mathbf{q}_{\|}) \rangle = \frac{1}{A} \frac{k_B T}{\Delta\rho g + \gamma \mathbf{q}_{\|}^2}, \tag{9.5}$$

represented in Fig. 9.2 left. This spectrum has been well characterised for many liquid surfaces by light scattering down to wavelengths in the micrometer range [6] and is valid is the limit of small in-plane momentum $\mathbf{q}_{\|}$. It describes thermally excited capillary waves, limited by gravity for lengthscales larger than the capillary length, and by surface tension for smaller lengthscales. The resulting surface structure is isotropic in plane. Then, the rms roughness of the interface is obtained by summing over all the modes:

$$\langle z^2 \rangle = \frac{1}{A} \sum_{q_{\|}>0} \frac{k_B T}{\Delta\rho g + \gamma q_{\|}^2}, \tag{9.6}$$

where the summation runs from $q_{\min} = 2\pi/L$ to $q_{\max} = 2\pi/a$ where a is a molecular length. In the continuous limit:

$$\langle z^2 \rangle = \frac{k_B T}{4\pi\gamma} ln \left[\frac{1 + q_{\max}^2 (\gamma/\Delta\rho g)}{1 + q_{\min}^2 (\gamma/\Delta\rho g)} \right]. \tag{9.7}$$

Generally, we may assume $q_{\min}^2(\gamma/\Delta\rho g) \ll 1$ and

$$\langle z^2 \rangle = \frac{k_B T}{4\pi\gamma} ln \left[1 + q_{\max}^2 (\gamma/\Delta\rho g) \right] \tag{9.8}$$

For $\gamma_{H_2O} = 73mN/m$, one obtains $\langle z^2 \rangle = 0.4nm$. Also interesting are the limits:

[1] For a rigorous calculation of the statistical average see for example Ref. [5].

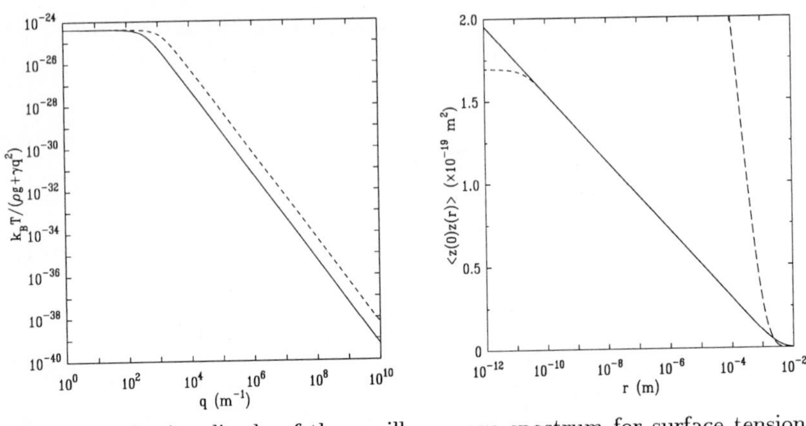

Fig. 9.2. left: Amplitude of the capillary wave spectrum for surface tension $\gamma = 73mN/m$ (continuous line) and $\gamma = 7.3mN/m$ (broken line). right: The height-height correlation function $\langle z(\mathbf{0})z(\mathbf{r}_{\|})\rangle$ for water (continuous line) and the surface of a liquid having the same density but a 10 times smaller surface tension (long dashed line). The dotted line is obtained by attenuating the spectrum for water at a molecular cut-off $q = 2\pi/10^{-10}m^{-1}$

$$A \to \infty \; \langle z^2 \rangle \sim -lng,$$
$$g \to 0 \quad \langle z^2 \rangle \sim lnA.$$

These logarithmic divergences do not imply that there is no interface. The interface does exist, but is not localised in space [7], in agreement with the fact that the divergence is due to small wavevector modes $q \to 0$.

An open question is that of the corrections to the surface energy at very small lengthscales. This is an important question since, as shown by Eq. (9.5) the (deformation) surface energy determines the surface structure. Taking into account the coupling between capillary modes [8] would lead to a renormalisation of the surface tension equivalent to a larger effective surface tension at very small lengthscales. The opposite trend is displayed when we take into account the effect of long-range dispersion forces which lead to a smaller effective surface tension [9].

The height-height correlation function can be obtained by Fourier transforming the spectrum Eq. (9.5):

$$g(\mathbf{r}_{\|}) = \langle \left[z(\mathbf{r}_{\|}) - z(\mathbf{0}) \right]^2 \rangle = 2\langle z^2 \rangle - \frac{k_B T}{\pi \gamma} K_0 \left(r_{\|} \sqrt{\Delta\rho g/\gamma} \right), \tag{9.9}$$

and

$$\langle z(\mathbf{0}) z(\mathbf{r}_{\|}) \rangle = \frac{k_B T}{2\pi\gamma} K_0 \left(r_{\|} \sqrt{\Delta\rho g/\gamma} \right) \tag{9.10}$$

(figure 9.2 right). K_0 is the modified second kind Bessel function of order 0. $K_0(x)_{x\to 0} \simeq Log2 - \gamma_e - Logx$ and $lim\; K_0(x)_{x\to\infty} = 0$.

As indicated in Chap. 2, there is a specular reflection on a liquid surface only because gravity limits the logarithmic divergence. This logarithmic divergence with distance of the roughness, only limited by gravity or the finite size of the surface is a distinctive property of liquid surfaces.

Substituting Eq. (9.9) into Eq. (4.41), it is possible to find the following approximation for the scattering cross-section which is valid in the limit $q_{\|} \gg \sqrt{\Delta\rho g/\gamma}$ [10–13]:

$$\frac{d\sigma}{d\Omega} \approx \mathcal{A}\frac{k_0^4(1-n^2)^2}{16\pi^2}|t_{0,1}^{\mathrm{in}}|^2|t_{0,1}^{\mathrm{sc}}|^2\frac{k_BT}{\gamma q_{\|}^2}\left(\frac{q_{\|}}{q_{\max}}\right)^{\eta}, \tag{9.11}$$

where $\mathcal{A}$ is the illuminated area and $\eta = (k_BT/2\pi\gamma)|q_{z,1}|^2$.

9.1.2 Relation to Self-Affine Surfaces

Many solid surfaces are well described by correlation functions of the form:

$$\langle z(\mathbf{0})z(\mathbf{r}_{\|})\rangle = \sigma^2 e^{-\left(r_{\|}/\xi\right)^{2h}} \ ; \ 0 < h < 1. \tag{9.12}$$

Such surfaces are known as self-affine surfaces and h is the roughness or Hurst exponent (see chapter 2). σ is the roughness, and ξ the roughness correlation length. A self-affine fractal differs from a self-similar fractal in that all directions of space are not equivalent for the self-affine case. In contrast to self-similar fractals, self-affine fractals do not have a unique fractal dimension [14]. As a whole, they are surfaces, $d = 2$, but they can alternatively be described by a local fractal dimension $D = 3 - h$ [2]. A fairly good approximation of the roughness spectrum of such surfaces [15] is

$$\langle z(q)^2\rangle \ = \ \frac{A}{(2\pi)^5}\,\frac{\sigma^2\xi^2}{(1+aq_{\|}^2\xi^2)^{(1+h)}}, \tag{9.13}$$

which is an exact relation for $h = 0.5$ (see also section 6.2.4). The smaller h, the rougher the surface appears. The correlation function of a liquid surface is obtained in the limit $h \to 0$. As compared to a solid surface described by the correlation function Eq. (9.12) the range of the correlations of a liquid surface is much longer (logarithmic divergence). The correlation length is on the order of $1mm$.

[2] The local fractal dimension can be defined by mapping the object on a lattice of constant l. If $N(l)$ lattice sites are occupied, then $D = lim_{l\to 0}\left[ln(N(l)/ln(1/l)\right]$. Globally $h < 1$ and $lim_{x\to 0}h(x) = 0$ thus the dimension is 2. Locally however, $h(ax,y) \propto a^h h(x,y)$ and $N(l) \propto l^{-(3-h)}$; $D = 3 - h$.

9.1.3 Bending Rigidity

If a film is present at the interface, it will reduce the surface tension but also resist bending. The simplest treatment of these effects which will be most important for very low surface tensions, is due to Helfrich [16], and is described below. Very low surface tensions can be achieved for example in microemulsions composed of brine, oil (e.g. alkanes) and surfactants where interfacial tensions as low as a few thousandths of a mN/m can sometimes be obtained. Very low surface tensions are also obtained for self-assembled amphiphilic films (vesicles, lamellar phases) in water or brine. The molecular area in such systems results of a balance between attractive hydrophobic interactions and repulsive interactions (hydrophilic, steric, electrostatic) between headgroups. At equilibrium $\partial F/\partial A = 0$, and surface tension is therefore not a relevant parameter for describing the system.
In all these systems fluctuations play an important role. This for example is the case of lamellar phases which can be diluted down to very low concentrations where the separation between lamellae ($> 100nm$) is larger than the range of electrostatic forces. Such structures are stabilised by the so-called Helfrich entropic interaction (undulation forces): the fluctuations of a lamella are limited by the neighbouring lamellae, resulting in a repulsive effective interaction. For such systems having a very low surface tension, the fluctuations are no longer limited by the surface tension but by the bending stiffness.
The curvature is defined by two parameters independent of the surface parametrisation, the mean curvature C and the Gaussian curvature G. If R_1 and R_2 are the principle radii of curvature, then:

$$C = \frac{1}{2}\left(\frac{1}{R_1} + \frac{1}{R_2}\right), \tag{9.14}$$

$$G = \frac{1}{R_1 R_2}. \tag{9.15}$$

The deformation free energy can now be developed as a function of the mean curvature and the Gaussian curvature to the second order:

$$F = \int dA \left(F_0 + \lambda C + 2\kappa C^2 + \overline{\kappa} G\right). \tag{9.16}$$

κ is the bending rigidity modulus and $\overline{\kappa}$ Gaussian bending rigidity modulus. Eq. (9.16) can be alternatively written:

$$F = \int dA \left(F_0' + 2\kappa (C - C_0)^2 + \overline{\kappa} G\right). \tag{9.17}$$

C_0 is the spontaneous curvature. If the surface topology does not change, the integral of G is a constant (Gauss theorem). Methods similar to those previously used lead to:

$$\langle z(\mathbf{q}_{\parallel})z(-\mathbf{q}_{\parallel})\rangle = \frac{1}{A}\frac{k_B T}{\Delta\rho g + \gamma q_{\parallel}^2 + \kappa q_{\parallel}^4}. \tag{9.18}$$

Note that a comprehensive understanding of the bending rigidity in terms of molecular order and chain conformations is still lacking. This is a central problem in soft condensed matter physics where systems are often composed of films (monolayers or bilayers) and the role of fluctuations is dominant.

9.2 Experimental Measurement of the Reflectivity of Liquid Surfaces

9.2.1 Specific Experimental Difficulties

Considering the measurement of the reflectivity of liquid surfaces, we first note that the liquid surface must necessarily be horizontal. If the source is a sealed tube, it can be moved to probe the scattering vectors, whereas using a rotating anode or synchrotron radiation sources alternative solutions must be considered. The angular spread of a rotating anode source can be used to change the incidence on a fixed point by displacing a monochromator on a circle containing the source and the target point (Fig. 9.3).

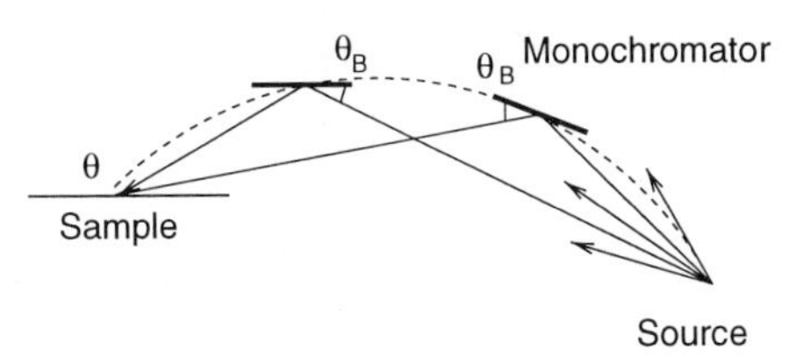

Fig. 9.3. Experimental setup for changing the angle of incidence at a fixed point using a divergent x-ray source, e.g. a rotating anode. The monochromator is moved on a circle containing the source and target point

Using synchrotron radiation, two different solutions have been found. For grazing incidence a mirror is generally used. For higher incidences, a crystal deflector can be used (Fig. 9.4) to deflect the beam [17], for example a very thin silicon crystal. By rotating the crystal around the incident beam, the diffracted beam describes a cone of opening angle $4\theta_B$ if θ_B is the Bragg angle. The whole diffractometer must then be used to keep a fixed point of impact. This crystal can be bent to fit the divergence requirements of the incident beam.

The other crucial experimental problem posed by reflectivity measurements on liquid surfaces is that of the background due to scattering in the bulk. If one is interested in specular reflectivity, the most efficient method consists in scanning in q_x for each q_z value in order to determine and subtract the background.

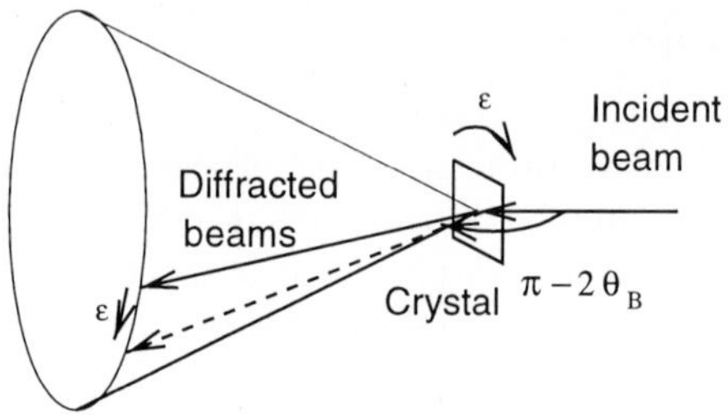

Fig. 9.4. Beam Deflector. By rotating the crystal around the incident beam, the diffracted beam describes a cone of opening angle $4\theta_B$ if θ_B is the Bragg angle

9.2.2 Reflectivity

The interpretation of reflectivity curves can be tricky because for liquid surfaces the diffuse intensity is generally peaked in the specular direction as shown in Fig. 9.5 [18]. This is of course a consequence of the long-range correlations. This point was already discussed in Chap. 4 where it was shown in particular that in a reflectivity experiment diffuse scattering will always eventually dominate the true specular (coherent) component for large wavevector transfers. A reflectivity experiment should therefore never only consist in the

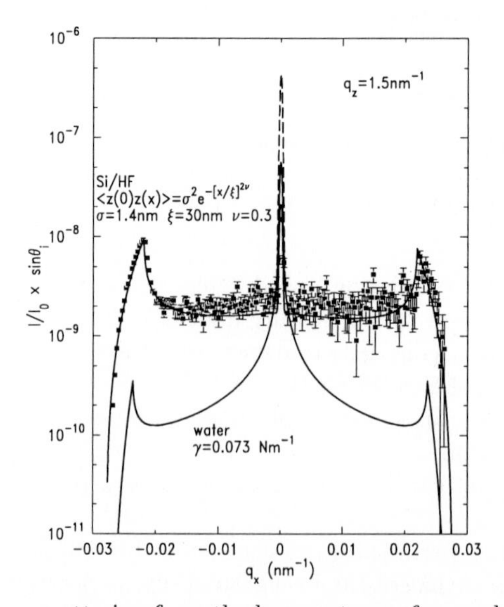

Fig. 9.5. Diffuse scattering from the bare water surface and a solid surface

measurement of the specular intensity, since the measured signal is always sensitive to the exact dependence of the height-height correlation function (see Sect. 4.8 for details).

This is illustrated in Fig. 9.6. In that case, we calculated that, using the spectrum Eq. (9.18), the intensity measured in the specular direction for a resolution Δq_x in the plane of incidence is smaller than the reflectivity of an equivalent smooth interface by a factor

$$\pi^{-1/2}\Gamma\left[\frac{1}{2}-\frac{k_BTq_z^2}{4\pi\gamma},\frac{1}{2}\Delta q_X^2\frac{\kappa}{\gamma}\right]\times\exp-\left[\frac{k_BTq_z^2}{2\pi\gamma}Log\left(\frac{e^{\gamma_E}}{\sqrt{2}}\frac{\sqrt{\gamma/\kappa}}{\Delta q_x}\right)\right], \tag{9.19}$$

where Γ is the incomplete Γ function and $\gamma_E = 0.577$ is Euler's constant. This factor is larger than $e^{-q_z^2\langle z^2\rangle}$ because diffuse scattering has been taken into account in addition to specular reflectivity. One can see on Fig. 9.6 that the diffuse intensity dominates the specular intensity for $q_z > 2nm^{-1}$. Note also that a data analysis not taking into account diffuse scattering leads to an erroneous estimation of the structural parameters.

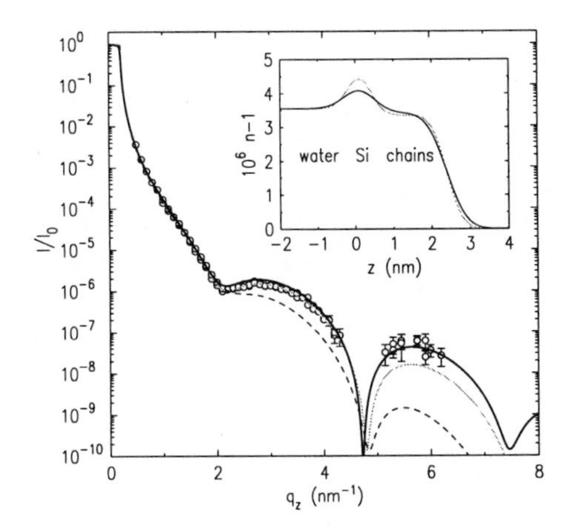

Fig. 9.6. Interference pattern resulting from the reflection of an x-ray beam on an octadecyltrichlorosilane monolayer at the air/water interface and its corresponding electron density model (inset, black curve). The broken curve represents the specular reflection, the long-dashed curve the diffuse intensity, and the thick line the total intensity. The grey curve in the inset is obtained when the data are analysed using a "box model" with error function transition layers, not taking into account diffuse scattering

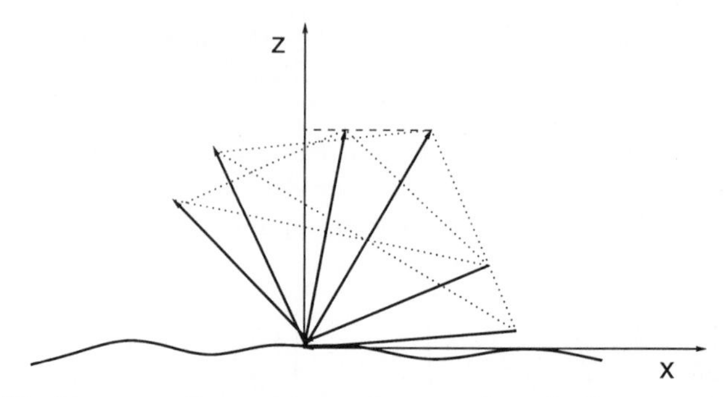

Fig. 9.7. "Rocking curve"geometry. q_z is approximately kept constant as q_x is varied by "rocking" the incident beam and the detector

9.2.3 Diffuse Scattering

The case of diffuse scattering is even more difficult. One would expect that the same kind of measurements that are successfully carried out on solid surfaces could be also applied to liquid surfaces. In particular, "rocking curves" (equivalent to q_x scans at fixed q_z, Fig. 9.7) would yield a very good resolution along q_x:[3]

$$\Delta q_x = \frac{2\pi}{\lambda} \left(\sin\theta_{\mathrm{in}} \Delta\theta_{\mathrm{in}} + \sin\theta_{\mathrm{sc}} \Delta\theta_{\mathrm{sc}} \right) . \tag{9.20}$$

In fact (Fig. 9.8, left) such measurements lead to "flat" spectra for $q_z \geq 2nm^{-1}$. This is because when the incidence angle becomes larger than the critical angle for total external reflection, bulk scattering dramatically increases at large q_z (Fig. 9.8, right). This example shows that it is in practice necessary to fix the incidence angle below the grazing angle for total external reflection θ_c. It is then possible to measure the scattered intensity either in the plane of incidence (projected on q_x) or in the horizontal sample plane. In the first case, q_x and q_z are varied together and it is possible to measure the normal structure of, for example, a film, and verify that surface scattering is indeed measured. However, one has to decouple the structural effects from the fluctuation spectrum.

Measuring diffuse scattering in the plane of incidence should be considered whenever one is interested in the determination of the normal structure of thin films using synchrotron radiation. This has two main advantages over reflectivity (see Fig. 9.11):

- The reduced background.
- The much lighter experimental setup (only a mirror is required instead of a beam deflector).

[3] The resolution function is in particular discussed in section 4.7.2 of this book and in Refs. [19,20].

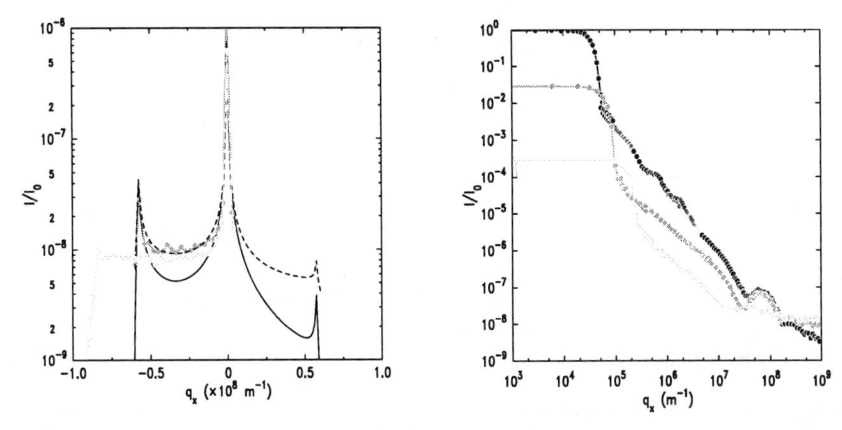

Fig. 9.8. left: Diffuse scattering (rocking curves) from the bare water surface for $q_z = 2.5nm^{-1}$ (dark grey circles) and $q_z = 3nm^{-1}$ (light grey circles) as a function of q_x. the reflected beam is at $q_x = 0$. Note the Yoneda peaks. Calculated surface signal at $q_z = 2.5nm^{-1}$ (black line). A constant background has been added to calculate the dashed line, giving a better agreement with the experimental data. Right: Diffuse scattering by an arachidic acid $(CH_3 - (CH_2)_{18} - COOH)$ film at the air/water interface. The fixed angles of incidence are respectively $2mrad$ (black symbols), $6mrad$ (dark-grey symbols) and $10mrad$ (light-grey symbols). $\theta_c = 2.4mrad$. Note that for a grazing angle of incidence equal to $10mrad$, the surface sensitivity revealed on the other curves by the constructive interference for $q_x \approx 10^8 m^{-1}$ is lost because of bulk scattering

When we are only interested in the roughness spectrum, a second kind of scan (in the horizontal sample plane) which directly yields a signal proportional to the roughness spectrum should be preferred (Fig. 9.10).

A last important point which is not specific to liquid surfaces is that the diffuse intensity is proportional to the resolution volume (Fig. 9.12). It is therefore necessary to precisely determine the resolution function as a function of slit openings and of the footprint of the beam on the surface to precisely determine the magnitude of this intensity.

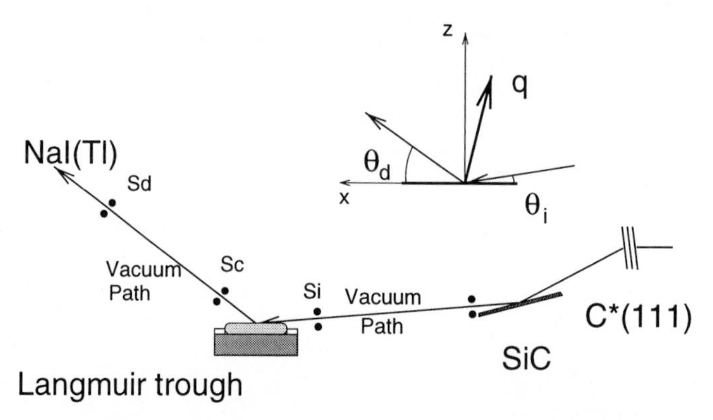

Fig. 9.9. Schematics of experiments (Troika beamline, ESRF). (q_x, q_z) plane of incidence geometry. in-plane q_y geometry. $C^*(111)$: diamond monochromator, SiC: mirror, NaI(Tl): scintillation detector. Typical distances are: Sample-to-S_d distance 700 mm, S_c-S_d distance 500 mm. Typical horizontal $\times$ vertical openings of the slits S_i, S_c and S_d are $w_i \times h_i$: 0.4 mm $\times$ 0.2 mm, $w_c \times h_c$: 2 mm $\times$ 2 mm, $w_d \times h_d$: 10 mm $\times$ 0.250 mm

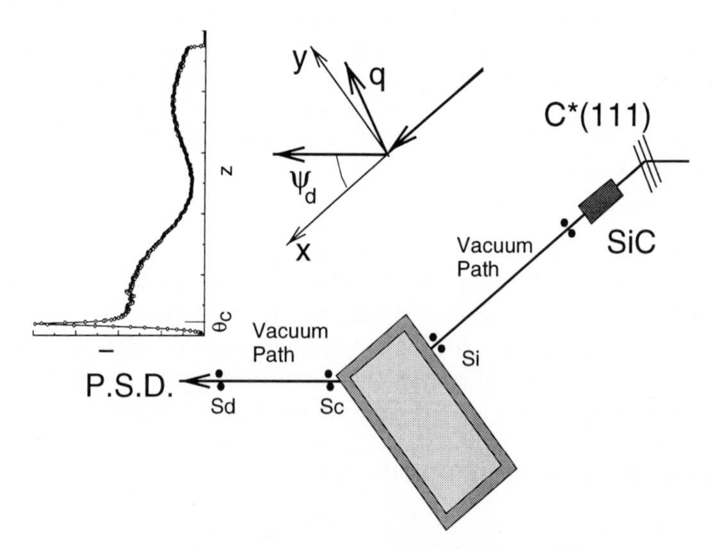

Fig. 9.10. Schematics of experiments (Troika beamline, ESRF). In-plane q_y geometry. $C^*(111)$: diamond monochromator, SiC: mirror, PSD: position sensitive gas-filled (xenon) detector. The experimental curve represents the scattered intensity (horizontal axis) as a function of the vertical position on the PSD. Typical distances are: Sample-to-S_d distance 700 mm, S_c-S_d distance 500 mm. Typical horizontal $\times$ vertical openings of the slits S_i, S_c and S_d are $w_i \times h_i$: 0.3 mm $\times$ 0.2 mm, $w_c \times h_c$: 0.3 mm $\times$ 100 mm, $w_d \times h_d$: 0.5 mm $\times$ 100 mm

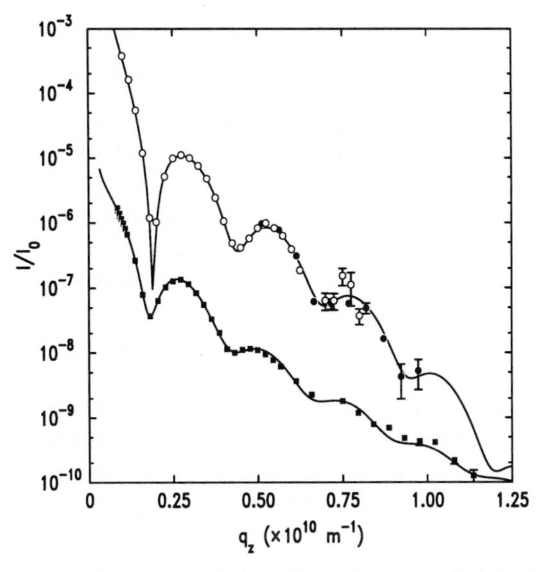

Fig. 9.11. Laboratory (empty circles) and synchrotron (filled circles reflectivity experiments (top). Diffuse scattering experiment in the plane of incidence (filled squares, bottom) for the same arachidic acid monolayer on a $CdCl_2$ subphase

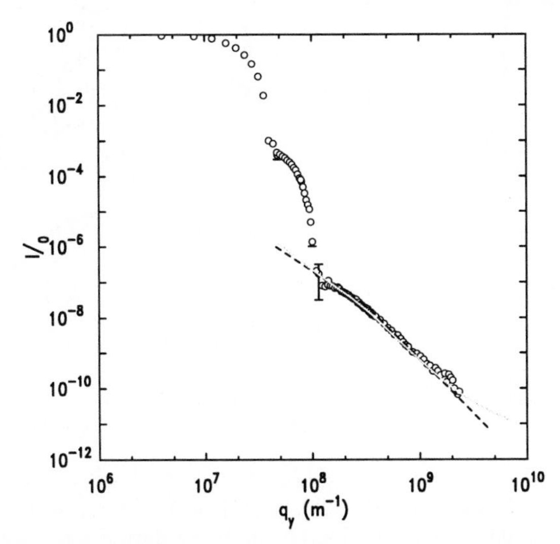

Fig. 9.12. Diffuse scattering from the bare water surface. The dependence of the intensity above $10^{-7} I_0$ results from the convolution of the reflected beam with the experimental resolution. Even the slope of the intensity scattered by the rough water surface is modified by the integration over the resolution volume (-3 instead of -2 for the scattering cross-section)

9.3 Some Examples

The first reflectivity experiments on liquid surfaces were carried out on liquid metals [21,22] in the 70's and the roughness of the free surface of water was demonstrated to be consistent with the capillary wave model in the mid 80's [23–25]. In reflectivity studies however, only density profiles averaged over the surface can be measured. From the beginning of the 90's, the analysis of diffuse scattering from the surface, which gives access to the roughness spectrum has been undertaken[26,11]. The spectrum could be measured up to wave-vectors of the order of $10^7 m^{-1}$. The method has been more recently extended down to molecular length-scales giving access to new phenomena [27]. After a short presentation of the method we will discuss the results pertaining to the bare water surface and give some examples for liquid metal surfaces, surfactant monolayers, and liquid-liquid interfaces. References [28–30] contain recent review articles on the scattering by liquid surfaces.

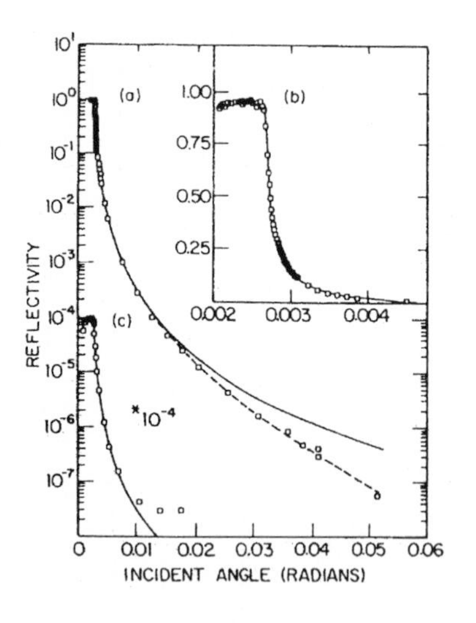

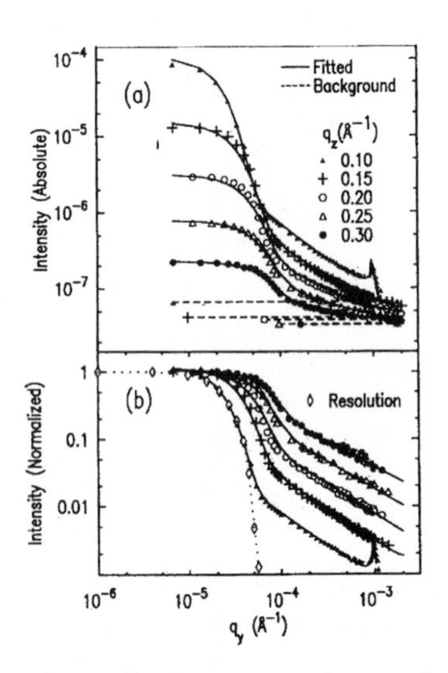

Fig. 9.13. Left: First reflectivity experiment on the bare water free surface by Braslau et al. at Hasylab in 1985 [23]. (a) Measured x-ray reflectivity (circles), calculated Fresnel reflectivity (continuous line); the dotted line takes the capillary wave roughness into account. (b) Expanded version around the critical angle for total external reflection. (c) same as (a) but measured with use of a rotating anode generator. With kind permission of A. Braslau and P.S. Pershan. Right: Diffuse scattering by the ethanol free surface (Sanyal et al. [11], NSLS beamline X22B). q_x scans at constant q_z; The background is subtracted in Fig. (b). With kind permission of M.K. Sanyal

9.3.1 Simple Liquids Free Surface

The bare water free surface was studied for the first time in 1985 by Braslau et al. at Hasylab, Hamburg (Fig. 9.13, left) [23]. Their results were successfully interpreted within the frame of the capillary wave model and they found a rms roughness $\langle z^2 \rangle = 0.32nm$. These results have been confirmed and improved in Refs. [24] and [25]. In Ref. [26] diffuse scattering from the surface was also measured, and it was demonstrated that the capillary wave model could be applied down to distances as small as $50nm$. This model was applied in Ref. [11] for the ethanol surface up to wavevectors on the order of $10^7 m^{-1}$ (Fig. 9.13, right).
In both cases, the limit was fixed by the source flux and background subtraction. This is no longer the case for the experiments of Ref. [27] carried out at the European Synchrotron Radiation Facility. The intensity scattered in the plane of incidence is presented in Fig. 9.14a.

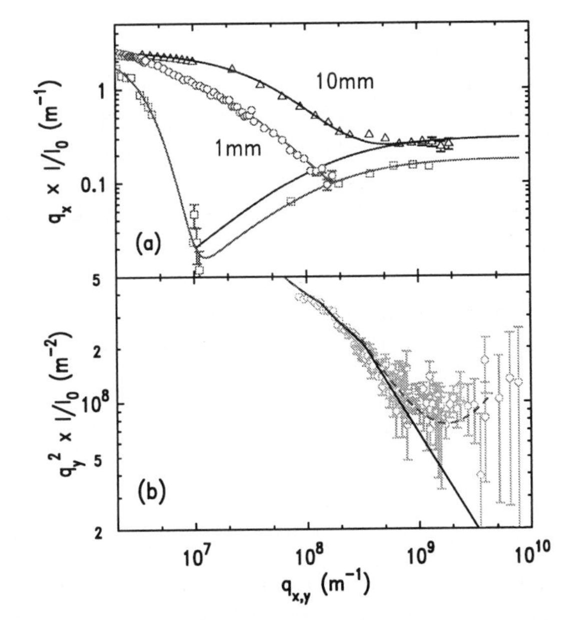

Fig. 9.14. Scattering by the bare water surface in the plane of incidence (a) and in the plane of the surface (b). For measurements in the plane of incidence q_x is the projection of the wave-vector transfer on the horizontal. "$1mm$" (circles) and "$10mm$" (triangles) is the opening of slit s_3 in Fig.9.9. Squares represent the background which was not subtracted. In (b) the background was subtracted following the procedure indicated in text. The full line is calculated by using the capillary wave spectrum and the acoustic wave scattering cross section Eq. (9.21) has been included to calculate the dashed line curve. Note that the great dynamic range of the y-scale has been compressed by a multiplication of the measured scattered intensity by q_x or q_y^2

The data extend to $q \geq 10^8 m^{-1}$ and are well described by the capillary wave spectrum. The data taken in the interface plane (q_y) extend to even larger wave-vector ($\approx 10^{10} m^{-1}$, note that this is more than two orders of magnitude than that of the best previous measurements [26]) but can only be described with the spectrum of Eq. (9.5) up to $q \approx 5 \times 10^8 m^{-1}$. The excess scattering at $q \geq 10^9 m^{-1}$ indicates either a smaller effective surface tension at such large wavelengths [9] or another source of scattering. This scattering has no measurable dependence on q (the apparent q^{-1} dependence on Fig. 9.14 is due to the resolution function), and can be attributed to density fluctuations (acoustic waves) within the penetration depth of the beam. The corresponding scattering cross-section [17] can be calculated from the density-density correlation function[4] at or near the interface which can itself be determined by using the linear response theory [6]:

$$\frac{d\sigma}{d\Omega} = \frac{k_0^4}{16\pi^2} \frac{A(1-n^2)^2 |t_{0,1}^{\mathrm{in}}|^2 |t_{0,1}^{\mathrm{sc}}|^2 k_B T \kappa_T}{2\mathcal{I}m(q_{z,1})}, \tag{9.21}$$

where n is the refractive index of water, t_{in} and t_{sc} are the transmission coefficients of the air/water interface for the incident and scattering beams, κ_T is the isothermal compressibility of water ($4.58 \times 10^{-10} m^2 N^{-1}$), and $\mathcal{I}m(q_{z1})$ is the imaginary part of the normal component of the wave-vector transfer in the liquid (inverse of the penetration length). Including this contribution gives a better agreement (Fig. 9.14b), without discarding the possibility of other corrections.

Alcohol and alkane surfaces have also been extensively studied [31]. In particular, partial wetting of the liquid phase by a crystal phase has been discovered.

9.3.2 Liquid Metals

The first reflectivity investigation of the mercury surface was carried out by Lu and Rice in 1978. A more recent experiment is that of Magnussen et al. (Fig. 9.15) [32]. The data extend to $25 nm^{-1}$ and the structuration of the interface is unambiguously demonstrated. Except for the first layer which is shifted towards the vapor, the distance between layers is smaller than the distance between atoms in the liquid, but larger than the distance between atoms in the solid. The decay length of the order is on the order of $0.35 nm$ and the peaks are smeared out by capillary waves.

[4] see Sect. 4.5 for details.

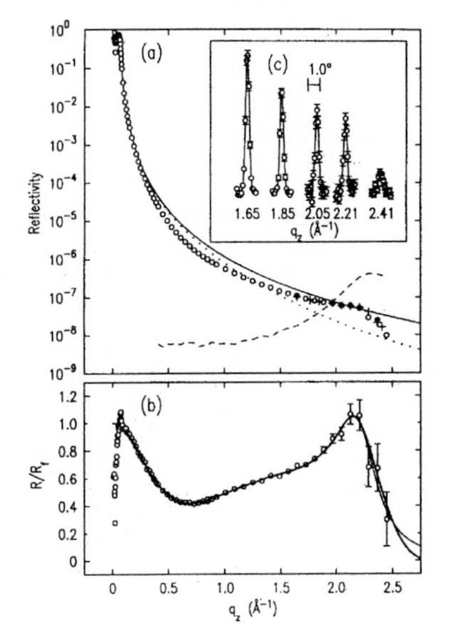

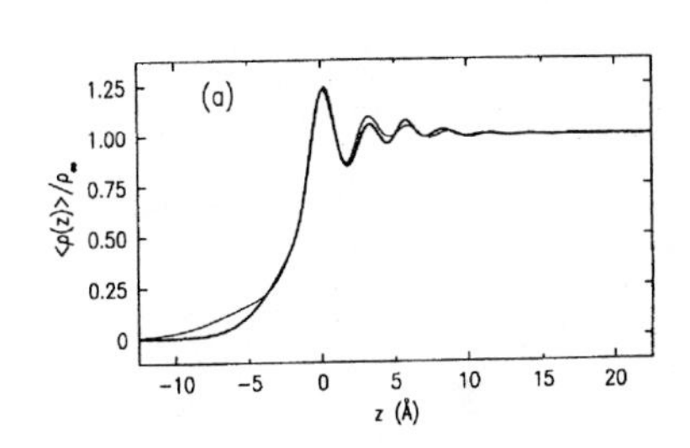

Fig. 9.15. Left: Reflectivity of the mercury surface according to Ref. [32] (a). In (b) the reflectivity curve has been divided by the Fresnel reflectivity of a smooth surface of a liquid that would have the bulk mercury density up to its surface in order to enhance the structures in the reflectivity profile. Different reflections are shown in (c). Note that the data extend up to $25nm^{-1}$. Right: Model profile of the density at the mercury free surface obtained from the previous reflectivity curves. With kind permission of P.S. Pershan

9.3.3 Surfactant Monolayers

Reflectivity Studies Surfactant monolayers have been the subject of many x-ray and neutron reflectivity studies. We shall only discuss one of them to illustrate the manner in which information can be extracted from a reflectivity experiment.

A very comprehensive study of long-chain alcohols ($C_{10} \rightarrow C_{16}$) at the air/water interface has been carried out by Rieu et al. [33] (Fig. 9.16). From their very precise measurements, the authors were able to measure the density of both the aliphatic chains and of the headgroups as a function of temperature. The data were sufficiently well resolved to evidence thickness and density changes at a two-dimensional liquid-solid (rotator phase) transition (Fig. 9.16). The bending rigidity modulus was also evaluated. Such a determination is necessarily very rough in a reflectivity experiment, but can now be performed by grazing incidence diffuse scattering.

Diffuse Scattering When a surfactant monolayer is present at the interface, its first effect is to reduce the surface tension: $\gamma = \gamma_{H_2O} - \Pi$, where Π is

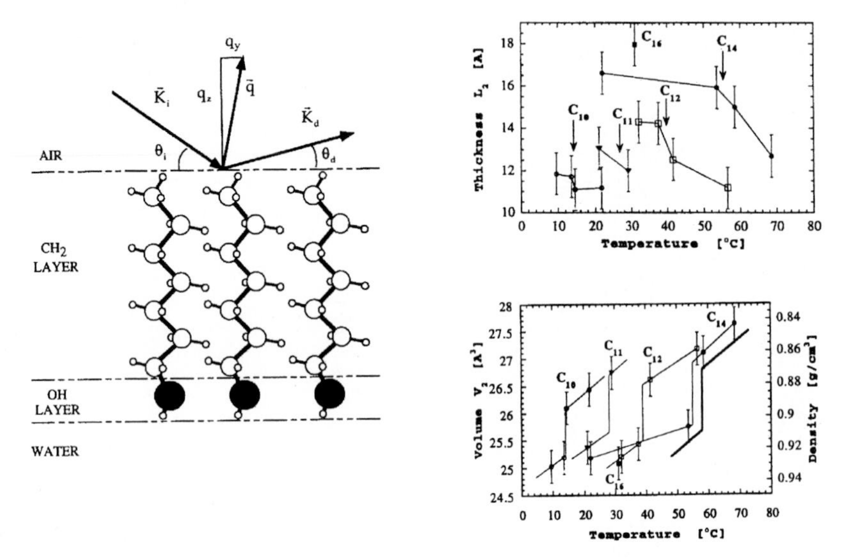

Fig. 9.16. Left: Schematics of the alcohol monolayer at the air/water interface. Top, right: Film thickness as a function of temperature. The arrows indicate the phase transition. Bottom, right: Volume per CH_2 as a function of temperature. Note the density jump at the 2-d liquid to solid (rotator phase) transition. With kind permission of B. Berge and J.P. Rieu

the surface pressure, as illustrated in Fig. 9.17 for an arachidic acid ($CH_3 - (CH_2)_{18} - COOH$) film.

Higher order corrections to the spectrum, i.e. effects of the bending stiffness of the film are also apparent. Results for a L_α di-palmitoylphosphatidylcholine (DPPC) film on pure water are presented in Fig. 9.18. Whereas at small q_y values the scattered intensity scales with the surface tension as expected, this is no longer true at large q_y due to the effect of bending stiffness. The data of Fig. 9.18 have been analysed using the spectrum Eq. (9.18) including the additional term Kq^4 in the denominator. For the more compressed film of Fig. 9.18 it is found that $\kappa = (5 \pm 2)k_BT$, smaller than generally expected in condensed DPPC films [34]. The observed wave-vector range is not large enough to allow the precise determination on the exponent 4. Smaller exponents are however found with the very rigid films[27] formed by fatty acids (here behenic acid $CH_3 - (CH_2)_{20} - COOH$) on divalent cation subphases ($5 \times 10^{-4} mol/l\ CdCl_2$) at high pH (8.9) and low temperature ($5°C$). Uncompressed, such films exhibit a $q^{-3.3}$ power law which has been attributed to the coupling between in-plane (phonons) and out-of-plane elasticity [27].

Finally, in systems with more than one interface, it is possible to measure the correlation between the interfaces (see Sect. 4.3.3). This is the case for soap films [35] and also for free standing smectic films for which the elastic constants can be measured [36].

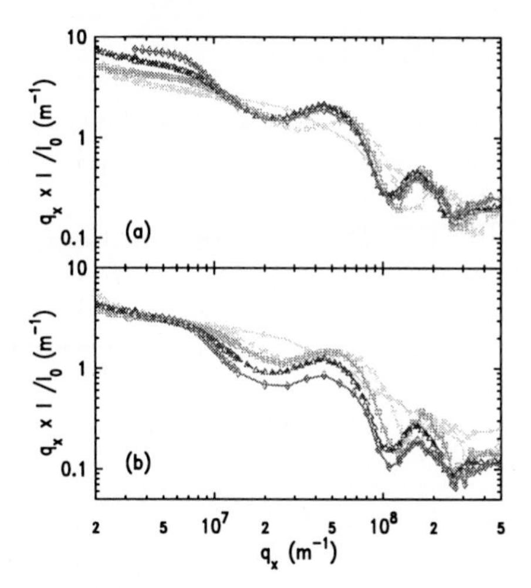

Fig. 9.17. (a): Intensity scattered by an arachidic acid film (black curves) and water (grey curves). The surface tensions are (top to bottom) 33 mN/m (diamonds), 43mN/m (triangles), 53mN/m (squares), 69mN/m (circles) and 73mN/m. (b): The same data normalised by γ/γ_{water} in order to illustrate the scaling $I \propto \gamma$ in the range $3\times10^6 m^{-1} \leq q_x \leq 8\times10^6 m^{-1}$ where capillary waves dominate the fluctuation spectra. The fringes are due to the normal film structure since q_z is not constant in the (x,z) configuration

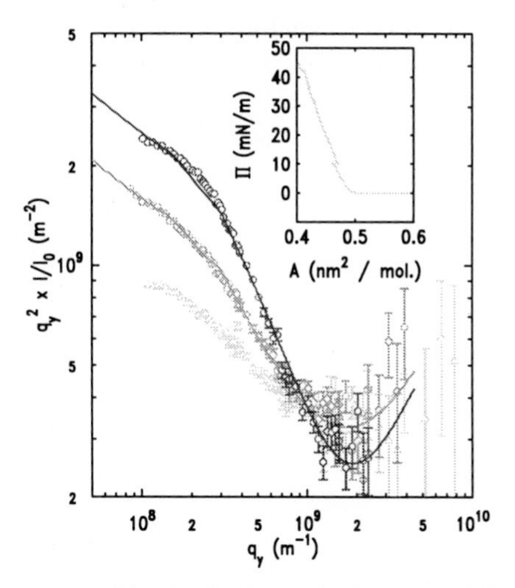

Fig. 9.18. Intensity scattered in the horizontal plane by a bare water surface (grey triangles) and a DPPC film at $3°C$ compressed at $20mN/m$ (grey circles) and $40mN/m$ (black circles). Lines are the best fits as indicated in text. Note that the scattered intensity scales with the surface tension at low q_y but that this is no longer true at large q_y due to the effect of bending stiffness (the black curve passes below the grey curves). Inset: corresponding molecular area - surface pressure isotherm of the DPPC film

9.4 Liquid-liquid Interfaces

Only a few x-ray reflectivity experiments have been attempted up to now (neutron reflectivity experiments will not be discussed here). In Ref. [37], the possibility of measurements using the high energy bremstrahlung of a tungsten tube was demonstrated. Two geomeries are possible with the incident beam coming either through the top or the side (Fig. 9.19). The experiment was done at a fixed angle with an energy sensitive detector. The main difficulty is related to the transmission through the $7cm$ wide cell (Fig. 9.20, left). Quite surprisingly, this classical setup allowed the measurement of a very nice reflectivity curve at the water-cyclohexane interface (Fig. 9.20, right).

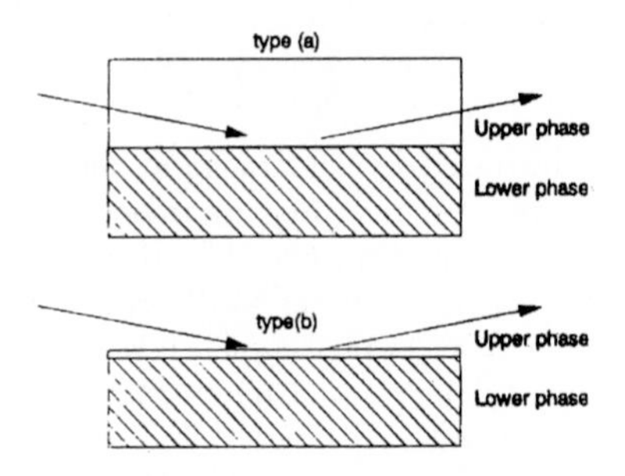

Fig. 9.19. Possible geometries for liquid-liquid surface scattering (a) coming from the side of the upper liquid. (b) beam coming from the top of the upper liquid. With kind permission of S.J. Roser

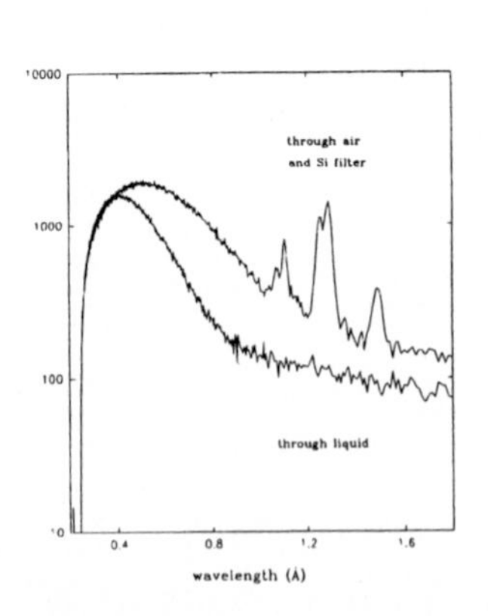

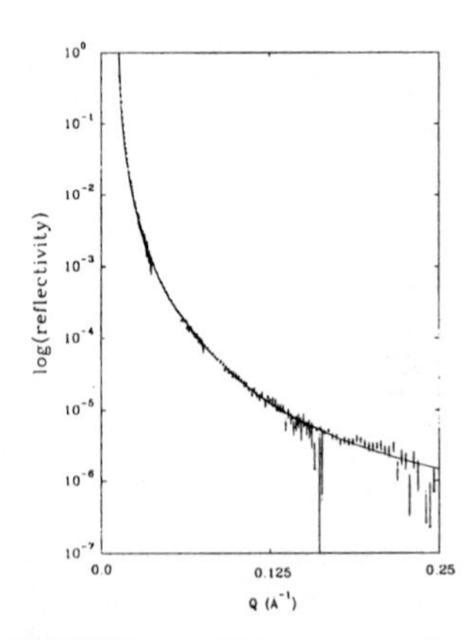

Fig. 9.20. Left: Absorption through air and liquid. The curves represent the transmitted intensity as a function of the wavelength. Right: Reflectivity of the water/cyclohexane interface and model fit using a Fresnel reflectivity profile with additional roughness. With kind permission of S.J. Roser

McClain et al. [38] studied a surfactant film in contact with the microemulsion middle phase in a decane, water, triethylene glycol monooctyl ether (C_8E_3) ternary mixture at $17keV$ using synchrotron radiation (NSLS, beamline X20B). They measured both reflectivity (Fig. 9.22, left) and diffuse scattering (Fig. 9.22, right). The rms roughness extracted from the reflectivity experiment was $\langle z^2 \rangle^{1/2} = 8.5nm$. In addition, the diffuse scattering experiment yielded $\gamma = 0.11mN/m$ and $K < 0.5k_BT$. Note however that the curves of Fig. 9.22 suffer from background subtraction problems at large q_z values which represent in fact the main difficulty of such experiments.

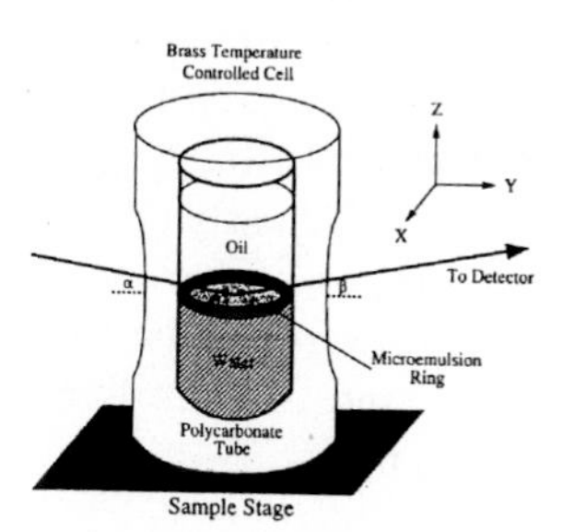

Fig. 9.21. The experimental cell of McClain et al. at beamline X20B, NSLS. α in the angle of incidence. β is the angle of reflection. The decane-water-C_8E_3 microemulsion middle phase is in equilibrium with decane and water. With kind permission of B.R. McClain

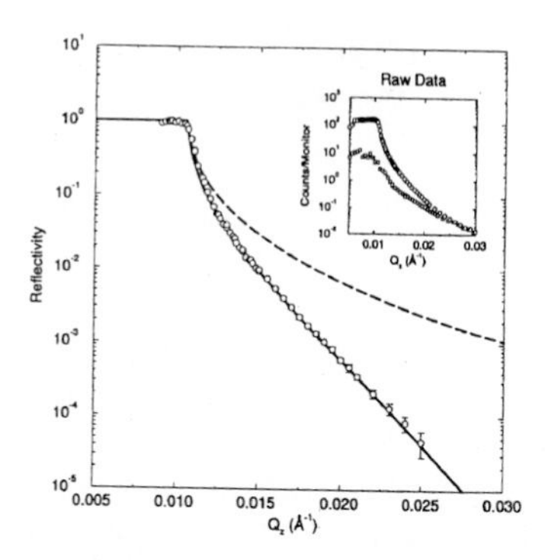

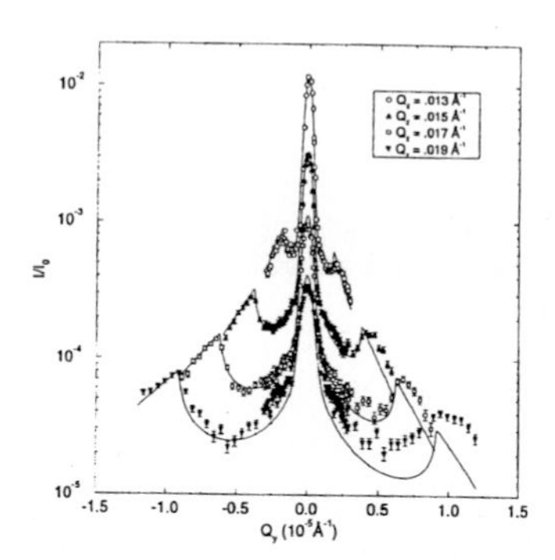

Fig. 9.22. left: Specular reflectivity from the water-microemulsion interface. The dashed line is the Fresnel reflectivity of the corresponding flat interface, and the solid line is calculated with a r.m.s. roughness of $8.5nm$. Raw data (circles) and background (squares) are displayed in the inset. Right: Diffuse scattering from the water-microemulsion interface (q_x scans at fixed indicated q_z value). The theoretical lines are calculated with a surface tension $\gamma = 0.11mN/m$. With kind permission of B.R. McClain

Another illustration of a liquid-liquid interface measurement is given in Fig. 9.23 [39]. A dedicated two-barrier Langmuir trough made of glass (lower part containing water) and teflon (upper part containing oil) to avoid leakage, equipped with very thin ($50\mu m$) teflon windows for the x-ray beam was used. The optimal cell width resulting of a balance between absorption and the requirement of a flat meniscus is $7cm$. A high energy has to be used ($18keV$, $\lambda = 0.068nm$, for which the transmission through the $7cm$ wide film of hexadecane is ≈ 0.1). Detector scans in the plane of incidence are represented Fig. 9.23. The amphiphile used is the phospholipid di-palmitoyl-phospshatidyl-choline (DPPC) which forms very stable films at the water/oil interface and can therefore be compressed to high pressures (i.e. low surface tensions), thus giving rise to a large diffuse scattering signal. The fluctuations of this amphiphilic film ($\gamma = 10mN/m$) at the oil water interface could be measured up to wave-vectors $\approx 10^8 m^{-1}$. The background is very large but the subtraction procedure is sufficiently efficient and reliable to allow the measurement of very small signal to background ratios. The structural parameters used to analyse the data were similar to those of compressed DPPC films at the water/air interface, and the film fluctuations could be analysed using Eq. (9.5) with $\gamma = 10mN/m$.

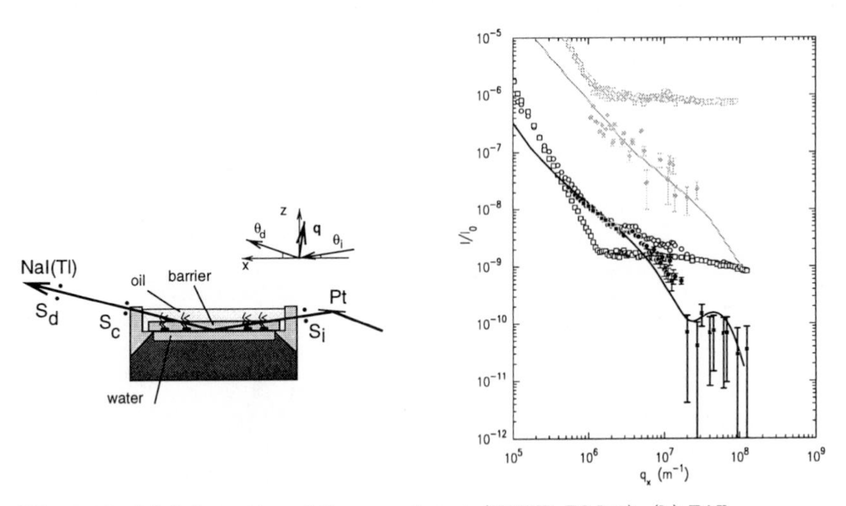

Fig. 9.23. (a) Schematics of the experiment (ESRF, BM32). (b) Diffuse x-ray scattering at the hexadecane-water interface. Detector θ_{sc} scans in the vertical plane of incidence. Grey symbols and curve: bare hexadecane-water interface. Black symbols and curve: compressed L-α-dipalmytoilphosphatidylcholine film ($\gamma = 10mN/m$). These curves are divided by a factor of 1000. Empty circles: signal, empty squares: background mainly due to bulk hexadecane scattering, filled cirles: signal minus background

References

1. J. D. van der Waals *Verhandel. Konink. Akad. Weten. Amsterdam* **1**, 8 (1893).
2. J.S. Rowlinson and B. Widom, "Molecular Theory of Capillarity" Oxford, Clarendon Press,1982.
3. J.W. Gibbs, *On the equilibrium of heterogeneous substances* in *The scientific papers of J. Willard Gibbs* reprinted by Dover Publications, New-York, 1961.
4. F.P. Buff, R.A. Lovett and F.H. Stillinger Jr. *Phys. Rev. Lett.* **15** 621 (1965).
5. R.F. Kayser *Phys. Rev. A* **33**, 1948 (1986)
6. See for example R. Loudon, "Ripples on liquid interfaces" in "Surface excitations" edited by V.M. Agranovich and R. Loudon, Modern Problems in Condensed Matter Science, vol. 9, North-Holland, Amsterdam, 1984.
7. R. Evans, *Molecular Physics* **42** 1169 (1981).
8. J. Meunier, *J. Physique* **48** 1819 (1987).
9. M. Niapórkowski, S. Dietrich, *Phys. Rev. E* **47** 1836 (1993).
10. S.K. Sinha, E.B. Sirota, S. Garoff, and H.B. Stanley, Phys. Rev. B **38**, 2297 (1988).
11. M.K. Sanyal, S.K. Sinha, K.G. Huang, B.M. Ocko, *Phys. Rev. Lett.* **66** 628 (1991).
12. M. Fukuto, R.K. Heilmann, P.S. Pershan, J.A. Griffiths, S.M. Yu, and D.A. Tirrell, *Phys. Rev. Lett.* **81**, 3455 (1998).
13. H. Tostmann, E. DiMasi, P.S. Pershan, B.M. Ocko, O.G. Shpyrko, M. Deutsch, *Phys. Rev. B* **59**, 783 (1999).
14. "Dynamics of fractal surfaces" edited by F. Family and T. Vicsek, World Scientific, Singapour, 1991.
15. G. Palasantzas, *Phys. Rev. B* **48** 14472 (1993).
16. W. Helfrich, *Z. Naturforschung* **28 c** 693 (1973).
17. J. Daillant, K. Quinn, C. Gourier, F. Rieutord, *J. Chem. Soc. Faraday Trans.*, **92** 505.
18. L. Bourdieu, J. Daillant, D. Chatenay, A. Braslau, and D. Colson, *Phys. Rev. Lett.* **72**, 1502 (1994).
19. J. Daillant, O. Bélorgey *J. Chem. Phys.* **97** 5824 (1992).
20. W.H. de Jeu, J.D. Schindler, E.A.L. Mol, *J. Appl. Cryst.* **29** 511 (1996).
21. B.C. Lu, S.A. Rice, *J. Chem. Phys.* **68** 5558 (1978).
22. L. Bosio, M. Oumezine, *J. Chem. Phys.* **80** 959 (1984).
23. A. Braslau, M. Deutsch, P.S. Pershan, A.H. Weiss, J. Als-Nielsen, J. Bohr, *Phys. Rev. Lett.* **54** 114 (1985).
24. A. Braslau, P.S. Pershan, G. Swislow, B.M. Ocko and J. Als-Nielsen, *Phys. Rev. A*, **38**, 2457 (1988).
25. J. Daillant, L. Bosio, B. Harzallah, J.J. Benattar, *J. Phys. France II* **1** 149 (1991).
26. D.K. Schwartz, M.L. Schlossman, E.H. Kawamoto, G.J. Kellog, and P.S. Pershan, *Phys. Rev. A* **41** 5687 (1990).
27. C. Gourier, J. Daillant, A. Braslau, M. Alba, K. Quinn, D. Luzet, C. Blot, D. Chatenay, G. Grübel, J.F. Legrand, G. Vignaud, *Phys. Rev. Lett.* **78** 3157.
28. S. Dietrich, A. Haase, *Physics Reports*, **260** 1 (1995).
29. R.K. Thomas, J. Penfold *Current opinion in colloid and interface science* **1** 23 (1996).
30. S.K. Sinha *Current opinion in solid state and material science* **1** 645 (1996).

31. Many references concerning this work can be found in M. Deutsch, B.M. Ocko, X.Z. Wu, E.B. Sirota, S.K. Sinha, in "Short and long chains at interfaces" edited by J. Daillant, P. Guenoun, C. Marques, P. Muller, J. Tran Thanh Van, Editions Frontières, Gif-sur-Yvette 1995, p.155.
32. O.M. Magnussen, B.M. Ocko, M.J. Regan, K. Penanen, P.S. Pershan, and M. Deutsch, *Phys. Rev. Lett.* **74** 4444 (1995).
33. J.P. Rieu, J.F. Legrand, A. Renault, B. Berge, B.M. Ocko, X.Z. Wu, M. Deutsch, *J. Phys. II France* **5** 607 (1995).
34. E. Sackmann in "Handbook of biological physics", vol. 1A edited by R. Lipowsky and E. Sackmann, Noth-Holland, Amsterdam, 1995.
35. J. Daillant, O. Bélorgey *J. Chem. Phys.* **97** 5837 (1992).
36. E.A.L. Mol, J.D. Schindler, A.N. Shalaginov, W.H. de Jeu *Phys. Rev. E* **54** 536 (1996).
37. S.J. Roser, S. Felici, A. Eaglesham, *Langmuir* **10** 3853 (1994).
38. B.R. McClain, D.D. Lee, B.L. Carvalho, S.G.J. Mochrie, S.H. Chen, J.D. Litster, *Phys. Rev. Lett.* **72** 246 (1994).
39. C. Fradin, D. Luzet, A. Braslau, M. Alba, F. Muller, J. Daillant, J.M. Petit, F. Rieutord, *Langmuir* **14**, 7329 (1998).

10 polymer Studies

Günter Reiter

Institut de Chimie des Surfaces et Interfaces CNRS, 15 rue Jean Starcky, B.P. 2488, 68057 Mulhouse, France

10.1 Introduction

In this chapter I would like to present some examples for the great success of neutron and x-ray reflectometry in polymer science. These techniques are unique for the determination of interfacial density profiles, even of buried interfaces. The vertical resolution of these techniques is at least comparable with SFM (scanning force microscopy) but at the same time they take averages over large enough areas to give a representative and characteristic information of the system. Other techniques may be more direct (like SFM or NRA (nuclear reaction analysis)) but they have severe disadvantages concerning either vertical resolution or lateral sampling. While SFM may provide detailed information on small sample sizes this may not be representative for the whole sample. NRA and other ion beam techniques are certainly more direct as they work in direct space (and not in Fourier space as neutron and x-ray reflectivity). However, their vertical resolution is in many cases insufficient to detect all important features of polymeric interfaces. Comparing the vertical depth resolution of neutron and x-ray reflectometry (of the order of Angstroms) to the typical size of a polymer (some hundreds of Angstroms) clearly shows the possibility to measure changes at a submolecular level.

Here I will give several examples where neutron and x-ray reflectivity have been successfully used to investigate interfacial problems in polymer science. It should be noted that samples generally need to be quite large (several cm^2) and homogeneous over this area. In some cases this may present a major difficulty concerning sample preparation. For stratified systems the interfaces need to be extremely parallel. Otherwise averaging of the large areas illuminated by the incident beam will lead to smearing effects. As a consequence, the high vertical resolution of this technique would be lost.

I have selected mostly examples regarding problems from polymer sciences which are demonstrating the many possibilities x-ray and neutron reflectometry offer. They will show how powerful, versatile and unique these techniques are. Due to space limits, I have not been able to give an exhaustive survey and focused on some rather unique, successful and convincing setups. It was also not my intention to give a review on polymer physics and thus most of the explanations and interpretation are short and limited.

Note that k used in this chapter is defined as $k = k_z = (2\pi/\lambda) sin\theta = q_z/2$ for specular reflectivity. This convention is mostly used by the neutron community.

10.2 Thin Polymer Films

I will start with work on thin polymer films. Although there are simpler (and faster) techniques available to measure the thickness of thin polymer films (like e.g. ellipsometry) x-ray and especially neutron reflectometry are nonetheless used. This comes from the possibility to distinguish between film thickness and film density, as well as the density profile, even for films as thin as some nanometers. One can thus determine density and thickness changes separately if e.g. the film is measured at different temperatures.

In Fig. 10.1 one can see a typical reflectivity curve for a thin polymer film. It represents a 49.3 nm thick polystyrene film deposited by spincoating onto a glass slide and measured by x-ray reflectometry. The thickness can be determined quite precisely (better than 0.1 nm) from the well pronounced interference fringes. Assuming an error-function density profile for the substrate/film and the film/air interfaces one obtains roughness values of 0.6 and 0.3 nm, respectively. The specific density of the film (which shows up mostly at the critical wavevector for total reflection, see Sects. 3.1.2 and 5.3.2) corresponds well with the density of polystyrene in bulk samples (1.05 g/cm^3).

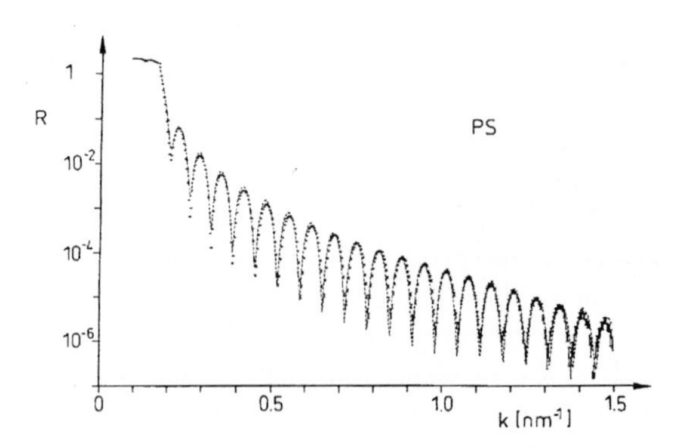

Fig. 10.1. A typical x-ray reflectivity curve for a polymer film: polystyrene spincoated from toluene onto float glass. Experimental data points are shown with error-bars. The dotted curve is the best fit yielding a thickness of 49.3 nm, and 0.3 and 0.6 nm for the roughness of the polymer-air and the polymer-substrate interface, respectively. The density of the film corresponds to the bulk density of polystyrene

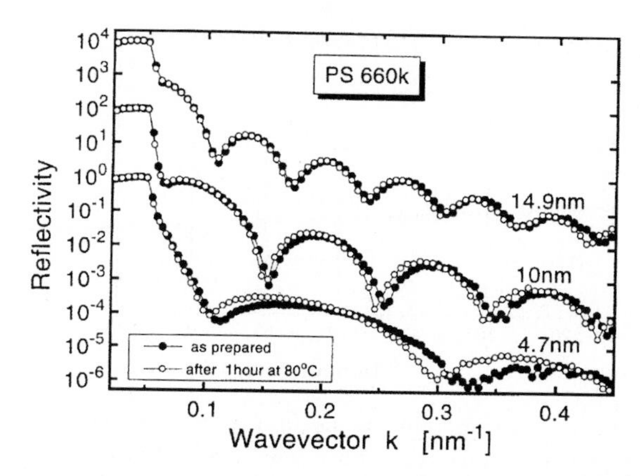

Fig. 10.2. X-ray reflectivity curves for 3 different polystyrene films ($M_w = 660k$) before and after annealing for one hour at $80°C$, i.e. below the bulk glass transition temperature. The thicknesses are indicated in the figure. (adapted from Ref. [1])

The high vertical resolution of x-ray reflectometry is extremely useful and necessary for the investigation of changes in polymer films after (or during) thermal treatments. In Fig. 10.2 reflectivity curves for three polystyrene films (differing in initial thickness) are shown. The curves were measured at room temperature but after the sample was annealed for one hour at 80° C. Each time the curve is compared with the result just after sample preparation. One can clearly see that the curves have changed after heating. These changes are attributed to relaxations of the polymers.

The high sensitivity of x-ray reflectometry has been used to measure the thickness of an ultrathin PS-film at different temperatures (see Fig. 10.3). Contrary to usual thermal expansion extremely thin films rather showed reversible contraction. The film thickness changed by more than 10%. X-ray and neutron reflectometry can be used not only to investigate featureless thin films but these techniques are also able to provide information on the internal structure of the films. Using specular reflection one may obtain results e.g. on the multilayer structure induced by surface directed orientation of self assembling block copolymers (see Figs. 10.4-10.6). These examples show polystyrene-poly(methyl methacrylate) block copolymers (P(S-b-MMA) deposited by spincoating onto silicon wafers. The second bounding medium may either be air (i.e. a FREE surface) or a deposited SiO2-layer (confined polymer layer).

In Fig. 10.4 one can see the reflected intensity (R) as a function of the incident wavevector for a 556 nm thick P(S-b-MMA) film. Several features can be noted. First, the intensity drops quite abruptly at about $0.01Å^{-1}$. This is due to passing the critical angle of total reflection for the copolymer film, i.e. the x-ray beam is penetrating the film. The slow increase of R below $0.01Å^{-1}$ is due to an increase of the fraction of the incident beam hitting the

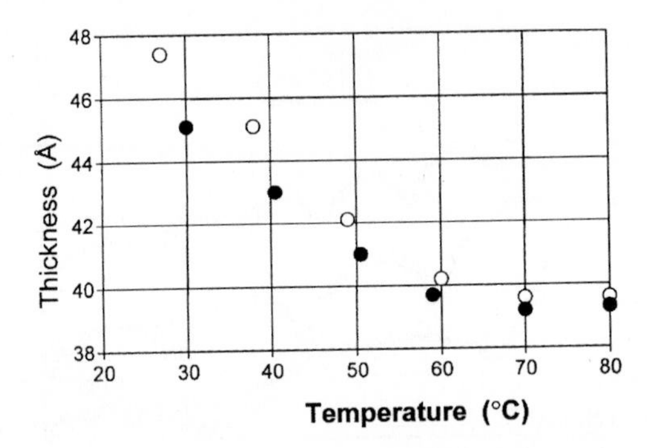

Fig. 10.3. Thickness of a thin polystyrene film as a function of annealing temperature. After a first annealing (to remove non-equilibrium conformations due to spincoating) the sample was heated incrementally to 80° *C (full circles), was cooled to room temperature, and re-heated to 80° C (open circles) (adapted from ref. [2])*

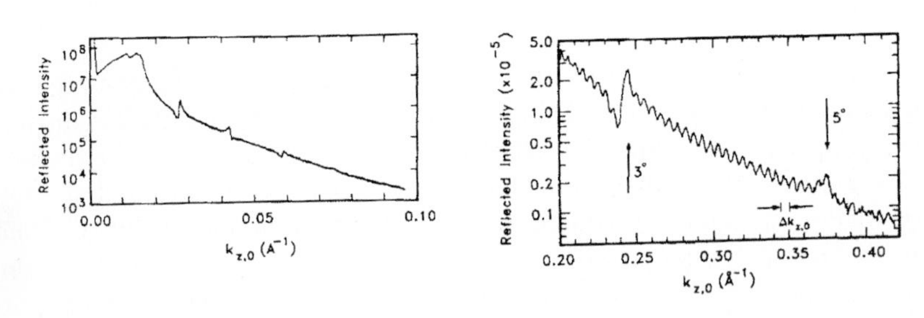

Fig. 10.4. X-ray reflectivity curve of a P(S-b-MMA) diblock copolymer that has self-assembled into a multilayer structure after annealing at 170° *C. The right curve shows an expansion of the curve on the left (adapted from ref. [3])*

sample. Around $0.015 Å^{-1}$ the intensity drops again as now the substrate is not totally reflecting anymore. Beyond $0.015 Å^{-1}$ the reflection curve consists of essentially two features:

- The interference fringes (separated by $\Delta k_{z,o}$) which are proportional to the total film thickness of 556 nm
- The multilayer Bragg reflections (see Chap. 8) corresponding to a lamellar period of 44.5 nm.

Thus this film consists of exactly 12.5 layers.

Fig. 10.5 shows a similar P(S-b-MMA) (150nm thick) film, now investigated by neutron reflectometry below and above the order-disorder transition of the system. Again, the clearly visible multilayer reflection peak around 0.02 $Å^{-1}$ reflects the layering of the lamellae parallel to the substrate surface. A good fit is obtained by using the model shown in the inset. Above the order-

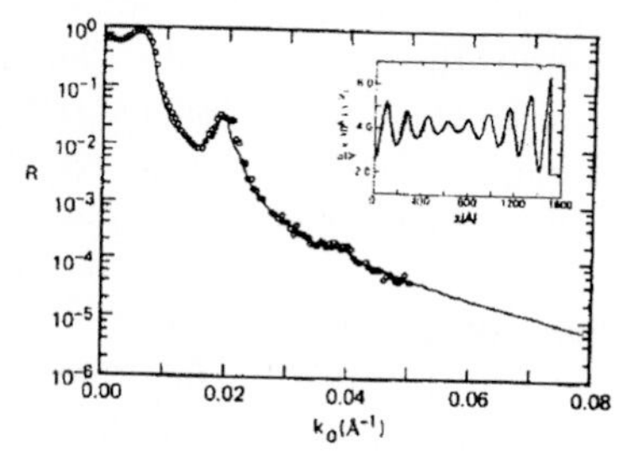

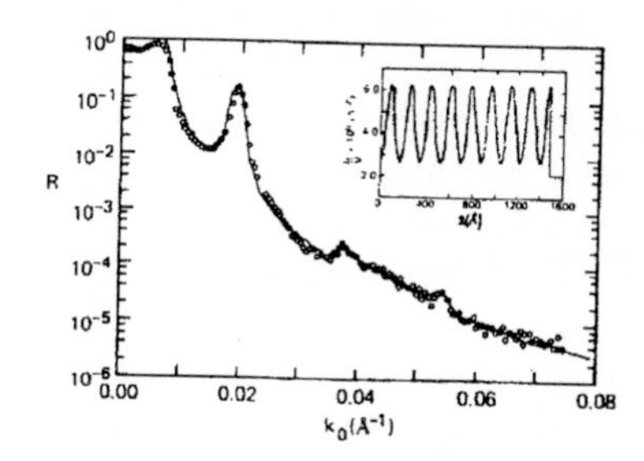

Fig. 10.5. Neutron reflectivity curve of a P(S-b-MMA) diblock copolymer film. The molecular weight of the copolymer is about 30k. The sample was annealed at 176° *C (left panel) and at 140° C (right panel), respectively. The insets show the scattering length density profiles that yielded the best fits, indicated by the solid lines (adapted from ref. [3])*

disorder transition (ODT) the lamellae further away from the interfaces are less developed. However, annealing this sample below the ODT (i.e. in the ordered region) leads to an increase of the Bragg peaks. Even higher orders can be seen. Due to the large contrast (see Sect. 5.3.2) of the scattering length obtained by deuteration of the PMMA block, details of the interfaces between the lamellae could be resolved. In particular, the interfacial width between PS and PMMA was found to be 5± 0.2nm.

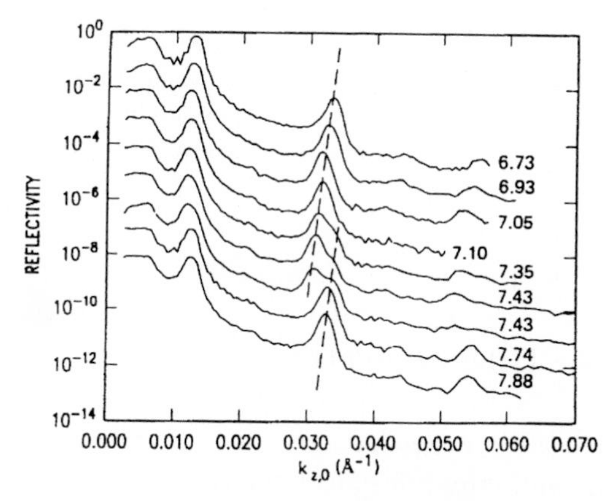

Fig. 10.6. Neutron reflectivity curves of a P(S-b-MMA) diblock copolymer confined between two surfaces where the separation distance has been changed. This distance is indicated by the t/L_0 values (where t is the film thickness and L_0 is the equilibrium lamellar period of the block copolymer) on the right side of each curve (adapted from ref. [3])

Polymer films can also be confined between two solid walls. The second wall may be produced by evaporating SiO_2 onto the polymer film. The only technique which is able at present to measure the density distribution in such confined thin films with the necessary precision is neutron reflectometry. Fig. 10.6 shows an example for P(S-b-MMA) films of different thicknesses. For increasing film thicknesses (indicated at the right side of the figure by t/L_0, where L_0 is the thickness of an unperturbed lamella) the shift of the third order reflection peak corresponds to an increase of the period. The occurrence of a double peak indicates two distinct lamellar thicknesses. Thin films may also prepared from solutions of polymer mixtures. In Fig. 10.7 one can see the results of neutron reflectometry from thin PS/PB films containing 44% and 12% of deuterated PS. In this case the interference fringes are due to segregation of PS to the substrate interface creating a sharp PS-PB interface. The interface is the sharper the lower the amount of PS is.

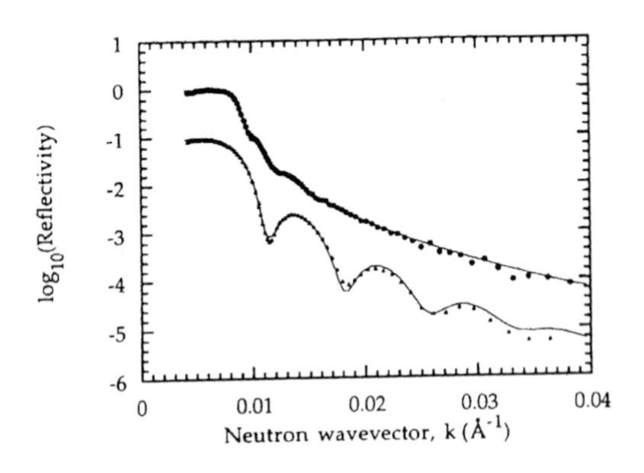

Fig. 10.7. Neutron reflectivity curves from films spincoated from a toluene solution of a mixture of deuterated polystyrene and polybutadiene (d-PS/PB). The volume fraction of d-PS was 0.44 for the upper curve and 0.12 for the lower curve, respectively (adapted from ref. [4])

10.3 Polymer Bilayer Systems

X-ray and neutron reflectometry have been extensively used to study polymer-polymer interfaces and interdiffusion. For this purpose double layer samples have been prepared. Usually, a first film is spincoated onto a silicon wafer. Using a highly selective solvent, the second film may be spincoated directly onto the first one. Alternatively, one may prepare the second film on a different substrate and then float off this film onto a clean water surface. This floating film now can be picked up by the substrate coated with the first film.

Based on this technique double layer samples up to some $10 \times 10cm^2$ have been prepared. Although the technique appears to be rather crude the width (or the roughness) of the polymer-polymer interface can be as low as 1nm.

In most cases neutron reflectometry is more favorable due to the much larger scattering length density (see Sects. 1.2, 1.3.1, 3.1.2, 5.3.2) contrast as compared to x-rays. Fig. 10.8 gives a calculated example for an incompatible system of PS and brominated PS (=PBrS). Due to the many electrons of the Br-atoms this system has already a quite measurable contrast for x-rays. But deuteration allows to improve this contrast for neutrons by almost an order of magnitude. Consequently, the reflectivity curves (via their fringe spacing, see Chap. 3) mainly reflect the thickness of the overall system (in the case of x-rays) or the thickness of the deuterated layer (in the case of neutrons).

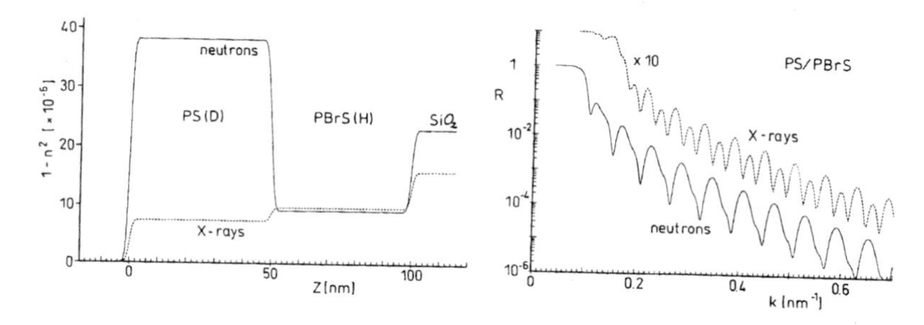

Fig. 10.8. Calculated reflectivity curves (right panel) for the refractive index profiles shown on the left panel

If one is interested in the interdiffusion between identical polymers neutron reflectometry is the only possible reflection technique. Deuteration of the molecules of one film does not strongly modify the system (unless the polymers are extremely long, where deuteration may lead to incompatibility). A system studied by several groups is polystyrene interdiffusing into polystyrene. The most important question in this context is how do polymers diffuse across an interface. The reptation model by de Gennes predicts significantly different interfacial profiles with respect to the interdiffusion of simple (small) molecules. As the polymer chains are supposed to cross the interface first via their ends, the interface should stay rather sharp at its center for the characteristic "reptation time". Only few molecules or parts of the molecules are able to cross initially. With progressing time more and more chain segments will be able to diffuse across the interface and the density profile eventually can be described by an error-function, typical for Fickian diffusion.

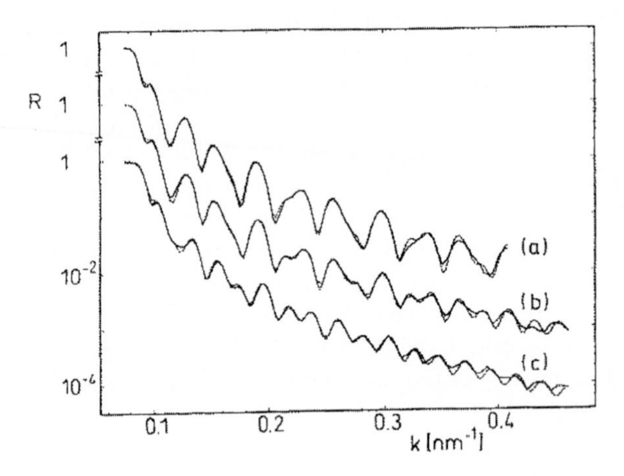

Fig. 10.9. Measured neutron reflectivity curves (full lines) for a bilayer system of deuterated and protonated polystyrene on glass. Results are for a) the unannealed, b) 2 min at 120° *C, and c) 3900 min at 120° C annealed sample. The broken lines represent the best fits using the refractive index profiles shown in Fig. 10.10 for the polymer-polymer interface (adapted from ref. [5]).*

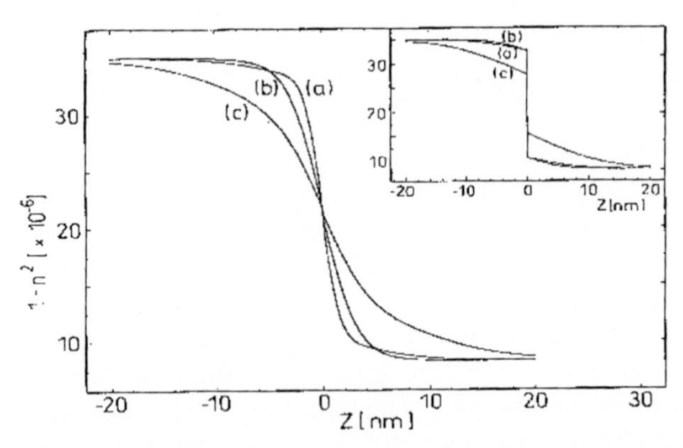

Fig. 10.10. Refractive index profiles for the polymer-polymer interface used to fit the data of Fig. 10.9. The inset shows the contribution due to reptation alone (adapted from ref. [5]).

Figures 10.9-10.11 show typical results for the early stages of interdiffusion, where reptation is "visible" [5]. The system consists of a deuterated and a protonated film of about the same thickness. In curve a) (just after preparation) the fringe spacing (Chap. 3) is mainly determined by the deuterated

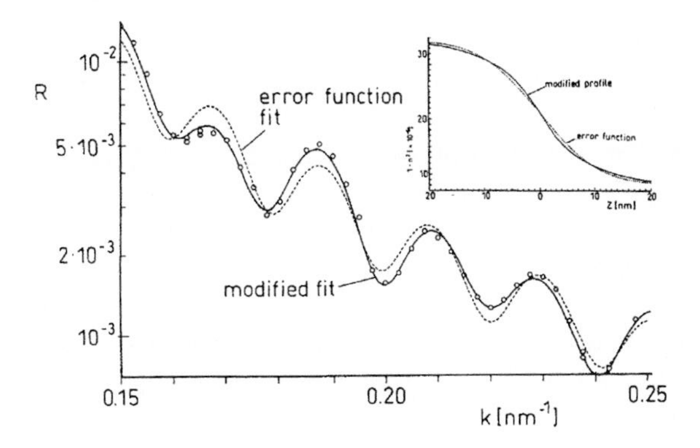

Fig. 10.11. Comparison of the best fits from an error-function (broken line) and the modified profile (full line) as described in the text for a section of curve c of Fig. 10.9

layer on top. The protonated layer is only visible through the distortions of the interference fringes. As the interface between the two polymer layers gets smeared out (a consequence of interdiffusion) the fringe spacing gets smaller (the number of fringes doubles). The air/polymer and the polymer/substrate interfaces stay sharp during the whole annealing procedure and eventually will dominate the reflection of neutrons. Thus, the fringe spacing corresponds to the overall thickness of the double layer system. These reflectivity curves have been fitted to a model which used the following refractive index profile $n(z)$ of two superposed error-functions for the polymer/polymer interface:

$$n(z) = n_o + \Delta n\{(1-p)(2-\mathrm{erfc}(z/\sigma_c)) + p(2-\mathrm{erfc}(z/\sigma_t))\} \qquad (10.1)$$

with $\mathrm{erfc}(z) = 1 - \mathrm{erf}(z)$, n_o being the refractive index of the first layer, Δn the maximum difference in refractive index between the two layers, σ_c and σ_t are the widths of the error functions and p described the relative weights. (The indices c and t represent core and tail contribution). Such a profile is based on the possibility of restricted local movements between the entanglement points (responsible for the core part) and the pure reptation contribution which shows up in the tails of the profile (Figure 10.10 shows the profiles used to fit the curves of Fig. 10.9. The inset shows the "pure" reptation part).

In order to show how powerful neutron reflectometry is, we enlarged a section of curve (c) of Fig. 10.9. The broken and the full lines represent the best fit using a simple error-function profile and the profile given above, respectively. Although the modified profile is deviating only slightly from an error-function profile (see inset of Fig. 10.11) it gives a significantly better fit. In reference [5] interdiffusion has been followed as a function of time and the results (in particular the profiles) have been successfully compared to the reptation theory.

At this point, I want to add a word of caution. In order to be able to resolve characteristic features of an interfacial profile the reflectivity curve has to be measured in the appropriate k-range. In Fig. 10.12 simulated reflectivity curves for step (broken lines) and linear (full lines) interfacial profiles of different widths are compared. The system and the k-range chosen are the same as for the example of interdiffusion shown in Figs. 10.9-10.11. One can clearly see that fine details of a profile can only be determined if the k-range is sufficiently large, e.g. a linear profile of 4 nm width and a step profile of 2 nm width can hardly be distinguished if the reflectivity curve is limited to $k = 0.6nm^{-1}$.

10.4 Adsorbed Polymer Layers

In contrast to the problems mentioned above where the interfaces have been rather sharp, adsorbed polymer layers are usually diffuse. Nonetheless, neutron reflectometry can be rather sensitive and, especially for liquid systems, is a unique technique for determining characteristic features of such layers. Maximum contrast or an improved contrast can be achieved by contrast matching the environment of the adsorbed polymer layer to the bounding medium (air or the substrate). In many cases, one has also the possibility to invert the contrast (deuterated polymer layer in protonated environment and vice versa). Results of two complementary curves have to be described by a single model. This helps to reduce ambiguities in analysing the data.

Due to the monotonic and featureless decay of the reflectivity such curves need significant theoretical support (predicting a model for the interfacial profiles) to enable a satisfying analysis. In Fig. 10.13 a result from one of the first successful experiments in this context is shown. The full lines represent the model fits based on a powerlaw decay over a distance of 60nm (for further details, see ref. [6]). X-ray reflectometry has also been used to measure self-assembled surface micelles of end-functionalised AB diblock copolymers. For this purpose a polymer solution has been spread onto a water surface on a Langmuir trough. The entire film balance system was placed in a Plexiglas container with kapton windows for the x-ray beams. Figure 10.14 shows a schematic drawing of the setup and Fig. 10.15 gives some typical results (including least-squares fits based on the model shown in right part of Fig. 10.15) for polystyrene-alkylated polyvinylpyridine terminated by iodine (PS-P4VP-C8I-). The different curves are for different compression (i.e. different areas per molecule) of the surface layer.

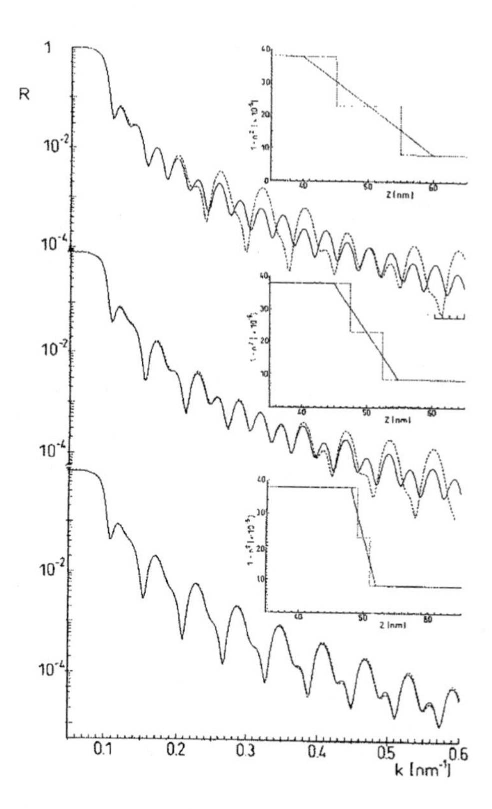

Fig. 10.12. Calculated reflectivity curves for a double layer system, representing 50 nm deuterated on 50 nm protonated polystyrene on glass. The interfacial profile between the two polymers is assumed to be a step function (dotted line) or a linear profile (full line). The three sets compare widths of 2 nm/4 nm, 5 nm/10 nm, and 10 nm/20 nm for the step and the linear profile, respectively

10.5 Polymer Brushes

A major area in polymer physics which has been extensively investigated by x-ray and neutron reflectometry is the formation and the properties of polymer brushes, either at a solid or a liquid interface. A brush is formed if many polymers are anchored with one endgroup at an interface. The molecules arriving first are not yet interacting with each other. Increasing the number of grafted molecules (per unit area) leads to overlap between these molecules and eventually to a stretching of the chains if more than one chain per cross-section of an unperturbed polymer chain is grafted onto the substrate. Fig. 10.16 gives a schematic drawing of this process.
The degree of stretching depends (for a given number of grafted molecules) on the quality of the solvent. In Fig. 10.17 one can see how this behavior was observed by neutron reflectometry. Increasing the temperature of a polystyrene

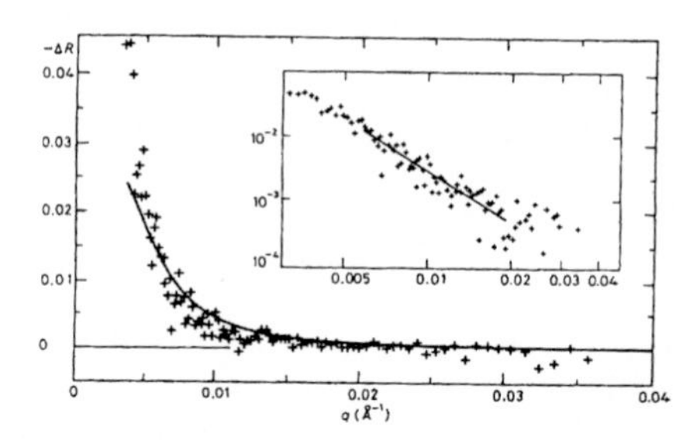

Fig. 10.13. Experimental (crosses) and calculated (full lines) values of $\Delta R = (R(q) - R_{Fresnel})$ for polydimethylsiloxane (Mw=4200k) adsorbed at an air/deuterated toluene interface. q is the momentum transfer normal to the surface with total reflection stopping at q=0 (adapted from ref. [6])

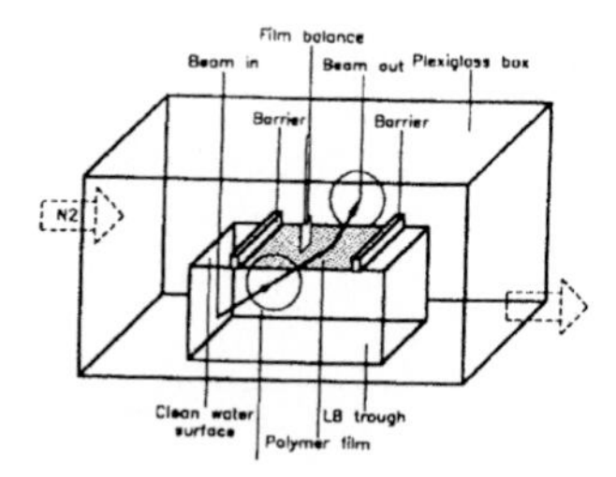

Fig. 10.14. Schematic drawing for the system used to measure block copolymers adsorbed at a liquid/air interface, allowing to compress the monolayer (adapted from ref. [7])

brush in deuterated cyclohexane increases the interaction between polymer and solvent and thus leads to swelling of the brush. It should be noted that in this example the neutron beam passed through the substrate (silicon single crystal, entered at one edge at an angle of about 90°) and then is reflected at the interfaces of the brush. The solvent on top of the brush is sufficiently thick (of the order of mm), deuterated and the solvent-glass interface is not necessarily parallel to avoid contributions from reflection from this interface.

Brush formation can also be followed as a function of time allowed for the grafting. For this purpose one can, for example, start from a thin film containing a mixture of deuterated non-functionalised and protonated functionalised molecules. As the brush forms the deuterated and the protonated molecules are progressively separated. This leads to a distinct two layer system which can be clearly seen in the reflectivity curves shown in Fig. 10.18. One starts with a film containing homogeneously distributed deuterated and protonated molecules. As the protonated molecules are grafted one obtains a layer of mainly deuterated molecules with a thickness which is less than the initial

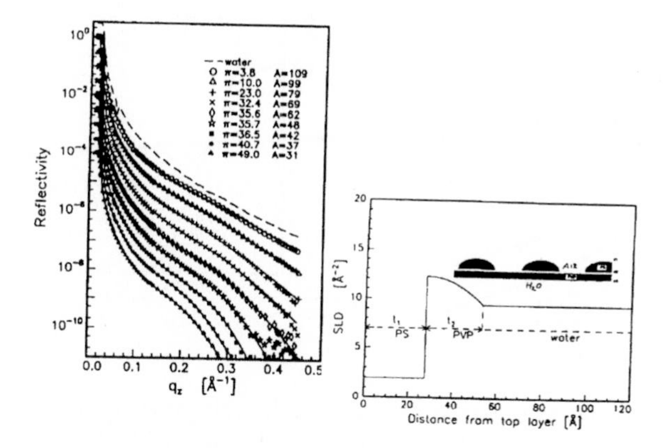

Fig. 10.15. Measured x-ray reflectivity curves (symbols) for PS-P4VP-C8I- adsorbed on the water surface for different surface pressures. The lines represent best fits based on the two-layer model shown on the right. The dashed line represents the specular reflectivity from the pure water surface (adapted from ref. [7])

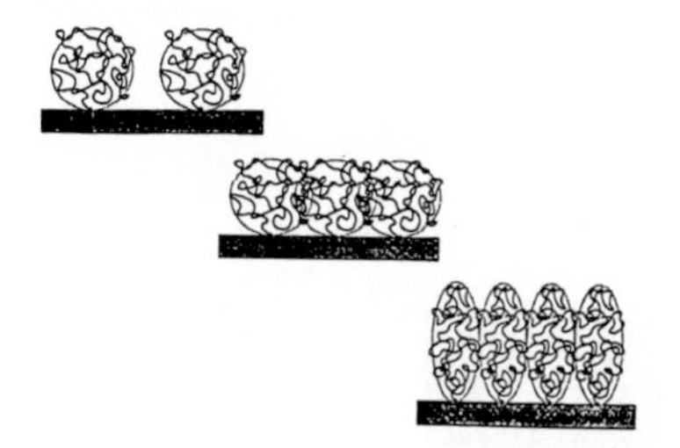

Fig. 10.16. Drawing showing the transition from isolated to overlapping and stretched grafted polymers during grafting (adapted from ref. [8])

film thickness on top of a mainly protonated layer. Thus the fringe spacing increases because the protonated layer is not contributing significantly to the reflected intensity. At the same time the critical wavevector for total reflection shifts to higher values due to an increase of the volume fraction of deuterated molecules in the top layer. It is possible to use a single sample to measure the kinetics because neutrons are not affecting (e.g. degrading) the polymer. This is advantageous as different samples are always slightly different due to preparation etc. (e.g. in their film thickness). Using a single sample furthermore allows to compare reflectivity curves directly. All differences between the curves are necessarily due to the grafting process and not to differences of the samples. Thus, one can directly see, without having to use a model

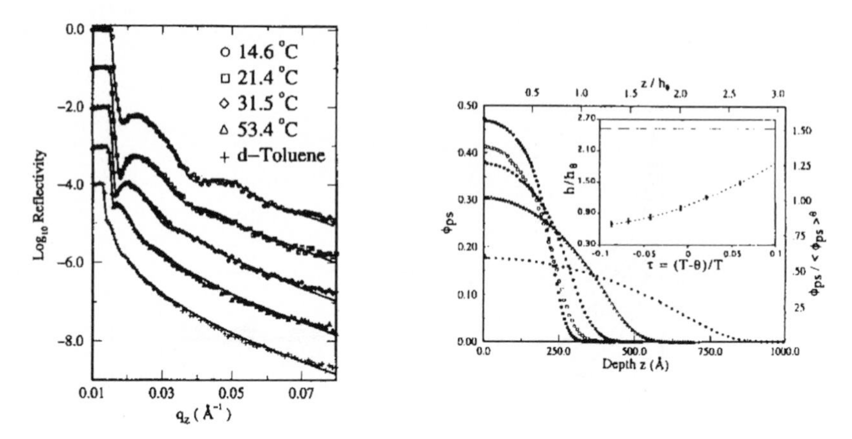

Fig. 10.17. Neutron reflectivity from a polystyrene brush grafted onto a silicon wafer and dissolved in deuterated cyclohexane for different temperatures. The lines give the best fits using the brush profiles shown on the right side (adapted from Ref. [9])

for the fit, if significant changes have occured. The fit is only necessary to quantify these changes.

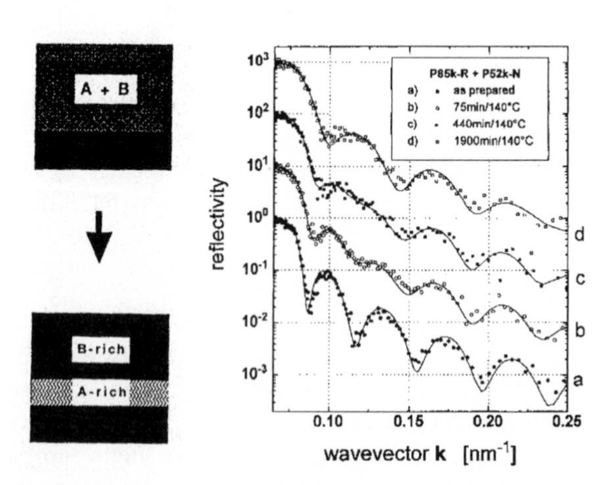

Fig. 10.18. Neutron reflectivity from a thin film containing a mixture of functionalised and non-functionalised (deuterated) polystyrene molecules. In the course of annealing the functionalised molecules graft onto the substrate. This leads to a separation of deuterated and protonated molecules as indicated by the schematic on the left

10.6 Polymer-Metal Interfaces

A major disadvantage of x-rays for the investigation of polymers is the low difference in the refractive index between various polymers. Usually the contrast is of the order of 10^{-6} for most polymers. The refractive index for x-rays depends mostly on the electron density. Thus it is quite obvious that the contrast can be improved by more than an order of magnitude if one deals with polymer-metal interfaces.
Here I want to give an example where the increased contrast has been used to investigate polymer interdiffusion between (protonated) polystyrene molecules of different lengths. As the molecular weight does not affect the refractive index the interface between the two layers of polymers was marked by evaporating a thin layer (about 5 nm) of gold. The surface tension of the polymer is much lower than the one of gold. Thus, gold only partially wets the polymer. Consequently, one obtains tiny droplets or particles of gold on top of the polymer film. The individual particles cannot be resolved by the x-ray beam as its lateral coherence length (Sects. 2.3, 4.7.2) is of the order of some microns. The reflectivity curve from such a thin layer of gold on top of a polymer film clearly represents two fringe periods (see Fig. 10.19). The well pronounced interference pattern originating from the gold layer allows one to obtain an average density (or equivalently a mean coverage, in the present example it is about 50%) and a density profile of the gold layer indicating the shape of the particles.

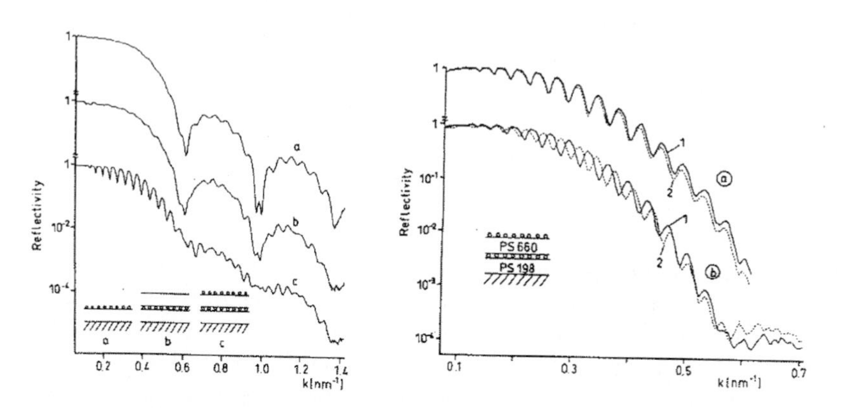

Fig. 10.19. Left panel: x-ray reflectivity curves from various steps of preparation of a four-layer gold-polystyrene (Mw=198k)-gold-polystyrene (Mw=660k) system as indicated in the inset. Right panel: Comparison of curves from the system shown in the inset after different annealing steps ($T = 120°C$): curves a): 1=675sec, 2=6830sec, curves b): 1=245 400sec, 2=658 000sec (adapted from ref. [10])

Putting a second polymer layer on top of this system does NOT much change the reflectivity curve. It would be impossible to measure changes e.g. of the thicknesses of the polymer layers. Thus, a second gold layer has been evaporated onto this tri-layer system. Now the two gold layers mainly determine the reflectivity of the system and thus the distance between these two layers is clearly visible from the interference fringes below about 0.6 nm^{-1}. This four-layer system has been used to investigate the "fast" interdiffusion of the shorter polymer into the layer of longer molecules. The imbalance in the fluxes led to a "swelling" of the top layer. This is clearly visible in a change of the spacing of the interference fringes (see Fig. 10.19). However, such a swelling only occured after an induction period which could be related to the reptation time of the shorter molecules.

10.7 Spreading of Polymers

In most cases of x-ray and especially neutron reflectometry, it is necessary to have rather large samples of at least several cm^2. It is, however, also possible to collimate the x-ray beam. The remaining intensity is still sufficient to measure much smaller areas.

Such a collimated system (see Fig. 10.20) has been used to measure final

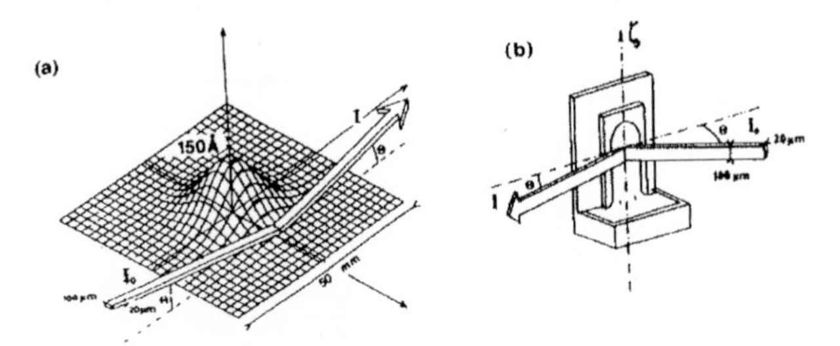

Fig. 10.20. Schematic view of (a) a microscopic droplet and (b) a capillary rise experiment studied by x-ray reflectometry (adapted from ref. [11])

stages of spreading of polymer droplets. As the beam could be moved across the droplet this setup also allowed to detect the shape of the droplet and to compare it with theoretical predictions. In particular, the existence of a "pancake", i.e. a homogeneous, quite dense and flat film of molecular thickness ($= 0.8nm$) could be detected. It has to be noted, however, that the illuminated area depends on the angle of incidence. For a correct interpretation this has to be taken into account by a film thickness which may vary with the angle. The agreement between such a model and the measured data is very satisfying (see Fig. 10.21).

A similar setup allows also to measure a thin film climbing up a solid substrate in a capillary rise experiment. In both cases, the well collimated beam

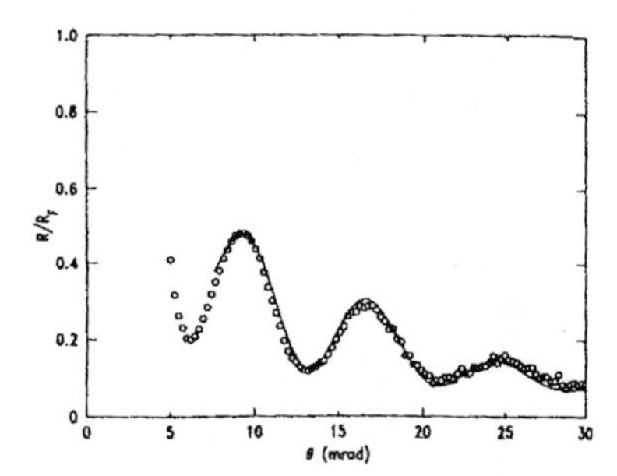

Fig. 10.21. Reflectivity curve recorded at the edge of a droplet. The mean thickness is 10nm, the slope is 2nm/mm, and the beam size is $20 \times 100\mu m$ (adapted from ref. [11])

and the possibility to translate the sample provide a means to measure height variations with a lateral resolution of the order of mm. The main advantage of this technique (in comparison with ellipsometry, which has a better lateral resolution, of the order of $50\mu m$) is the sensitivity for density-variations in addition to the high vertical resolution which can be better than $0.1nm$. For the case of spreading of polymers one thus can distinguish if the final stage is a dense and homogeneous film or an incomplete layer covering only partially the substrate.

10.8 Dewetting of Polymers

Dewetting may be understood as the opposite process to spreading. One starts from a smooth and homogeneous film and ends up with many droplets. In some cases these droplet may sit on top of a remaining thin layer (e.g. a monolayer of adsorbed or grafted polymers). Under certain conditions, x-ray reflectometry is extremely well suited for a) demonstrating that the film has become unstable and holes have been created and b) that the final stage (=droplets) has been reached but a thin and compact layer remains on the substrate.

In the first case, one takes advantage of the fact that the lateral coherence length (Sects. 2.3, 4.7.2) of the incident beam allows to average over areas of several μm^2. Thus, the fact that the film contains holes is reflected in a density profile schematically shown in Fig. 10.22. The decrease of the average density of the film is proportional to the fraction of the film which is now replaced by holes. The material removed from the holes is deposited in rims around the holes. These rims lead to tails of the density profile which have an effect similar to roughness. It should be noted however, that the film between the holes remains unperturbed as can be seen from the fringe spacing representing this thickness. Some typical examples for unstable films of end-functionalised polystyrene are shown in Fig. 10.23. As dewetting proceeds

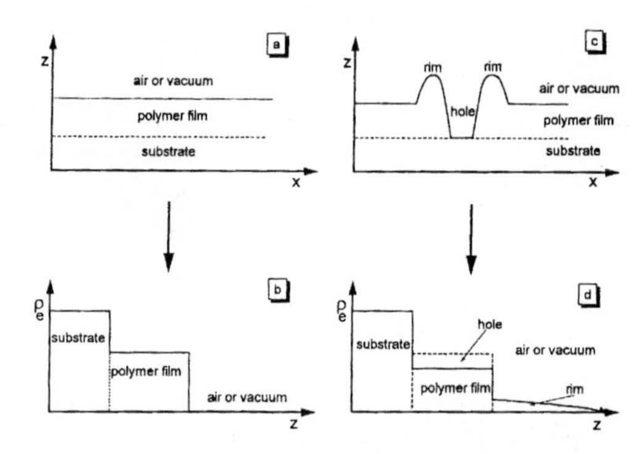

Fig. 10.22. Schematic representation of how the density profile used in the case of dewetting was deduced

this spacing disappears and, because a grafted layer is formed, a new larger spacing evolves giving the thickness of this layer. At the final stage where only droplets remain on a monolayer one only detects this monolayer. The droplets are not significantly contributing. Firstly, because they are occupying only a minor fraction of the surface (typically 10%, but this depends sensitively on the initial film thickness) and secondly, because they scatter the x-rays in directions off the specular direction. (Off-specular scattering can be used to detect the size and the distribution of the droplets)

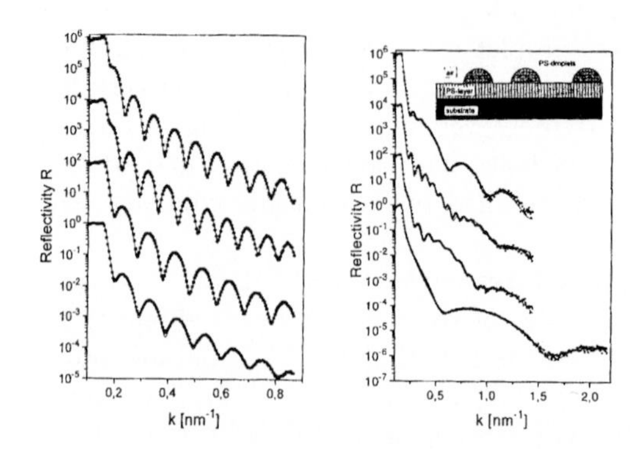

Fig. 10.23. X-ray reflectivity from thin films of ω-barium sulfonato polystyrene of 30 - 40 nm thickness. The molecular weight varies from bottom to top: Mw=2.8k, 13k, 18k, 19.5k. Left panel: BEFORE annealing, right panel: AFTER annealing for 200h at 175° *C. The inset shows the model used for the fits, shown by the solid lines (adapted from ref. [12])*

In conclusion, x-ray and neutron reflectometry are extremely versatile, powerful and in certain cases, unique techniques for the investigation of thin films, or interfacial problems in general, in polymer science.

References

1. G. Reiter, *Macromolecules*, **27**, 3046 (1994).
2. W.J. Orts, J.H. van Zanten, W.-L. Wu, S.K. Satija, *Phys. Rev. Lett.*, **71**, 867 (1993).
3. T.P. Russell, *Physica B*, **221**, 267 (1996). see also: T.P. Russell, *Materials Sci. Rep.* **5**, 171 (1990).
4. M. Geoghegan, R.A.L. Jones, A.S. Clough, J. Penfold, *J. Polym. Sci.: B: Polym. Phys.*, **33**, 1307 (1995).
5. G. Reiter, U. Steiner, *J. Phys. II*, **1**, 659 (1991).
6. X. Sun et al., *Europhys. Lett.*, **6**, 207 (1988).
7. Z. Li et al., *Langmuir*, **11**, 4785 (1995).
8. R.A.L. Jones et al., *Macromolecules*, **25**, 2352 (1992).
9. A. Karim, S.K. Satija, J.F. Douglas, J.F. Ankner, L.J. Fetters, *Phys. Rev. Lett.*, **73** 3407 (1994).
10. G. Reiter, S. Hüttenbach, M. Foster, M. Stamm, *Macromolecules*, **24**, 1179 (1191).
11. J. Daillant, J.J. Benattar, L. Léger, *Phys. Rev. A*, **41**, 1963 (1990). J. Daillant, J.J. Benattar, L. Bosio, L. Léger, *Europhys. Lett.*, **6**, 431 (1988).
12. G. Henn, D.G. Bucknall, M. Stamm, P. Vanhoorne, R. Jérôme, *Macromolecules*, **29**, 4305 (1996).

Main Notation Used in This Book

z	Direction normal to the surface
x, y	Directions in the plane of the surface
$\|\|$	Used to describe a component parallel to the interface plane
xOz	Plane of incidence
j	Label of layer. Numbering of layers goes from 0 (upper medium) to N the last layer. s is the substrate
Z_j	Average location of the $j-1, j$ interface
$z_j(x, y)$	Fluctuations of the interface location around Z_j
$\mathbf{k}$	Wave vector
$\mathbf{k}_\mathrm{in}$, $\mathbf{k}_\mathrm{r}$, $\mathbf{k}_\mathrm{tr}$, $\mathbf{k}_\mathrm{sc}$	Incident, reflected, transmitted and scattered wavectors
$k_{\mathrm{in}\ z,j}$	z component of the incident wavevector in the jth layer
$k_{z,j}$	when unambiguous
$\mathbf{q}$	Wave-vector transfer
q	Modulus of the wave-vector transfer
q_x,$q_\parallel$,q_z	Components of the wave-vector
$\mathbf{u}$	Scattering direction
r, t	Reflection and transmission coefficients in amplitude
R, T	Intensity reflection and transmission coefficients
$r_{j-1,j}$	Reflection coefficient in amplitude when passing from medium $j-1$ to medium j
$t_{j-1,j}$	Transmission coefficient in amplitude when passing from medium $j-1$ to medium j
$\mathbf{E}$	Electric field
$\hat{\mathbf{e}}_\mathrm{in}$, $\hat{\mathbf{e}}_\mathrm{sc}$	Polarisation vectors of the incident and scattered fields
$\mathbf{B}$	Magnetic field
$\mathbf{j}$	Current density
$\mathbf{P}$	Electric polarisation
$\mathbf{A}$	Vector potential
$\mathbf{S}$	Poynting's vector
$A_\mathrm{j}^\pm$	Amplitude of the upwards and downwards propagating electric fields in layer j
$U(\pm k_{\mathrm{in}\ z,j}, z)$	$A_j^\pm \mathrm{e}^{\pm \mathrm{k}_{\mathrm{in}\ z,j} z}$
$\mathcal{M}$	Transfer matrix
p_n	n-point probability distribution
σ	r.m.s. roughness. $\sigma^2 = \langle z^2 \rangle$

$C_{zz}(x_1, x_2, y_1, y_2)$	Height-height correlation function Also denoted $\langle z(x_1, y_1)z(x_2, y_2)\rangle$
$g(r)$	$2\sigma^2 - 2C_{zz}(x_1, x_2, y_1, y_2)$
G	Green function
$\overline{\overline{\mathcal{G}}}$	Green tensor (electromagnetic case)

$e^{i(\omega t - \mathbf{k}.\mathbf{r})}$ waves are used except in Chap. 5 devoted to neutron reflectivity (see Sect. 1.2.1 for details related to the conventions used in this book, and Sect. 5.1 for the notation used in Chap. 5).

Table 1. Typical length scales for x-ray reflectivity experiments

	Definition	Value
Wavelength λ		1 Å
Scattering length	b	$r_e = 2.81810^{-15}$ m for 1 electron
Extinction length	$L_e = \frac{\lambda}{2\pi\|n-1\|}$	1 μm
Longitudinal coherence length	$\lambda^2/\delta\lambda$	1 μm
Incidence slit opening		0.1 mm
Detector slit opening normal to the plane of incidence (y)	h_y	10 mm
Detector slit opening in the plane of incidence (x)	h_x	0.1 – 1 mm
Sample-to-detector distance	L	1 m
Transverse coherence length normal to the plane of incidence (y) (when fixed by the detector)	$\lambda/\Delta\theta_y$ with $\Delta\theta_y = h_y/L$	10 nm
Transverse coherence length in the plane of incidence projected on the surface (x) (when fixed by the detector)	$\lambda/(\theta\Delta\theta)$ with $\Delta\theta_x = h_x/L$	100 μm for $\theta = 10$ mrad
Illuminated area (length × width)		$(0.1\,\text{mm}/\theta)\times$ $(1 - 10\,\text{mm})$
Absorption length	$\mu = \lambda/4\pi\beta$	0.1 mm – 1 mm for $\beta = 10^{-7} - 10^{-8}$

Index

Printing: Druckhaus Beltz, Hemsbach
Binding: Buchbinderei Schäffer, Grünstadt